Withdrawn

Advances in Downy Mildew Research

Advances in
Downy Mildew
Research

Edited by

P.T.N. Spencer-Phillips
University of the West of England,
Bristol, U.K.

U. Gisi
Syngenta Crop Protection Research,
Basel, Switzerland

and

A. Lebeda
Palacký University in Olomouc,
Olomouc-Holice, Czech Republic

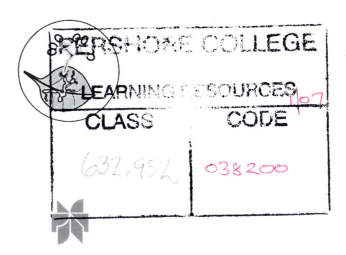
KLUWER ACADEMIC PUBLISHERS
DORDRECHT / BOSTON / LONDON

A C.I.P. Catalogue record for this book is available from the Library of Congress.

ISBN 1-4020-0617-9

Published by Kluwer Academic Publishers,
P.O. Box 17, 3300 AA Dordrecht, The Netherlands.

Sold and distributed in North, Central and South America
by Kluwer Academic Publishers,
101 Philip Drive, Norwell, MA 02061, U.S.A.

In all other countries, sold and distributed
by Kluwer Academic Publishers,
P.O. Box 322, 3300 AH Dordrecht, The Netherlands.

Printed on acid-free paper

Front cover:
Logo of the Downy Mildew Research Group, University of the West of England,
Bristol, UK, designed by Jeremy Clark and Peter Spencer-Phillips.

TABLE OF CONTENTS

PREFACE

P. T. N. SPENCER-PHILLIPS
Co-ordinator, Downy Mildew Working Group of the International Society for Plant Pathology
University of the West of England, Coldharbour Lane, Bristol BS16 1QY, UK
Email: peter.spencer-phillips@uwe.ac.uk

It is a very great privilege to write the preface to the first specialist book on downy mildews since the major work edited by D. M. Spencer in 1981.

The idea for the present publication arose from the Downy Mildew Workshop at the 7th International Congress of Plant Pathology (ICPP) held in Edinburgh in August 1998. Our intention was to invite reviews on selected aspects of downy mildew biology from international authorities, and link these to a series of related short contributions reporting new data. No attempt has been made to cover the breadth of downy mildew research, but we hope that further topics will be included in future volumes, so that this becomes the first of a series following the five year ICPP cycle.

The emphasis here is on evolution and phylogeny, control with chemicals including those that manipulate host plant defences, mechanisms of resistance and the gene pool of wild relatives of crop plants. The value of these contributions on downy mildews has been broadened by comparison with other plant pathogenic oomycetes, especially *Phytophthora* species. In addition, lists of binomials and authorities prepared by Dick provide a key reference source. Readers requiring an introduction to the biology of downy mildews are referred to the review by Clark and Spencer-Phillips (2000), part of which was originally intended for the present book.

As with many of these publishing projects, there has been a long and often frustrating gestation period. However, the editors have ensured that the book is as current as possible by giving authors the opportunity to update their contributions to the end of 2001, immediately prior to submission to the publishers. We are indebted to all for their perseverance and commitment. I also wish to give special thanks to my co-editors Ulrich Gisi and Ales Lebeda for their work; without them this project would not have been completed.

The next ICPP is in Christchurch, New Zealand in 2003. Potential contributors to the Downy Mildew Workshop and authors of review articles for the next volume are invited to contact me with their proposals. We are particularly keen to include progress on genomics, the biology of compatible interactions, control through non-chemical means and the epidemiology of downy mildew diseases.

Clark, J.S.C. and Spencer-Phillips, P.T.N. (2000) Downy Mildews, In J. Lederberg, M. Alexander, B.R. Bloom, D. Hopwood, R. Hull, B.H. Iglewski, A.I. Laskin, S.G. Oliver, M. Schaechter, and W.C. Summers (eds), *Encyclopedia of Microbiology, Vol. 2*, Academic Press, San Diego, pp. 117-129.

Spencer, D. M. (1981) *The Downy Mildews*, Academic Press, London.

TOWARDS AN UNDERSTANDING OF THE EVOLUTION OF THE DOWNY MILDEWS

M. W. DICK
Centre for Plant Diversity and Systematics, Department of Botany, School of Plant Sciences, University of Reading, 2 Earley Gate, READING RG6 6AU, U.K.

1. Introduction

The present review is a revision and expansion of the latter part of a discussion by Dick (1988), much of which has also been incorporated in *Straminipilous Fungi* (Dick, 2001c). New data provided by molecular biological techniques and the resultant data analyses are critically assessed. The strands of the widely disparate arguments based on molecular phylogenies and species relationships, morphology, biochemistry and physiology, host ranges, community structures, plate tectonics and palaeoclimate are drawn together at the end of this chapter.

The downy mildews (DMs) are fungi (Dick, 1997a, 2001c; Money, 1998) but they do not form part of a monophyletic development of fungi within the eukaryote domain. While the closest branches to the Ascomycetes and Basidiomycetes are animals and chytrids, the sister groups to the DMs and water moulds are chromophyte algae and certain heterotrophic protoctista (Dick, 2001a, b, c). The fundamental characteristic of fungi is that of nutrient assimilation by means of extracellular enzymes which are secreted through a cell wall, with the resultant digests being resorbed through the same cell wall. This physiological function has usually resulted in the familiar thallus morphology of a mycelium composed of hyphae.

The unifying structural feature of the chromophyte algae (which include diatoms, brown seaweeds, chrysophytes, yellow-green algae and other photosynthetic groups - see Preisig, 1999), the labyrinthulids and thraustochytrids, some vertebrate gut commensals and free-living marine protists, and the biflagellate fungi (including, by association, certain non-flagellate DMs and a few uniflagellate fungi) is the possession of a distinctively ornamented flagellum, the straminipilous flagellum (see Dick, 1997a, 2001c). Molecular sequencing has confirmed that this diverse group of organisms is monophyletic (Cavalier-Smith, 1998; Cavalier-Smith, Chao and Allsopp, 1995). The group certainly warrants its kingdom status (Dick, 2001c), being more deeply rooted within the eukaryotes than either the kingdoms Animalia or Mycota, but there is debate as to whether or not the photosythetic state is ancestral (discussed below) and therefore

1

P.T.N. Spencer-Phillips et al. (eds.), Advances in Downy Mildew Research, 1–57.
© 2002 *Kluwer Academic Publishers. Printed in the Netherlands.*

TABLE 1. Downy mildews and related taxa. Synopsis of the current ordinal and familial classification of part of the sub-phylum (or sub-division): PERONOSPOROMYCOTINA (Class PERONOSPOROMYCETES). For full synoptic classification, see Dick (2001c).

Sub-class: **Peronosporomycetidae**

Thallus mycelial, rarely monocentric or with sinuses; asexual reproduction diverse; oosporogenesis centripetal; mostly mono-oosporous (exceptions in Pythiales); oospores with a semi-solid, hyaline or translucent ooplast, lipid phase dispersed as minute droplets; able to use $SO_4^=$, ability to metabolize different inorganic N sources variable. Basal chromosome number x = 4 or 5. Two orders, one order including downy mildews.

Peronosporales

Obligate parasites of dicotyledons (very rarely monocotyledons). Thallus mycelial and intercellular with haustoria; zoosporogenesis, when present, by internal cleavage, otherwise asexual reproduction by deciduous conidiosporangia; conidiosporangiophores well-differentiated, persistent; oogonia thin-walled, oospore single, aplerotic with a well-defined exospore wall layer derived from persistent periplasm. Possibility that all are dependent on exogenous sources of sterols.

Peronosporaceae: Myceliar fungi with large, lobate haustoria. Asexual reproduction by deciduous conidiosporangia or conidia borne on conidiosporangiophores; conidiosporangia pedicellate; conidiosporangiophores dichotomously branched, monopodially branched, or unbranched and clavate; conidiogenesis simultaneous; zoosporogenesis, when present, internal within a plasmamembranic membrane, zoospore release by operculate or poroid discharge.

Genera: *Basidiophora, Benua, Bremia, Bremiella, Paraperonospora, Peronospora, Plasmopara, Pseudoperonospora.*

Albuginaceae: Myceliar fungi with small, spherical or peg-like haustoria. Asexual reproduction by deciduous sporangia, conidiosporangia or conidia borne on unbranched conidiophores. Conidiogenesis sequential and percurrent. Zoosporogenesis internal with papillate discharge.

Genus: *Albugo.*

Pythiales

Parasites or saprotrophs; parasites mostly in axenic culture. Some members parasitic on fungi and some on animals. Thallus mycelial with little evidence of cytoplasmic streaming; zoosporangium formation terminal, less frequently sequential, then percurrent or by internal or sympodial proliferation; sporangiophores rarely differentiated; oogonial periplasm minimal and not persistent; oospore usually single, plerotic or aplerotic. Evidence of partial dependence on exogenous sterol precursors.

Pythiaceae: Thallus mycelial or monocentric and pseudomycelial; hyphae 6-10(15) μm diameter; zoosporogenesis either by internal cleavage without vesicular discharge or with a plasmamembranic vesicle or by external cleavage in a homohylic vesicle; oogonia (with very few exceptions) thin-walled; oospores never strictly plerotic. Aerobic metabolism. Freshwater or marine.

Genera: *Cystosiphon, Diasporangium, Endosphaerium, Halophytophthora, Lagenidium sensu strictissimo, Myzocytium sensu strictissimo, Peronophythora, Phytophthora, Pythium, Trachysphaera.*

Pythiogetonaceae: Thallus mycelial, with or without sinuses, perhaps rhizoidal; hyphae <5 μm diameter; zoosporogenesis by external cleavage in a detached homohylic vesicle, or absent; oogonia thick-walled; oospore plerotic. Probably with anaerobic metabolism.

Genera: *Medusoides, Pythiogeton.*

TABLE 1, *continued.*

Sub-class: Saprolegniomycetidae

Thallus mycelial, coralloid or monocentric; zoosporogenesis and oosporogenesis centrifugal; oogonia sometimes poly-oosporous; oospores with a fluid, more or less granular ooplast and variable degrees of lipid coalescence; unable to utilize $SO_4^=$, NO_3^-. Basal chromosome number x = 3. Four orders, only one order including downy mildews.

Sclerosporales

All species are known only as parasites of Poaceae. Mycelium of very narrow (<5 μm diameter) hyphae, with granular cytoplasm, cytoplasmic streaming visible where wide enough; zoosporogenesis by internal cleavage; discharge vesicles not formed; oogonia very thick-walled, often verrucate, with a single, often plerotic oospore; periplasm minimal or absent; distribution of oil reserves as minute droplets. Two families.

Sclerosporaceae: Parasitic, not culturable. Mycelium with peglike or digitate haustoria; sporangiophores grossly inflated; more or less dichotomous; zoosporangium formation sequential or more or less simultaneous on inflated sporangiophores, zoosporangium/conidium maturation more or less simultaneous. Zoospore release, if known, by operculate discharge.

    Genera: *Peronosclerospora, Sclerospora.*

Verrucalvaceae: Parasitic but culturable. Mycelium without haustoria; sporangiophores poorly differentiated; sporangium formation sequential, either by internal or sympodial renewal. Zoospore release, if known, by papillate discharge.

    Genera: *Pachymetra, Sclerophthora, Verrucalvus.*

---

the kingdom may be referred to as the Chromista (photosynthetic endosymbiont ancestral) or the Straminipila (heterotrophy ancestral).

The fungal component of the Straminipila has been named as the sub-phylum (sub-division) Peronosporomycotina, class Peronosporomycetes, using suffixes familiar to mycologists (Dick, 1995), but since the nomenclature within the kingdom spans the zoological and botanical codes, these suffixes may change (cf. Labyrinthista instead of Labyrinthulomycetes, Labyrinthulales, previously included within mycological works). The class is divided into three sub-classes, one of which, the Peronosporomycetidae, at present contains two orders, the Peronosporales and the Pythiales. It has been postulated that the DMs are polyphyletic within the straminipilous fungi (Dick, 1988); the graminicolous DMs were placed in a different sub-class from the dicotyledonicolous DMs, the Saprolegniomycetidae, in the order Sclerosporales (Dick, Wong and Clark, 1984; Dick 2001c; Spencer and Dick, in press) (Tables 1 and 2). The relationships among the families and genera of the Peronosporomycetidae are in a state of flux (discussed below) so that any discussion of the evolution of the DMs must make reference to taxa not regarded as DMs. For general morphological and taxonomic reviews of the DMs and some of the related genera such as *Phytophthora* and *Pythium*, see de Bary (1863); Gäumann (1923); Gustavsson, (1959a, b); Waterhouse (1964, 1968); Kochman and Majewski (1970); Plaats-Niterink (1981); Spencer (1981); Waterhouse and Brothers (1981); Dick (1990b); Constantinescu (1991a); Lebeda and Schwinn (1994) and Erwin and Ribeiro (1996); Dick (2001c).

Downy mildews (DMs) and some related or comparable genera in the Peronosporomycetidae and Saprolegniomycetidae are necrotrophic to biotrophic obligate parasites. 'Biotrophic obligate parasitism' is not always fully developed, so that a limited range of host/parasite relationships may be covered by this phrase. Biotrophic obligate parasites, such as DMs, have advanced genetic and biochemical attributes often, but sometimes unjustifiably, equated to an evolutionary status. Biotrophic obligate parasitism certainly requires a degree of specialization and a constraint to variation: there must be elements of genome protection or conservation in both partners. The basis for this harmony probably lies in unique pairings of 'metabolic packages', the principal components of which may differ from parasite to parasite, or host to host, or both (Dick, 1988). Such 'pairings' are probable between the DMs and their hosts. Dependence might be based upon an 'empathy' between certain crucial metabolic pathways of host and parasite, so that the catabolism and anabolism of both are in accord, rather than there being a determining demand for a particular chemical. It should be noted here that the straminipilous fungi have unique biochemical requirements and metabolic products, many of which are under-rated and some of which will be of significance to the establishment of parasitic relationships.

From the coevolutionary viewpoint, there are distinctions to be drawn between obligate parasitism, species-specific parasitism, and special-form relationships. A discussion on infra-specific differences could, in time, illuminate the processes of speciation compared with population diversity, but the data are too fragmentary at present. Whereas obligate parasitism merely requires the presence of a regular (but possibly periodic) and renewable (but possibly highly transient) nutrient availability from living protoplasm, species-specific parasitism implies a much more restricted range for potential complementary metabolisms. The concept of a 'tolerance range', probably much narrower *in planta* than *in vivo* and thus analogous to the ecological ranges of saprotrophs *in situ* in soils (Dick, 1992), might provide a better model than a search for a package of absolute metabolic requirements.

Different host pathways may be pre-eminent for different parasites. Because of these differences, individuals of a single host species may be infected by several parasites (see Sansome and Sansome, 1974) and the parasites may, by the same token, also encompass different degrees of host specificity.

The systematic range of hosts known to be parasitized by DMs is both taxonomically diverse yet at the same time very limited. But the outstanding characteristic of this distribution is that it is not primarily the more primitive or ancient orders of angiosperms that are affected (Dick, 2001c). Angiosperms parasitized by DMs are mostly in highly specialized taxa, or in recently evolved families, or in taxa that may have a propensity to produce high levels of secondary metabolites. The biochemistry of secondary metabolic pathways, and their importance, has been fundamental to the biotrophic phyto-parasitic coevolution of the DMs in these hosts. In order to understand this non-phylogenetic coevolution, it is essential for this review to outline angiosperm evolution from the Cretaceous through the Tertiary, including a summary of plate tectonic movements, orogeny, resultant climatic life-zones and climatic change over this span of geological time. The stimuli for the development of secondary metabolic pathways may be sought in the exposure of angiosperms, which had evolved in sub-optimum light, to

TABLE 2. Summary of ultrastructural, developmental and morphological differences between the DMs of dicotyledons and the DMs of grasses (from Dick, 2001*c*)

Criterion	DMs of dicotyledons (Peronosporaceae)	DMs of monocotyledons (Sclerosporales)
Flagellar concertina	long, without struts (Barr & Désaulniers, 1987: Peronosporales)	short, struts possibly present (pers. comm., L. W. Olson, decd. re: Lange, Olson and Safeeulla, 1984; cf. Barr & Désaulniers, 1987: Saprolegniales)
Zoosporangial vesicle	plasmamembranic if present	precipitative (or homohylic?) if present
Conidiosporangiophore persistence	found on herbarium specimens of hosts	evanescent, not preserved in exsiccatae
Conidiosporangiophore basal septation	no basal septum	biconvex septum present
Hyphal diameter	broad *c*. 10-12 μm diameter	very narrow *c*. 3-5 μm diameter
Haustoria	Peronosporaceae: large, lobate [Albuginaceae: peg-like]	Sclerosporaceae: small, spherical
Oogoniogenesis sequence	swelling of initial; followed by plug formation, then wall thickening	wall deposition continuous with swelling of initial; followed by septation
Oosporogenesis	centripetal, fundamentally aplerotic	centrifugal, fundamentally plerotic
Periplasm	usually present and contributing a prominent, often ridged, exospore	usually absent; exospore wall minimal or non-existent
Ooplast	translucent, not thoroughly described	translucent, sometimes condensing, not thoroughly described
Oospore wall	relatively thin	usually very thick

pressures for herbaceous development in open canopy. Here, photosynthetic activities would lead to excess photosynthate and high exposure would require UV protection. The development of secondary metabolites would have been responsible for further ramifications of the angiosperm/animal coevolution. Straminipilous fungi, previously adapted to high protein/hydrocarbon/carbohydrate nutrition (perhaps primarily provided by animal substrata), might have been stimulated to colonize roots and crowns which had accumulated excess photosynthate.

2. What are the downy mildews?

The downy mildews (DMs) are parasitic in highly restricted groups of angiosperms. No evidence for parasitism of other vascular, but non-angiospermic, plants exists. The DMs are typically confined to the stem cortex and leaf mesophyll, but some species may be systemic, with the mycelium ramifying throughout the host plant. The long conidio-sporangiophores which emerge from stomata are responsible for the downy appearence of the mildew. The assimilative stage of the dicotyledonicolous DMs is usually a restricted intercellular mycelium with haustoria which penetrate the host cell walls (cf. rust fungi). However, systemic infections are known to occur in the Peronosporaceae (Goosen & Sackston, 1968; Heller, Rozynek and Spring, 1997) and Albuginaceae (Jacobson et al., 1998). Infections caused by the DMs of panicoid grasses may also be systemic (Kenneth, 1981). Not all species are fully biotrophic. Significant cell damage is caused, for example, by *Peronospora tabacina* and *Plasmopara viticola* (Lafon and Bulit, 1981): the plasmamembranes of the host mesophyll cells become excessively leaky, resulting in a distinctive greasy or wet appearance to the infected part of the leaf. This is essentially a moderated manifestation of the symptoms associated with wet rots caused by certain species of *Phytophthora* (Keen & Yoshikawa, 1983) and probably resulting from a similar biochemical interaction. Biphasic culture has been achieved for several genera and species (Ingram, 1980; Lucas et al., 1991, Lucas, Hayter and Crute, 1995), but none is yet in axenic culture.

Any discussion of the systematics and evolution of the DMs must involve some related pathogens in the orders Peronosporales (including *Albugo* in the monogeneric Albuginaceae), Pythiales (*Phytophthora* and *Pythium* in the Pythiaceae) and Sclerosporales (*Pachymetra* and *Verrucalvus* in the Verrucalvaceae). The white blister rusts (*Albugo* species) are also obligately biotrophic parasites of dicotyledons, commonly recorded from stems and leaves. *Pachymetra*, parasitic on roots of sugar cane, and *Verrucalvus*, parasitic on roots of *Pennisetum*, are known only from eastern Australia. Both of these genera are monotypic and can be maintained, with difficulty, in axenic culture (Dick et al., 1984, 1989).

In the Pythiales, the sister order to the Peronosporales, *Phytophthora* is known as a pathogen of a wider range of woody and herbaceous angiosperms and conifers; different species may parasitize roots, hypocotylar regions, leaves or fruits. A few species are saprotrophic. *Pythium* has a still wider host range, embracing invertebrate and vertebrate animals, marine red and freshwater green algae, charophytes and vascular plants, and fungi. (Animal parasitism by straminipilous fungi has been reviewed by Dick, 2001*b*; algal parasitism by Dick, 2001*c*.) Almost all species of *Pythium* are readily culturable and possibly because of this, the extent of the truly saprotrophic habit is unknown. *Phytophthora* species are also culturable, but require more care.

The DMs have been classified within the family Peronosporaceae since de Bary (1863, 1866) first coined the family concepts "Saprolegnieen" and "Peronosporeen". An annotated list of genera, and binomials therein, for the DMs and Pythiales is published elsewhere in this volume. The hierarchy of classification has changed in the intervening years, but Dick et al. (1984), and more recently Dick (1990*a*, 1995, 2001*a*, 2001*c*; Table 1) has proposed the class Peronosporomycetes, including the subclasses

Peronosporomycetidae and Saprolegniomycetidae. Support for these subclass divisions is now available from molecular data (Dick *et al.*, 1999; also discussed below).

The DMs comprise two distinct groups; those almost exclusively associated with herbaceous dicotyledons and those parasitic on grasses, particularly the panicoid grasses. In the classifications of Dick (1995, 2001*a*, 2001*c*; Dick *et al.*, 1984) the graminicolous DMs have been placed in a separate order, the Sclerosporales, in the Saprolegniomycetidae. The host differences can be correlated with morphological characters (Dick *et al.*, 1984, 2001*c*; Table 2), of which flagellar ultrastructure may be the most significant. Other criteria worthy of note are: the persistence of the branched conidiosporangiophore in the Peronosporales and its ephemeral nature in the Sclerosporales; the different haustorial morphologies of the Albuginaceae, Peronosporaceae and Sclerosporaceae; the degree to which oosporogenesis is periplasmic (Albuginaceae, Peronosporaceae; but see Vercesi *et al.*, 1999) or whether the oospore is plerotic (Sclerosporaceae); and differences between the obligately parasitic families (including the Albuginaceae) with respect to the morphogenesis of asexual propagules.

The DMs and *Phytophthora* have received attention because of the damage caused to yields from crop plants. Together these organisms have been responsible for socio-economic-political change (Large, 1940; Smith, 1884; Woodham-Smith, 1962), the birth of plant pathology (de Bary, 1863, 1876), and the first developments of the agricultural chemical industry instigated by Millardet (Ainsworth, 1976; Schneiderhan, 1933). It is essential to acknowledge that distributions of many of the *Phytophthora* parasites were provincial until recently, when trade movements disseminated these species, often with an eventually dramatic new pathogenic impact (Late Blight of potatoes and the Irish Famine).

The importance of the DMs in agriculture has only arisen within the last 500 years, brought about by significant intercontinental trade and movement in grain, root and fruit crops. There was almost certainly previous limited movement around the Mediterranean; across the Panamanian isthmus; and between the great south-east Asian river delta systems, but this trade did not, as far as is known, cause problems with pathogen introductions, nor did it impinge on hosts vulnerable to DMs. Despite this extremely recent, in eco-evolutionary contexts, movement of the potential host plants, there has been time for the world-wide spread of DM diseases associated with the crops of millet, sugar cane, corn, potato, squashes and grape (*Pennisetum, Sorghum, Saccharum, Zea, Solanum, Cucumis, Vitis*), and for the evolution of intraspecific geographic differences in the fungal populations. Distribution by man has resulted in the parasitism of *Zea* (host from central America and parasite, *Sclerospora*, from south-east Asia); ploidy differences in *Phytophthora infestans* in Europe following the exotic introduction of both host and parasite, which originate from the Equador (Boussingault, 1845) - Central American region (Lucas *et al.*, 1991; Daggett, Knighton and Therrien, 1995) and *Peronosclerospora* in India (originating from Africa, Ball and Pike, 1984; Idris and Ball, 1984).

In contrast to these recent developments, it should be recalled that cereal cultivation was the original crop of agriculture, with at least four independent and long-localized origins: the 'Fertile Crescent', the region including parts of Asia Minor and Mesopotamia (*Hordeum*); South East Asia (*Oryza*); north-east Africa (*Sorghum*) and central America (*Zea*). The earliest cultivation of any these crops (*Hordeum*) can be confidently dated to more than 10,000 years B.P., followed by *Zea* (>7000 years B.P.) and *Sorghum* (>3000 years B.P.) (Clayton and Renvoize, 1986). Comparable dates for pulses are: 10,000 years B.P. (*Pisum*, south west Asia); 10,000 years B.P. (*Lens*, Fertile Crescent); 7,000 years B.P. (*Phaseolus*, central and south America); 5,000 years B.P. (*Vigna*, west Africa). The potato has been cultivated for 7,000 years, with wild potatoes being used in southern Chile for 11,000 years (Vaughan and Geissler, 1997).

3. Evolutionary origins of the Straminipila, including the Peronosporomycetes

It is possible that the straminipilous *fungi* themselves (as a component of the undoubtedly early-originating monophyletic line of straminipilous *organisms*) have an ancient origin (Pirozynski and Malloch, 1975; Stanghellini, 1974), but the fossil evidence is equivocal (Pirozynski, 1976*a*, *b*). No fossils of downy mildews have been reported: the establishment of various fossil genera such as those of Duncan (1876: *Palaeoachlya*); Seward (1898: *Peronosporites*); Pampaloni (1902: *Peronosporites*, *Pythites*); Elias (1966: *Propythium*, *Ordovicimyces*); Douglas (1973: *Peronosporoides*); Stidd and Consentino (1975: *Albugo*-like oospores) can be discounted (Dick, 1988, 2001*c*). Fossil angiosperm leaves with diseased tissues are well-known (Dilcher, 1965), but none is obviously a DM association.

There are now sufficient ultrastructural and molecular biological data for it to be as near certain as possible that the DMs are part of a very diverse monophyletic lineage, the kingdom Straminipila (diagnosis in Dick, 2001*c*; name very commonly mis-spelt 'stramenopiles'). The monophyletic origin, probably prefungal, is ancient. This kingdom is separate, on the one hand, from green and red plants, and on the other hand, from animals and fungi (Mycota) all of which are now commonly accorded kingdom status. This straminipilous lineage is extraordinarily diverse, including photosynthetic (chrysophyte, diatomaceous and fucoid); heterotrophic (free-living marine and gut comensal protoctist, bicosoecid and labyrinthuloid), and osmotrophic (fungal) organisms (Gunderson *et al.*, 1987; Leipe *et al.*, 1994, 1996; Potter, Saunders and Andersen, 1997; Silberman *et al.*, 1996). These organisms are characterized primarily by the possession of a straminipilous flagellum (Dick, 1990*a*, 1997*a*, *b*, 1998, 2001*a*, *c*).

The straminipilous flagellum possesses two rows of tubular tripartite hairs (TTHs) which reverse the thrust of the flagellum, so that this flagellum is anteriorly directed (Dick, 1990*a*, 1997*a*, 2001*a*, *c* and references therein). The unique mode of motility conferred by the straminipilous flagellum has been discussed by Jahn, Landman and Fonseca (1964) using the model of Taylor (1952). However, the complexity of the morphology and the functional significance of each part of the TTH are still not explained (see Dick, 1990*a*, 2001*a*, *c*). It has been calculated that the anterior flagellum is about ten times more powerful as a 'motor' than the whiplash flagellum (Holwill, 1982). The evolution and possession of the straminipilous flagellum involved a transfer of receptor sites from the cell surface to this anterior flagellum, and a more efficient endogenous energy reserve and mobilization. The structural complexity and function of the straminipilous flagellum are such that it is unlikely to have evolved more than once

(Dick, 1990a, 2001c; Van Der Auwera & De Wachter, 1998). The straminipilous flagellum is thus advanced (derived *after* the evolution of the standard 9 + 2 flagellar axoneme) and evolutionarily conserved (Leipe *et al.*, 1996), occupying a place as significant as the evolution of chlorophylls *b* and *c* (both derived from chlorophyll *a*).

The possession of flagellation indicates an aquatic origin. However, the evidence for an aquatic photosynthetic precursor to the osmotrophic fungi is equivocal (for opposing views see Cavalier-Smith, 1986, 1989, 1998; Cavalier-Smith *et al.*, 1995; Nes, 1990) and for this reason I prefer a kingdom diagnosis based solely on the straminipilous flagellum (Dick, 2001c) rather than a diagnosis which includes the photosynthetic endosymbiont as a fundamental component, followed by its subsequent loss (the kingdom Chromista). Cavalier-Smith *et al.* (1995) argue that the Peronosporomycetes evolved from a photosynthetic ancestor. This argument, inferred from ultrastructural observations, depends on an rDNA analysis which used the very few peronosporomycetous data then published and the data for *Hyphochytrium*. All sequences were deeply rooted, but the position of *Hyphochytrium* was more deeply basal than those for other straminipilous fungi (Van der Auwera *et al.*, 1995). However, this basal position for *Hyphochytrium* may be challenged because none of the representatives of some critical taxa has yet been sequenced: recent unpublished data appear to suggest that *Halophytophthora* may also be ancestral. Neither of these genera is known to have a sexual phase, so data purporting to show ancestry to the teleomorphic straminipilous fungi are inevitably weak. The biochemical data of Nes (1990), based on the analyses of sterols and sterol synthetic pathways, suggested that the Peronosporomycetes did *not* have a photosynthetic antecedent (lanosterol is formed from squaline oxide cyclization via cycloartenol in photosynthetic lineages, but directly in non-photosynthetic lineages).

Flagellar loss, or partial flagellar loss (including loss or partial loss of straminipilous ornamentation), has probably occurred several times in the straminipilous fungi, as it has in the straminipilous algae (Leadbeater, 1989; Cavalier-Smith *et al.*, 1995) and 'straminipilous' protoctists (see Silberman *et al.*, 1996). Loss of the zoospore, and therefore flagellation, is a feature of both the Peronosporales and Sclerosporales and has minor phylogenetic significance. Until the data-base for the Peronosporomycetes is much larger, involving a wider range of peronosporomycetous fungi and more *straminipile* outgroups, the deposition of the heterotrophic orders viz-à-viz the photosynthetic orders must remain debatable (Leipe *et al.*, 1996; Potter *et al.*, 1997).

Coupled with the straminipilous flagellum, within the kingdom Straminipila, are: the possession of a mitochondrion with tubular cristae (as opposed to the plate-like cristae of animals and plants); a DAP lysine synthesis pathway (α, ϵ-diaminopimelic acid pathway) which they share with the angiosperm hosts (Vogel, 1964); and, if photosynthetic, a plastid with a second chlorophyll, chlorophyll *c* (not chlorophyll *b* as in green plants). Information on the nature of the phosphate storage mechanism (possibly the DBVs - dense body vesicles - in the straminipiles) is again unbalanced (see Chilvers, Lapeyrie and Douglass, 1985), with more information available from the fungal components of the kingdom. Metabolism is generally hydrocarbon-based, with high levels of non-cellulosic glucans.

In addition to the straminipilous characters listed above, the fungal class Peronosporomycetes is characterized by a combination of five characters not found in

any of the other major groups of straminipilous organisms (Dick, 2001*a, c*):
    haplomitotic B ploidy cycle (mitosis confined to the diploid phase)
    cruciform meiosis in a persistent nuclear membrane
    multiple synchronous meioses in coenocytic (paired) gametangia (meiogametangia)
    gametes without flagellation (donor gametes without cellular identity)
    formation of zygotic resting spores (oospores) in oogonia (receptor gametangia)

The diversity, extant lineages of, and genetic distances between the straminipilous organisms are such that the *kingdom* must have originated, and evolved initially, in the marine ecosystem, most probably in littoral and lagoon or estuarine environments (the photoendobiont was probably a red alga, strengthening the marine provenance; Potter *et al.*, 1997) and at an early geological Period (Cambrian? Precambrian?). Nevertheless, all existing evidence points to a freshwater or terrestrial origin for the straminipilous fungi, or Peronosporomycetes. The warm temperate lagoon ecosystem would rapidly have become world-wide during the early tectonic movements and sea level changes of the Gondwanaland, Laurentian, European and Siberian plates, which were all equatorial and separate during the Precambrian and Cambrian Periods, but which became united to form the supercontinent Pangaea during the Permian (Tarling, 1980). The principal questions remaining are:
    did the fungal organisms evolve from a *freshwater heterotrophic* ancestor (*Saprolegnia*- or *Pythium*-like) or from a *marine heterotrophic* ancestor (*Halophytophthora*-like)?;
    did the freshwater fungal straminipilous organisms evolve from a *freshwater photosynthetic* ancestor, or from an *originally and fundamentally heterotrophic* ancestor in *freshwater*?
In contrast to the oceanic margin ecosystems, the freshwater systems would not have been physically confluent, and therefore different communities could have evolved in isolation, possibly from estuarine habitats. Most species of the Saprolegniaceae and Pythiaceae are now cosmopolitan, with very little evidence of provincialism, and this might be taken as evidence for an early origin. On the other hand, there is evidence within the Saprolegniaceae that the Atlantic Ocean has provided a barrier for the evolution of separate species of *Aphanomyces* and for the distribution of *Aplanopsis terrestris* and *Newbya spinosa* ($\equiv$ *Aplanopsis spinosa*): *Aphanomyces astaci* from North America, now causes the crayfish disease in Europe (Dick, 2001*b*), while *Aplanopsis terrestris* and *Newbya spinosa*, both very abundant terrestrial saprotrophs in northern Europe, are not found in North America (Voglmayr, Bonner and Dick, 1999). The possibility of very rare events of transcontinental movement, and opportunistic saprotrophism, would still allow the hypothesis that *most* of the extant Peronosporomycetes could have evolved in the very recent (late Tertiary) past. Thus, although the origins of the straminipiles (Kingdom Straminipila) were probably Cambrian or Precambrian, the straminipilous fungi (class Peronosporomycetes) might have evolved *at any time* between the Early Palaeozoic (Ordovician?) some 438-488 million years Before Present (m.y.B.P.) and the late Tertiary (*ca* 20 m.y.B.P.), either as saprotrophs of dead aquatic animals, as animal parasites, or as saprotrophs and parasites of dead and dying phytoplankton. There remains an enormous gap between such a postulated early

origin (whether photosynthetic or fundamentally heterotrophic) and the distributions and diversity of present-day taxa.

Hypotheses for the evolution of the straminipilous fungi must therefore be based on circumstantial evidence of structure, morphogenesis and biochemistry (including molecular biology) of extant taxa. The high energy requirements of the straminipilous flagellum and the bacterially-induced anaerobic environment that would surround potential aquatic substrata appear to be mutually exclusive. The evolution of mechanisms for shifting the site of zoospore discharge from the site of zoosporogenesis would therefore have been beneficial and probably developed on more than one occasion. Similarly, substrata yielding readily available nutrients would also be favoured; concommitant tendencies to develop anaerobic metabolic pathways would follow, as shown by the Rhipidiales (Emerson and Held, 1969; Held, 1970) and Pythiogetonaceae (Winans in Emerson and Natvig, 1981; Voglmayr et al., 1999). The sub-cuticular coenocytium of nematodes and the ecdysic fluids of aquatic arthropods are obviously also such nutrient-rich substrata, and these substrata would have existed in the Palaeozoic. The freshwater/terrestrial origin of heterotrophs was probably coevolutionarily linked to arthropods, nematodes and the animal food chain (cf. Saprolegniales and Myzocytiopsidales), with freshwater algae (green algae and certain straminipilous algae) being the primary producers. The development of freshwater green algae would have been well advanced, since Charophytes (with their calcified fossils) of shallow brackish water are known from as early as the upper Silurian (Feist & Grambast-Fessard, 1991; Kenrick & Crane, 1997). Relationships with angiosperms, perhaps initially as saprotrophs of nutrient-rich substrata (seeds and fruits), comparable with animal substrata, must have occurred very much later, probably in the late Cretaceous. Most extant saprotrophic straminipilous fungal species are associated with animals or seeds and fruits. Links with vascular plant substrata may have started with detrital decay in water by transfer from animal substrata to fruit and seed decay with fermentative metabolism. The increasing availability of pollen and fruits (due to the coevolution between angiosperms and animals) in water systems in the late Triassic and early Cretaceous would have provided a novel source of nutrients. Twigs and leaves are less common substrata, but are utilized by *Phytophthora gonapodyides* and species of *Dictyuchus*, *Sapromyces* and *Apodachlya*. Leaf- and twig-decaying fungi in aquatic ecosystems are normally hyphomycetes which have a much older fossil history (Dilcher, 1965). It is noteworthy that evidence of associations of Peronosporomycetes with bryophytes, ferns, gymnosperms and early divergent angiosperms (Nymphaceae, Ceratophyllales, Laurales, Magnoliales and Piperales) is all but non-existant (but see *Albugo tropica* and compare with *Phytophthora* on Lauraceae, Erwin and Ribeiro, 1996).

3.1. THALLUS DIVERSITY

Diversity in the Peronosporomycetes is found in mycelial characteristics: the mycelial habit has probably developed on several separate occasions (Dick, 1995, 2001*a, c*). The origin of hyphae in the Peronosporomycetes was probably recent, either from a sporangio-gametangiophore with indeterminate tip growth (wider hyphae), or from a narrow germ tube developed from an infection peg (narrower hyphae). Phylogeny inferred from vegetative and asexual morphology is not acceptable, although subtle differences in morphogenesis might be invoked.

Morphological elaboration in the asexual system is also found in the continuum of sporangial forms in *Pythium*; the development of sporangiophores in *Phytophthora*; caducous zoosporangia; conidiosporangiophores and conidia. Differences exist in zoosporogenesis within both subclasses of the Peronosporomycetes with DMs. The ability to produce zoospores from conidia is usual in *Albugo*, variable in *Plasmopara* (see Wilson, 1907, re: *Rhysotheca* and *Plasmopara*) and has been lost in *Peronospora*, *Peronosclerospora* and *Pachymetra*. It could be inferred that morphology, particularly in relation to zoospore production, is an unreliable indicator of phylogenetic age or relationships among these parasites.

3.2. INTERCELLULAR HYPHAE

One of the striking features of the Peronosporomycetes (Peronosporales) is the development of biotrophy from necrotrophy. Savile (1968, 1976) has suggested that the first step towards phytoparasitism would have been the development of *systemic* (whole plant) myceliar parasitism to protect the hyphae from desiccation (note the extremely narrow and vulnerable hyphae of the Sclerosporaceae), and that lesions of limited mycelial extent would have evolved later. Another most important step would have been the development from mixed intra- and inter-cellular hyphae to mycelia solely of intercellular hyphae and haustoria (Fraymouth, 1956; Peyton and Bowen, 1963; Berlin and Bowen, 1964; Davison, 1968; Coffey, 1975). Parallel evolution of intercellular hyphae and haustoria (biotrophic parasitism) is manifest by the occurrence of these features in both the DMs and the phylogenetically unrelated Uredinales (*Puccinia*). Spencer-Phillips (Clark and Spencer-Phillips, 1993; Spencer-Phillips, 1997) has shown that the intercellular hyphae of the DMs retain the capacity for assimilation in the presence of haustoria. Differences could exist between the functions of haustoria in the nutrition of unrelated taxa. Thus, there is no reason to consider that this biotrophic development, even within the DMs, represents a monophyletic line. Indeed, the fact that the *morphology* of the haustoria is different in *Albugo*, *Peronospora*, and *Sclerospora* could point to independent origins, each possibly with a characteristic physiology.

4. Parasitism by the downy mildews

Parasitism by the downy mildews must be contrasted with the parasitoidal associations of the Myzocytiopsidaceae with nematodes and algae (Dick, 1997*b*, 2001*c*). These endobiotic parasites are always necrotrophic. Similarly, endobiotic Saprolegniaceae (*Aphanomyces parasiticus*), root-parasitic Saprolegniaceae (*Aphanomyces euteiches*) and Pythiaceae (*Pythium* species) are necrotrophic.

Developmental (evolutionary?) steps in parasitism can be traced at the assimilative and reproductive levels in the Peronosporomycetidae and Saprolegniomycetidae. Assimilation by means of necrotrophic intracellular root parasitism, systemic growth, development of intercellular hyphae, development of haustoria, nutrition by intercellular hyphae without haustoria and symptomless parasitism all occur.

Potentially interacting organisms must be able to come into contact, and there must be sufficient compatibility for nutritional requirements to be satisfied. Frequently, this will be because new hosts are phylogenetically close to former hosts. Host populations at the frontiers of their realizable niches are more liable to become involved in new coevolutionary initiatives, but the development of a stable relationship will depend on the generation cycles of the parasite and its capacity for genetic change. The critical factors for the nutritional environment of the parasite, the pathways, or the specific metabolites produced, may occur in organisms of differing phylogeny; or, they may only become evident in certain populations because of environmental circumstances. Two facets interconnect: the coevolutionary reliance by the parasite on a host species, and the restrictive nature of this reliance to particular metabolitic pathways. The critical factors involved may require subtle definition. Obvious basic carbon and nitrogen sources are unlikely to be crucial, but sulphur and combined forms of carbon and nitrogen may be so for DMs.

There must be physical or chemical similarities or analogues that enable an appropriate degree of association between previously separated populations. Too great a vulnerability will lead to an unstable and ephemeral (necrotrophic) relationship. The essence of coevolution is adaptive change in balanced relationships. It is possible that chance associations may lead to new relationships, as has been proposed by Baum and Savile (1985) for certain rusts. This may be more possible for parasites that produce a limited mycelium and for which physical rather than chemical environmental factors are more important. Chance associations leading to coevolution must be less likely for fungi that are essentially systemic, because there would be less likelihood that either host or parasite would survive long enough to reach reproductive maturity.

An obligate parasite that cannot be grown apart from its living host either requires particular metabolites that have not yet been identified, or the organism is intolerant of arbitrary levels of fluctuations in the concentrations and rates of supply of nutrients, or some other *in planta* factor is necessary. There are no suggestions that nutritional requirements are invariably linked to host range restriction in the DMs. The efficiency of waste removal may be a contributory factor. There is little evidence to support or refute any of these contentions. Moreover, extrapolations made from studies of related fungi that can be grown axenically could be misleading.

If parasite dependence is not based on a demand for particular chemical units, the dependence must have a different origin. I have suggested (Dick, 1988, 2001c) that this could be based upon an 'empathy' between certain crucial metabolic pathways of host and parasite, so that the catabolism and anabolism were in harmony. Different host pathways may be pre-eminent for different parasites, whether these are taxonomically related or not. Thus, individuals of a single host species may be infected by several parasites. The most notable example for DMs is the suggested synergism between *Peronospora* and *Albugo* in Brassicaceae (Sansome & Sansome, 1974). However, my hypothesis of critical pathway differences would not only explain the occurrence of simultaneous parasitism of a host by different, but systematically related biotrophic obligate parasites: it would also allow for the possibility that these parasites may have different degrees of host specificity.

Whatever the biochemistry underlying attraction to a particular host, and stimulation to germination and colonization by the parasite, there are well-documented examples of parasite-mediated modification of host physiology after establishment.   Green ear hyperplasia of pearl millet caused by *Sclerophthora* (Williams, 1984), hypoplasia of sunflower by *Plasmopara* (Sackston, 1981), and the well-known hypertrophy of crucifer stems by *Albugo* are three of the clearest examples relating to growth substance induction.   The precise mechanisms of the biochemical modifications have not been researched.

Symptomless occurrence of Peronosporales and Pythiales in angiosperms suggests that the evolution of parasitism has achieved the ultimate balance in some associations. Haustoria are not essential. *Pachymetra* in *Imperata cylindrica* var. *major* in Queensland (pers. comm., R. C. Magarey, Bureau of Sugar Experiment Stations, Queensland), *Phytophthora* in roots of raspberry and strawberry in Scotland (pers. comm., J. M. Duncan, Scottish Crops Research Institute), and *Pythium* in grass and herbaceous roots are all good examples of such symptomless associations. Symptomless association does not imply a 'no yield loss' situation.

The boundaries between obligate parasitism, species-specific parasitism, and special-form relationships are unclear: more research and discussion (cf. Skalický, 1964; Skidmore and Ingram, 1985) should elucidate the processes of speciation as opposed to different levels of infraspecific (population) diversity. Species-specific parasitism implies a much more restricted range for potential complementary metabolisms. This can be viewed as a *tolerance range* rather than a package of absolute metabolic requirements. The breadth of this tolerance range may well be extremely narrow *in planta*, in much the same way that saprotrophic *Pythium* species may co-exist in soil, but have very different patterns of relative frequency of occurrence *in situ* than might be predicted from growth studies *in vitro* (Dick, 1992).   The endpoint of this progression is the race concept of the special form for which biochemical compatibility is presumed to be the only apparent distinguishing feature. This may be merely the result of extremely narrow tolerance ranges for a number of factors.   But it may be, as with race induction in response to resistance cultivar production, a gene-for-gene evolution that may function through a variety of biochemical, physiological or morphological requirements.   An hypothesis for absolute metabolite requirement in the absence of strong selective pressure might require an improbably large number of genetic lesions to explain race-specific parasitism (*formae-speciales*) between related parasites and related hosts.

Discussions of single-gene host resistance in different systems of host resistance and pathogen virulence (e.g., Keen and Yoshikawa, 1983) ignore the attraction and stimulation that enables both species to coexist. It is unlikely that studies concentrating on intraspecific differences will reveal underlying coevolutionary factors.   There is a long-standing inverse relationship between the outlook and research momentum for plant pathology and the quest by mycologists for an understanding of species-specific coevolution.

The genetic bases for these distinctions may be diverse. Brasier (1992) and Brasier and Hansen (1992) have reviewed the evolution of *Phytophthora* from a genetic standpoint. Genome synteny (the presumption that syntenic loci are carried on the same chromosome) is now viewed somewhat differently with the demonstration that while most of the genes in the genome are similar, they may be distributed differently between the chromosomes, so that, as in the grasses, considerable differences in chromosome size and number conceal an underlying similarity (Moore *et al.*, 1995). Genome similarity should be assumed between genera, but it may involve chromosome inversions, chromosomal sections moved from one chromosome to another, with or without changes in chromosome length or number. Ploidy levels may be different, and here the breeding systems of the straminipilous fungi need to be taken into account, particularly when selfing and automictic sexual reproduction may be involved (Dick, 1972, 1987, 1995; Win-Tin and Dick, 1975). It is also possible that differences in virulence could be attributed to Simple Sequence Repeats (SSRs ≡ microsatellite DNA).

The diversity of genome variation, resulting in species-complexes in terms of chromosome sizes, chromosome numbers and genome size in angiosperms (see Vaughan, Taylor and Parker, 1996: *Scilla*), needs to be considered when reviewing DNA quantification (e.g., Martin, 1995*a*; Voglmayr and Greilhuber, 1998) and species based on a karyotype (*Phytophthora megakarya* - Brasier and Griffin, 1979).

From the systematic viewpoint, the above environmental/host distinctions of the parasite rest uneasily with the infra-specific categories of variety and form, together with *formae speciales* which are not governed by the rules of the International Code of Botanical Nomenclature (ICBN).

5. Molecular systematics, evolutionary origins and taxonomy, including a critique of available data

The advantages and disadvantages of Linnaean classifications need to be evaluated, since alternative systems, based on molecular phylogenies, have been proposed and these challenge the nomenclatural hierarchy (Hibbett and Donoghue, 1998). Despite considerable research activity, molecular phylogeny is still in its infancy: a number of considerations, in addition to questions of translating molecular phylogeny into classifications (outlined below), have yet to be fully addressed by mycologists. There is a tension between Linnaean/ICBN taxonomy and phylogenetic systematics (Brummitt, 1996; de Quiroz and Gauthier, 1994). Nevertheless, molecular phylogeny will provide information about relationships even if these relationships are not resolved into classifications. The following numbered points should be noted:

(i) To what extent should a clade node correspond to a 'classical' hierarchical level? Diversity within an ancient lineage may coexist with a more recently evolved, but fundamental attribute which so changes the evolutionary potential that the erection of a higher taxon is of practical value. Computer-generated similarity indices will reflect probable lineages, but these will not negate intra-subclass diversity in higher taxon concepts. Some higher taxa will encompass several nodes. Because of the progressively bifurcating nature of the cladogram, or lack of resolution for the origins of several lineages, phylogenetic approaches are not always best suited for establishing correlations (ie. discontinuities) with currently recognized hierarchies in systematics. It is not always possible to distinguish between derived (apomorphous) and ancestral (plesiomorphous) character states. At ultimate branches of phylogenetic trees single cladistic characters may be insufficiently diagnostic, so that a 'suite' of characters is necessary for separation at species (and sometimes genus) level (see Donoghue, 1985). With finger-printing techniques separation proceeds through infraspecific taxa all the way to populations, clones and individuals (Lévesque *et al.*, 1994; Liew *et al.*, 1998; Panabières *et al.*, 1989).

(ii)   The type concept is fundamental to systematics.   Genera are defined by historically determined *type species*, irrespective of whether the type species is uncharacteristic of the taxa presently included in the genus. The type species is based upon a *type specimen*, which again may deviate from the central tendency of the population from which it came. Although the *type material* may no longer be extant, or if extant no longer suitable for molecular analysis, it remains essential for the type species to be characterized before systematic changes can be justified. When the type material is not available, more recent isolates of the fungi (determined on morphological criteria) have to be used. These precepts are most pertinent to the systematics of the DMs. The type species of *Plasmopara* (Peronosporales) and both *Phytophthora sensu lato* and *Pythium sensu lato* (Pythiales) occupy extreme positions in the genera they characterize.

(iii)   There is no possibility of obtaining information from extinct taxa to qualify probabilities. In any systematic and phylogenetic (evolutionary) molecular reconstruction it is essential to recall that only relationships between *extant* species will be displayed.

(iv)   The basis for phylogenetic placement and relationships within the straminipiles depends, very largely at present, on long sequences of nucleotides in the gene encoding for ribosomal RNA. It is possible that one part of one gene is sufficient to establish a robust cladistic framework, but justification and support is normally required (see Doyle, 1992). In angiosperm phylogeny three independent genes are being used (Soltis, Soltis, Chase, *et al.*, 1998*a, b*; Soltis, Soltis and Chase, 1999; The Angiosperm Phylogeny Group (APG), 1998).   For entirely understandable reasons, the independent, endosymbiont genes most studied in straminipiles are *either* in the photoendobiont *or* in the mitochondrial endosymbiont (heterotrophs), so that comparability is lacking across the whole kingdom. Other genes have not yet been studied in sufficiently large samples of straminipilous fungi or other straminipiles to enable a robust phylogenetic hypothesis, similar to that for angiosperms, to be constructed.

(v)   For long sequences the number of informative, variable sites within the sequences that are necessary to give adequate characterization and separation within a particular group of related taxa should be noted: the region for data analysis must contain sufficient differences in sequences to allow closely related species to be separated; these differences should be the result of a single base change and be free of length mutations. Berbee *et al.* (1998) has shown, with ascomycetes, that while shorter sequences are sometimes adequate, there are some taxa for which much longer sequences are essential. It will be necessary to characterize the DMs and other straminipiles in this respect. Shorter sequences such as pertain to the ITS region are frequently used, but in *Pythium* there are length mutations in this region so that analysis becomes highly dependent on sequence editing.   In spite of this complication, the ITS region is effective in distinguishing between closely related species; other sections of the gene (the D2 region of the 28S rDNA gene) appear to be less suitable (pers. comm., F. N. Martin,

U.S.D.A., Salinas, Ca.).

(vi) Evolution is on-going. Species concepts (both real and postulated) vary widely, even in a single genus. Incipient speciation will occur. Isolation of, and modification of, the gene pool may not, initially, be correlated with or represented by morphological attributes. Population diversity and formally defined intra-specific taxa require reassessment; genetically controlled host/parasite associations will be characterized by *formae speciales*. Similarity, even identity, in nucleotide sequences with respect to one gene may be yoked to variation in another gene which codes for such host-specific functional associations.

(vii) There is no absolute time-scale for rates of molecular evolution, but eventually the molecular phylogeny should be integrated with geological time. The molecular clock for the straminipiles will be influenced by generation times and population sizes (the 'sloppy' clock hypothesis). The diatoms and other marine straminipilous unicells have enormous populations, short generation times and sexual reproduction is rare. *Hyphochytrium* (and all described members of the Hyphochytriales) and *Halophytophthora* are anamorphic; almost nothing is known of chromosomal or genetic stability in these genera. If a molecular clock cannot be determined, the apparent evolutionary distance, as represented by nucleotide sequence changes, will not necessarily be the same as the absolute evolutionary time-scale for all organisms. The rates of evolution of mitochondrial genes and nuclear genes may differ by a factor of 10 in other organisms. For a robust phylogeny of straminipiles, the factorial differences between the genes selected should be clarified.

(viii) The stability of the cladistic arrangement has yet to be established. The placement of some ordinal branches within the Peronosporomycetes, such as the Leptomitales (Dick, *et al.*, 1999; Riethmüller, Weiss, and Oberwinkler, 1999; Hudspeth, Nadler, and Hudspeth, 2000; Cook, Hudspeth, and Hudspeth, 2001), is still equivocal, even after analysis of long (>1800) nucleotide sequences from 18S rDNA. A comparable situation holds for the photosynthetic straminipiles (Potter *et al.*, 1997). Association depends on the algorithm used. Additional, independent data are needed. Positioning of so few, deeply rooted taxa in cladograms can also be influenced by the size of the data base and the outgroups used. Divergent orders with very few known species, such as the Leptomitales, present problems when interpreting cladograms. It must also be recognized that the addition of new information may affect the branching of the cladogram. In all cases it is desirable to rationalize cladogram differences with structural features (the 'common sense' factor).

Most deep phylogeny relies on the sequences of the small subunit (18S) of the rDNA gene. Phylogenies based on 18S rDNA are well-established for straminipilous organisms and largely confirm prior taxonomic conclusions from kingdoms down to orders: relationships between families and genera are more open to debate. No other sequences can compare, in the numbers and diversity of organisms assessed, with 18S rDNA at this stage.

More information (mainly restricted to shorter sequences) is known for other straminipilous fungi, especially *Phytophthora* and *Pythium* (Briard *et al.*, 1995; Lévesque *et al.*, 1993, 1994, 1998; Herrado and Klemsdal, 1998; Cooke *et al.*, 1996, 1999; Cooke *et al.*, this volume). The molecular data support, only in part, the hierarchical classification within the Peronosporomycetes (compare Grosjean, 1992 [pers. comm., J. M. Duncan, Scottish Crops Research Institute], Panabières *et al.*, 1997, and Ristaino *et al.*, 1998). The robustness of the cladogram branching order, as supported by Bootstrap and Jackknife procedures, is not always secure (values <75% should be viewed with caution; for the very much larger angiosperm database, this value could be set at <50%, pers. comm., M. W. Chase, Royal Botanic Garden, Kew).

Genetic relatedness as assessed by ITS1 sequences may indicate centres of speciation but not necessarily the evolutionary phylogeny. The same may also apply to the position of the 5S rDNA relative to the NTS of the rDNA repeat. Complexity between and within genera of both subclasses is found in the arrangement of 5S rRNA sequences; both tandem and inverted orientations are known, but both the Verrucalvaceae (Sclerosporales) and the filamentous-sporangiate *Pythium* species have the inverted orientation (Belkhiri, Buchko and Klassen (1992). Alignments in other unresolved regions may be problematic elsewhere in the entire rDNA gene.

The deep phylogenetic divide between the Peronosporomycetidae and the Saprolegniomycetidae within the Peronosporomycetes (de Bary, 1866; Dick *et al.*, 1984) has now been confirmed with 18S rDNA data (Dick *et al.*, 1999) and 28S rDNA data (Riethmüller *et al.*, 1999; Petersen and Rosendahl, 2000). This divide is supported by the mitochondrially encoded cytochrome oxidase (*cox* II) data of Hudspeth *et al.* (2000); see also Cook *et al.* (2001) for further support from their comparable study of *Lagenidium* and marine taxa. However, the placement of *Sapromyces* (Rhipidiales, Rhipidiomycetidae) may fall either in the Saprolegniomycetidae with the Leptomitales (Petersen and Rosendahl, 2000) or with the Peronosporomycetidae (Hudspeth *et al.*, 2000) depending on the molecular data used. Nevertheless, the Saprolegniomycetidae should be able to provide outgroups for phylogenetic analysis *within* the Peronosporomycetidae and *vice versa*. This will enable comparisons of longer nucleotide sequences, perhaps with additional variable and informative sites, than more distant outgroups. 18S rDNA-characterized type species which could be utilized now include: *Saprolegnia ferax*, *Leptolegnia caudata* and *Apodachlya brachynema* (Saprolegniomycetidae) and *Pythium monospermum* (Peronosporomycetidae).

The mitochondrial genome in straminipilous organisms is characteristically large. In some taxa the mitochondrial genomes are linear (Martin, 1995*b*) but this feature is probably not of phylogenetic importance. For straminipilous fungal genera such as *Pythium* and *Achlya* (i.e., in both subclasses), there is an inverted repeat in the mitochondrial genome (McNabb *et al.*, 1987). The length of this inverted repeat in *Pythium* (27-29 kilobase pairs) is quite different from that in *Achlya* (10 kb) or mycote fungal mitochondria (4-5 kb), but it is very similar to that in chloroplasts (20-28 kb) (Whitfield and Bottomley, 1983). *Phytophthora* lacks this inverted repeat although it is present in the Rhipidiales (McNabb *et al.*, 1987; McNabb and Klassen, 1988).

No complete 18S rDNA sequence has yet been published for any DM. The total molecular biological database for DMs is still fragmentary. Grosjean (1992), using ITS1, placed *Peronospora viciae* with *Phytophthora infestans*, and *Albugo candida* with *Pythium insidiosum* and *P. echinulatum*; both placements were at ultimate branches of the cladogram. Hudspeth (pers. comm., D. S. S. Hudspeth, Northern Illinois University, DeKalb), using mitochondrial *cox* II sequences, placed *Peronospora tabacina* and *Pe. nicotianae* with *Phytophthora megasperma*, also at ultimate branch points; in contrast, *Albugo candida* was basal to their phylogenetic tree, along with *Hyphochytrium* and *Sapromyces*. Cooke *et al.* (this volume) again using ITS sequences, have placed a small sample of *Peronospora* species parasitic in the Rosids and Asterids (see Dick, this volume, and Figure 1) with an intermediate branch of *Phytophthora* which includes *Ph. infestans* (the type species), *Ph. nicotianae* and *Ph. megakarya*. It is noteworthy that there is a measure of agreement with the molecular biological conclusions of Grosjean (1992) and Hudspeth (pers. comm., D. S. S. Hudspeth, Northern Illinois University, DeKalb) with respect to this particular group of *Peronospora* species with a similarly restricted group of *Phytophthora* species. On the other hand, there is no agreement concerning the placement of *Albugo*, although unpublished work generally places *Albugo* at some distance from *Phytophthora* and *Peronospora* (see below). However, when *Peronospora rumicis* (the type species, Corda, 1837) has been studied and shown to belong to this group, then the genus *Phytophthora sensu stricto* (type species the infamous *Ph. infestans*, de Bary, 1876) would become a synonym of *Peronospora*, heralding a nomenclatural nightmare! One solution might be to set the monophyletic generic concepts at a very low heirarchical level so that neither *Phytophthora* nor *Peronospora* would need to be abandoned, but this would necessitate the simultaneous erection of numerous other genera from *Phytophthora sensu lato* and *Pythium sensu lato*. Data from Cook *et al.* (2001) suggest that *Phytophthora sensu lato* is similarly nested within *Pythium sensu lato* with *Lagenidium*.

A further consideration must be the incongruence between the geographic evolutionary origins of *Peronospora* (Asia Minor? - see below) and the *Phytophthora* species listed above, most of which are thought to be of American origin. The other genera of the Peronosporaceae have yet to be investigated. Obviously, these data provide a totally inadequate framework for a robust DM phylogeny.

The genera *Phytophthora sensu lato* and *Pythium sensu lato* need to be retained, even though they may be paraphyletic, until a concensus of relationships has been established. Physiological traits and patterns of mycoparasitism (Pemberton *et al.*, 1990; Dick, 2000*c*) could provide independent supra-specific correlates. A meticulous taxonomic reassessment of historic generic names, and their type species, which have been placed in synonymy with *Phytophthora sensu lato* will be required (see Table 3). Stamps *et al.* (1990) separated *Phytophthora* into six morphological groups (identified by Roman numerals, see below) but molecular studies have not entirely endorsed this division (compare Cooke *et al.*, 1996, 1999; Ristaino *et al.*, 1998). Dick (in Klassen, McNabb and Dick, 1987; Dick, 1990*b*: Venn diagram) suggested that there were perhaps five major centres of speciation within *Pythium*, based on morphological criteria, and exemplified by (1) *P. monospermum*, *P. torulosum* and *P. diclinum*, (2) *P. anandrum*, (3) *P. ultimum*, (4) *P. irregulare* and (5) *P. ostracodes* and *P. oedochilum*: all of these groups are supported by deep clades in the ITS data of Grosjean (1992), but again, these data should not be regarded as sufficiently robust at this stage.

The symplesiomorphic trait in the *Phytophthora* line that gave rise to the apomorphies of the genera of the Peronosporales is yet to be defined. To put nomenclatural order into the phylogenetic classification of the Peronosporales, it will eventually be necessary to make a large number of name changes at genus and family levels using comprehensive

TABLE 3. The early chronology of taxonomic and plant-pathogenic studies of the downy mildews and related taxa up to Fitzpatrick (1930). *Sclerophthora* Thirumalachar, Shaw & Narasimhan (type species *S. macrospora*) was described in 1953.

date   fungus (only generic authorities given), citation and brief annotation

1807   *Albugo* zoospores described by Prévost (see Ainsworth, 1976: 62)

1821   *Albugo* (Pers.) Roussel (Gray, 1921: type species *A. candida*, and 2 other species)

1833   *Botrytis pygmaea* [type species of *Plasmopara*] named (Unger, 1833)

1837   *Peronospora* Corda (Corda, 1837: type species *P. rumicis* [holotype]) in Amaranthaceae

1843   *Bremia* Regel (Regel, 1843: type species *B. lactucae* [holotype]) in Asteraceae

1845   *Botrytis infestans* Mont. (Montagne, 1845) described as the causal agent of Late Blight of Potatoes

1845   Irish Famine - caused by Late Blight of Potatoes (symptoms also known as 'Potato Murrain'; 'Curl'; 'Rot') [N.B.: after *ca* 50 years of similar symptom reports; also of well-known occurrence in rainy years in Bogota (Boussingault, 1845)]

1846   Berkeley: "The decay [caused by Late Blight of Potatoes] is the consequence of the mould, and not the mould of the decay." (*contra* Lindley - see Smith, 1884; Large, 1940)

1851   *Peronospora pygmaea* [type species of *Plasmopara*] described with other species of *Peronospora* (Unger, 1847)

1851   *Botrytis viticola* [*Plasmopara*] described (Berkeley, 1851)

1858   *Pythium* Pringsh. described (Pringsheim, 1858; type species *P. monospermum*, with other species, all of which have subsequently been transferred to other genera)

1863   sexual reproduction in DMs described (de Bary, 1863): [modern nomenclature] *Bremia* (1 sp.), *Paraperonospora* (1 sp.), *Peronospora* (32 spp.), *Plasmopara* (5 spp.), *Albugo* (6 spp.), *Phytophthora* (1 sp) [the first synopsis of the classification of the DMs; still the most important comparative morphological account]

1866   the Peronosporeen and Saprolegnieen separated as the two major groups of biflagellate fungi (de Bary, 1863, 1866, 1887)

1869   *Basidiophora* Roze & Cornu described (Roze & Cornu, 1869; type species *B. entospora* [holotype])

1869   *Cystosiphon* Roze & Cornu described (Roze & Cornu, 1869; type species *C. pythioides* [holotype]; the first valid generic name for spherical-sporangiate *Pythium* species in *Pythium s.l.*)

1869   *Peronospora cubensis* [type species of *Pseudoperonospora*] described (Berkeley and Curtis, 1869: 363)

1876   *Protomyces graminicola* [type species of *Sclerospora*] described (Saccardo, 1876)

1876   *Phytophthora* de Bary described (de Bary, 1876; type species *P. infestans* [holotype] [group IV])

1876   *Plasmopara* [as *Peronospora*] found on vines (Farlow, 1876)

1878   *Plasmopara viticola* [as *Peronospora*] found in Europe (Millardet, 1885, in Schneiderhan, 1933)

1879   *Sclerospora* Schröter described (Schröter, 1879; type species *S. graminicola*)

1884   succinct accounts by Smith (1884) of plant pathology and morphology of DMs: [modern nomenclature] *Peronospora trifoliorum*, *P. destructor*, *P. parasitica*, *Plasmopara umbelliferarum*, *Bremia lactucae*, *Albugo candida*, *Phytophthora infestans*

1885   Bordeaux Mixture described, effective on *Plasmopara viticola* (Millardet, 1885, in Schneiderhan, 1933)

1886   *Plasmopara* Schroeter described (Schroeter, 1886; type species *P. pygmaea*)

1893   order Peronosporales described (Schröter, 1893)

1898   *Peronospora megasperma* [type species of *Bremiella*] described (Berlese, 1898)

TABLE 3, *continued.*

1899-1901 cytology of DM oosporogenesis by Stevens and others (summarized with illustrations in Lotsy, 1907, see Dick & Win-Tin, 1973)

1902   subgenus *Peronoplasmopara* Berl. described (Berlese, 1897-1902, type species *P. cubensis*)

1903   Cucurbit DM disease in Europe

1903   *Kawakamia* Miyabe described (Miyabe and Kawakami, 1903; type species *K. [Phytophthora s.l.] cyperi* [holotype] [group III])

1903   *Pseudoperonospora* Rostovsev described (Rostowzow, 1903; type species *P. cubensis*)

1905   *Peronoplasmopara* (Berl.) G. P. Clinton described (Clinton, 1905; type species *P. cubensis*)

1906   *Phloeophthora* Kleb. described (Klebahn, 1905; type species *P. [Phytophthora s.l.] syringae* [holotype] [group III])

1907   *Rhysotheca* G. W. Wilson described (Wilson, 1907; type species *R. umbelliferarum*, *Plasmopara* remaining monotypic)

1909   *Pythiomorpha* H. E. Petersen described (Petersen, 1909; type species *P. [Phytophthora s.l.] gonapodyides* [holotype] [group VI])

1913   *Nozemia* Pethybr. described (Pethybridge, 1913; type species *Nozemia (Phytophthora s.l.) cactorum* [group I])

1913   family Phytophthoraceae described (Pethybridge, 1913)

1913   subgenus *Peronosclerospora* S. Ito described (Ito, 1913; type species *P. sacchari*)

1914   *Bremiella* G. W. Wilson described (Wilson, 1914; type species *B. megasperma* [holotype])

1915   *Rheosporangium* Edson described (Edson, 1915; type species *R. aphanidermatum* [holotype])

1921   Jarrah dieback disease first noted (*Phytophthora cinnamomi* [group VI] on *Eucalyptus marginata*) (Podger, 1972)

1922   *Pseudoplasmopara* Sawada described (Sawada, 1922, type species *P. justiciae* [holotype])

1923   monograph on *Peronospora* by Gäumann (1923) - 243 species considered

1927   *Peronosclerospora* Hara raised to generic rank by Hara (Shirai & Hara, 1927; type species *P. sacchari*)

---

molecular, biochemical and morphological criteria. For example, this approach will be necessary for *Phytophthora undulata* (Dick, 1989; Mugnier and Grosjean, 1995) which is neither a *Phytophthora sensu stricto* nor a *Pythium sensu stricto*, while the *Pythium vexans* group almost certainly belongs to *Phytophthora sensu lato*, (cf. Dick, 1990*b*; Panabières *et al.*, 1997).

Other genes which might be suitable for providing data on the deeper phylogenies of the Peronosporomycetidae are actin coding regions (Hightower and Meagher, 1986; Dudler, 1990; Bhattacharya and Ehlting, 1995; Uncles *et al.*, 1997) and DNA dependent RNA polymerases (Klenk, Palm and Zillig, 1994). Complications arising from the use of actin gene sequences may arise because of gene duplication. In *Pythium irregulare* four 'copies' of the gene sequence occur. From sequence analyses, these 'copies' do not always fall in the same clade (one grouped with those for *Phytophthora* species) (pers. comm., F. N. Martin, U.S.D.A., Salinas, Ca.).

A different approach to phylogeny, using comparative DNA-based data (Feulgen Image Analysis) is that of Voglmayr and Greilhuber (1998) who have produced data suggesting that *Peronospora* and *Plasmopara* are probably not closely related (cf. Dick, 1988). Electrophoretic Karyotype (EK) polymorphisms (determined by CHEF analysis) occur in *Pythium* species (Martin, 1995*a*), but only one DM, *Bremia lactucae*, has been assessed (Francis and Michelmore, 1993) so again, evolutionary predictions would be premature. Analyses of EK whch show intra-specific heterozygosity are already known for several species of *Pythium* (pers. comm., F. N. Martin, U.S.D.A., Salinas, Ca.), therefore, a large database will be necessary to establish whether chromosome size and number also show consistent inter-specific differences.

5.1. PHYLOGENETIC TREES

The tacit acceptance (e.g., Sparrow, 1960; Karling, 1981) that obligately parasitic species with limited host ranges can be both closely related and have diverse hosts in widely disparate ecosystems must be rejected (Dick, 2001*c*). Similarly, well-known phylogenetic schemes for the Peronosporales (Shaw, 1981; Barr 1983) have been questioned by Dick *et al.* (1984) and Voglmayr and Greilhuber (1998). The simplistic linear evolutionary classification schemes of Shaw (1981) and Barr (1983), which have been largely based on subjectively selected morphological criteria, should be disregarded. The earlier suggestion (Skalický, 1966) that *Peronospora* and *Plasmopara* represent diverging lines rather than a progression along a single unbranching line from 'more primitive' to 'more advanced' characters is also supported by their different spectra of hosts, geographic centres and climatic zones, different conidiophore morphology and Feulgen analyses, but the points of divergence may have been earlier than the 'Peronosporales' and 'Peronosporaceae' of standard texts (cf. Dick, 1988: fig. 3, which may be a more representative hypothesis than that of Dick, 1990*a*: fig. 6). The currently-used genera of the Peronosporaceae may still be paraphyletic: one group of species may have a larger number of 'primitive' characters as well as other, more conspicuous 'advanced' characters.

FIGURE 1. Molecular phylogeny of angiosperms (after APG, 1998) with a summary of information for *Peronospora, Plasmopara, Albugo* and other downy mildews. A few families were not assigned to orders in APG (1998). The levels of Jackknife support, given in the original publication, have been omitted because they are not germane to this diagram. Unplaced families of no relevance to downy mildew parasitism have been excluded. Families from the Cretaceous include: Caryophyllales: Amaranthaceae; Malpighiales: Euphorbiaceae; Myrtales: Onagraceae; families from the Paleocene include: Solanales: Convolvulaceae; Malvales: Malvaceae; Gentianales: Gentianaceae.

The basal angiosperms, including the palaeoherbs (Ceratophyllales, Piperales), Laurales and Magnoliales have been omitted (but note that one species of *Albugo* occurs on Piperales). Monocotyledons have also been omitted, but note that one species of *Peronospora* occurs on Asparagales; one species of *Bremia* and two species of *Plasmopara* on Poaceae. In the absence of definitive check lists for DM genera, the numbers refer to binomials rather than taxa (see Dick, this volume); therefore there may be some redundancy within and between columns.

From *Straminipilous Fungi: Systematics of the Peronosporomycetes Including Accounts of the Marine Straminipilous Protists, the Plasmodiophorids and Similar Organisms* / by Michael W. Dick. © 2001 Kluwer Academic Publishers, ISBN 0-7923-6780-4. Table III: 3; pp. 128-129.

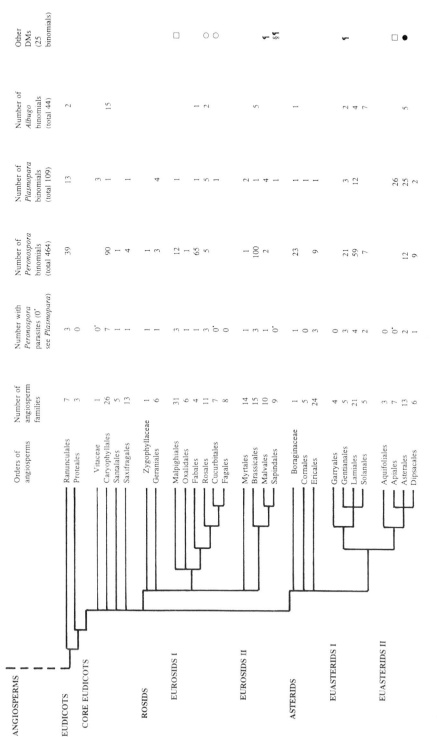

● = *Basidiophora* (2 spp), *Benua* (1 sp), *Bremia* (1 sp), *Paraperonospora* (9 spp);   □ = *Bremiella* (3 spp);   ○ = *Pseudoperonospora* (7 spp);   § = *Peronophythora* (1 sp) in *Litchi* (Sapindaceae)
¶ = *Trachysphaera* (1 sp) in *Theobroma* (Malvaceae, formerly Sterculariaceae), *Citrus* (Rutaceae - Watson, 1971),
Coffea (Rubiaceae) and *Persea* (Lauraceae - Spaulding, 1961) and *Musa* (Musaceae - Kader, 1999)

6. Taxonomic history of the downy mildews

The white blister rust, now known as *Albugo candida*, dating back to the writings of Persoon and observations of Prévost (Ainsworth, 1976) was listed, with two other species, by Gray (1821), although *Albugo* was not recognized as being closely related to the DMs for another 40 years, until the description of sexual reproduction in these genera by de Bary (1863).

The first DM to be formally and acceptably described and diagnosed was *Peronospora rumicis* by Corda (1837) from Poland; Unger (1833) had earlier noted what was to become known as *Plasmopara pygmaea*. Descriptions of the genera *Bremia* (Regel, 1843) and *Basidiophora* (Roze & Cornu, 1869) followed.

Phytophthora, the type species epithet of which was given by Montagne (1845), was not formally described until 1876 (de Bary, 1876), although the symptoms of Late Blight of Potato had been observed since the end of the eighteenth century in Europe (Woodham-Smith, 1962) and South America (Boussingault, 1845), and the disease was particularly prevalent in Europe during the 1840s (Berkeley, 1846; Smith, 1884). These and other early taxonomic landmarks are summarized chronologically in Table 3.

7. The angiosperm hosts: evolution in relation to tectonic movements, life-zones and the origins of major taxa

The hosts of the DMs are *angiosperms* (Dick, 1988: figs 4, 7). How did the broad sweep of angiosperm evolution result in such a restricted selection of angiosperms becoming vulnerable to parasitism by DMs? To answer this question it is necessary to summarize, as briefly as possible, the geological and climatic developments after the evolution of the angiosperm orders (Retallack and Dilcher, 1981*a*; Kenrick and Crane, 1997). This will enable the development of an argument for the coevolutionary development of biotrophic phyto-parasitism of these fungi (Figures 2-4).

7.1. THE CRETACEOUS

Retallack & Dilcher (1981*b*) have postulated that the angiosperms originated in the Gondwanaland rift valley that was eventually to separate the South American and African tectonic plates. At the beginning of the Cretaceous (144 m.y.B.P.), South America and Africa were united, only becoming completely separated, in southern latitudes, in the Maastrichtian, at the very end of the Cretaceous (65 m.y.B.P.), just before the 'Cretaceous Terminal Event'. The situation in southern Asia is palaeontologically obscure because of the India/Asia tectonic collision and Himalayan uplift, which did not occur until much later in the Tertiary, in the late Miocene (5-23 m.y.B.P.) and Pliocene (1.7-5 m.y.B.P.). By the early Cretaceous the angiosperms were merely a minor floristic component, consisting of streamside shrubs and palaeoherbs (Piperales, Nymphaceae and some aquatic Callitrichales). Subsequently, during most of the Cretaceous, the angiosperms contributed mainly to the lower tiers of the high-tiered tropical forest; thus most angiosperm photosynthesis was taking place in *sub-optimal light* conditions. It is apparent from the fossil record that, by the mid Cretaceous, angiosperm radiation and diversity were well-established and that their occurrence was primarily in the wet equatorial (pan-equatorial?) belt (Doyle *et al.*, 1982).

The combination of land masses, mountains and major meteorological systems results in the development of *life-zones*. Parrish (1987) has done much work on the prediction of palaeoclimates, but now these life-zones are being computer-modelled. The models are not yet definitive: in particular, alternating very wet and prolonged drought seasons (savannah country) and uniform high rainfall (tropical forest) may be included in the same computed life-zone (pers. comm., P. V. Valdes, University of Reading). This amalgamation may be an important consideration with respect to the evaluation of the evolution of the panicoid grasses, referred to below. Nevertheless, climatic models for the 50 m.y.B.P. period between the late Jurassic (150 m.y.B.P.) and mid Cretaceous (100 m.y.B.P.) are now available. During this period there was the break-up of the southern hemisphere super-continent, Gondwanaland, into South American, African, Indian, Australasian and Antarctic plates, but very little mountain orogeny was involved.

In contrast, in the northern hemisphere there were numerous barriers to plant dispersal. The changing plates of North America plus Laurentia (Greenland and northern Europe) gave rise to Laurasia (North America plus Europe); Angaraland (the continent east of the Urals), East Asia (Siberia) and South East Asia. Laurasia had the Laurentian (Scottish and Appalachian) and Variscan (Spanish and Moroccan) mountain ranges at this time, subsequently disrupted by the North Atlantic rift, with much of present-day southern Europe covered by epicontinental (continental shelf) seas. The Uralian Ocean disappeared as the eastern part of Laurasia collided with Angaraland, eventually becoming uplifted as the Ural mountains.

As the African plate drifted eastwards and northwards, the more or less equatorial circum-global ocean became bisected to give rise to the Central Atlantic Ocean and the Tethys Ocean. Monsoons would have been absent initially at the start of the Cretaceous (having been present in the Permian and Triassic). But, as the southern plates of Africa/Arabia and India moved north towards the Tropic of Cancer, monsoon climates would gradually have become reestablished, so that, by the late Tertiary, there would have been wet seasons on the southern edges of the northern plates.

Cretaceous sea levels were high, and there were epicontinental seaways north/south across North America and, perhaps intermittently, south-west/north-east across Africa from the nascent South Atlantic Ocean to the Tethys Ocean. Tropical forest probably covered most of northern South America and Africa, with separate tropical forest regions on the southern edges of the plates of East Asia and South East Asia together with island outcrop cover to the southern edge of Laurasia. Similar vegetation may have occurred on island outcrops, dependent on sea level fluctuations, in south western Asia and on the rising cordilleras of the western American plates. Thus, as a result of continental drift, by the end of the Cretaceous there would have been a more or less continuous band of tropical forest (with rain forest in western South America and South East Asia), isolated from the high latitude land by hot deserts on the Tropic of Cancer to the north, and the Tropic of Capricorn to the south. The southern deserts on both the American and African sides of the South Atlantic Ocean isolated the high latitude, warm temperate Gondwanaland flora from the equatorial forest.

26   M. W. DICK

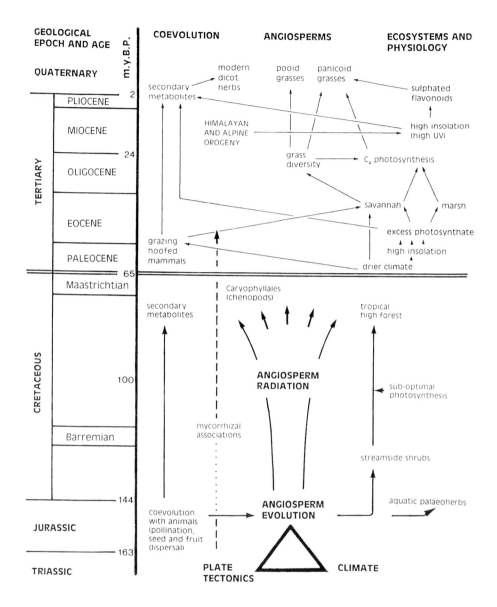

FIGURE 2: Summary diagram of angiosperm evolution through geological periods (e.g., CRETACEOUS), epochs (e.g., EOCENE) and ages (e.g., Maastrichtian), with climatic, ecological, physiological and coevolutionary landmarks (from Dick, 2001c). For further explanation, see text, also refer to Figure 3.

From *Straminipilous Fungi: Systematics of the Peronosporomycetes Including Accounts of the Marine Straminipilous Protists, the Plasmodiophorids and Similar Organisms* / by Michael W. Dick.
© 2001 Kluwer Academic Publishers, ISBN 0-7923-6780-4. Table III: 3; p. 131.

*Only at the end of the Cretaceous period of geological time is it possible to postulate a circum-global continuum of tropical/sub-tropical woody angiosperms and, by inference, pro-*Phytophthora *associations that could have evolved to give rise to the range of* Phytophthora *species and* Phytophthora *species distributions extant today and discussed below* (Figure 4).

The low latitude fossil megafloras are inadequate to support the postulation of this circum-global tropical forest continuum (Upchurch and Wolfe, 1987) but it is known that the mean global temperatures increased during the Maastrichtian, at the very end of the Cretaceous. The Cretaceous epoch ended with the 'Cretaceous Terminal Event', resulting in the extinction of 30% of the land plants, and rather higher percentages of some marine taxa such as diatoms (60% became extinct). Refugia on land were presumably able to provide for the biodiversity from the start of the Tertiary (65 m.y.B.P.).

7.2. THE TERTIARY

The early Eocene was probably warmer than at any time during Cretaceous (Wolfe and Upchurch, 1987). However, the Tertiary was to become increasingly cooler and drier, with a lowering of sea levels. Continental shelf was exposed for terrestrial colonization and the epicontinental seaways drained; a development which added to the drier climatic conditions of eastern North America and north-west Africa. The south Laurasian (Appalachian and Variscan) element of montane tropical forest was probably in rapid decline due to the cooling global climates and the loss of moisture as the epicontinental seaway of central North America drained and the Atlantic Ocean opened up. Further movement, this time northward, of the plates of Africa and Arabia and the Indian subcontinent reduced the Tethys Ocean to the Mediterranean Sea.

The first junction between North American and South American plates was via the Panamanian isthmus of ancient tectonic plates (Precambrian) and the Andean/Rockies cordilleras. This connection only became established in the late Tertiary (*ca* 20 m.y.B.P.), although there may have been volcanic island links earlier through the eastern Caribbean (present-day Jamaica/Haiti).

In the early Tertiary (Eocene) two climatic zones have been recognized throughout Eurasia (Takhtajan, 1969). These were a northern *temperate* zone and a southern *sub-tropical* zone (northern limits: southern Britain, Belgium, southern Baltic, south Urals, Kazakhstan, Korea, Honshu) dominated, in Europe, by a forest flora of Lauraceae (consider the genus *Cinnamomum* and the broad host-spectrum *Phytophthora cinnamomi*) and Fagaceae (with its host-restricted pathogens *Phytophthora quercina* on *Quercus* and *Ph. fagi* on *Fagus*), no doubt with appropriate root-associated fungi. In the Palaeocene the nature of the flora of Europe was sub-tropical due to maritime influences from the Indo-Tethys Ocean, the North Atlantic palaeogulf stream and the Ob Sea (the remnant of the Uralian Ocean separating the European from the Asian plates). Many of the sub-tropical families subsequently noted as minor host families for the DMs have been recorded from the late Eocene Baltic ambers, such as Apocynaceae, Cistaceae, Euphorbiaceae, Geraniaceae, Linaceae, Oxalidaceae and Rubiaceae (refer to Table 1 and the angiosperm-host classification for *Peronospora* in Dick, this volume). In the

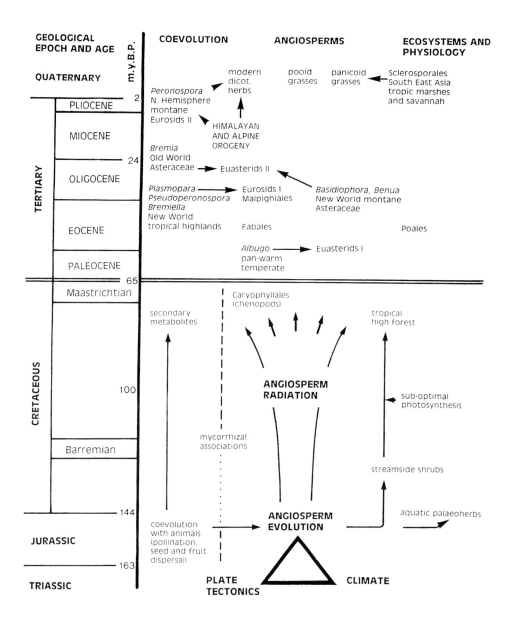

FIGURE 3: The summary diagram, Figure 2, substituted with the suggested downy mildew origins during the Tertiary. Figures 2 and 3 may be overlaid.

southern Urals, south-east Europe and Asia Minor the Eocene floras tend to be more xerophytic. Although the temperate flora extended southwards at the expense of the sub-tropical flora, all the regions named above from eastern Europe and western Asia lie *within* the northern boundary of the sub-tropical floristic zone (Takhtajan, 1969: fig. 30). These drier environmental influences resulted in the proliferation of endemics (see below) in these peri-montane parts of the temperate and sub-tropical vegetation boundary of the early Tertiary (Eocene).

The radiation of the angiosperms, which started in the Cretaceous, gradually resulted in the accumulation of morphological characters by which modern angiosperm orders can be recognized. *The pan-equatorial continuum of woody angiosperms and their possible pro-*Phytophthora *associates began to disintegrate and re-form as provincial communities.* Isolated, provincial communities, recovering from the Cretaceous Terminal Event, could evolve independently, allowing for biodiversity in hosts, parasites and host-parasite relationships: speciation would be expected to show adaptive radiation. The probable time of origin, based on palynological and fossil evidence, for orders of interest from the point of view of the DM parasites are given in Dick (1988: fig. 7).

By far the majority of named species in *Peronospora* are found in four clusters of angiosperm orders (refer to Soltis, Soltis and Chase *et al.*, 1998*a, b*; APG, 1998): Ranunculales and Caryophyllales; Eurosids I (especially Fabales); Eurosids II (especially Brassicales) and the Euasterids (especially various orders in Euasterids I) (see Dick, 2001*c* and elsewhere in this volume). It is probably no coincidence that Takhtajan's floristic regions 2 and 8 within the sub-tropical zone correspond to the main centres of distribution of these orders and families. In particular, the sub-region of group 8, the Armeno-Iranian region, is noted for many endemics in the families Amaranthaceae (synonym Chenopodiaceae), Caryophyllaceae, Brassicaceae, Rosaceae, Fabaceae, Zygophyllaceae, Scrophulariaceae, Lamiaceae, Campanulaceae, Asteraceae and Alliaceae. Floristically this sub-region stretches from central Turkey, around the Caspian sea and skirts the northern flanks of the Himalayan massif; it continues as the southern border of the Euro-Siberian floristic zone, which extends westwards though the Balkan mountains, the Alps and the northern Appennines and eastwards to the Kuril Islands and the Kamschatka peninsula. At this most eastern end there is continuity from the Kuril islands into Hokkaido and the other Japanese islands. Takhtajan (1969) also notes the floristic similarities between the Euro-Siberian and Canadian-Appalachian (4a) regions.

The tropical/sub tropical forest with a closed canopy gave way to more open vegetation. The circum-Mediterranean climate became 'mediterranean'. The mountain-building activity changed the topography and climate. The angiosperms, previously adapted to *sub-optimal light* conditions; *high ambient* temperatures and *high* humidity, were now exposed to *greater insolation*; *fluctuating diurnal* temperatures and *variable* (extreme seasonal) humidity. From trees, which had evolved in sub-optimum light, climatic pressures would have encouraged herbaceous development in open canopy: exposure to high levels of UV irradiation would have been deleterious; the high photosynthetic activities would have resulted in excess photosynthate. The consequent development of secondary metabolites from both of these causes would have produced further ramifications of the angiosperm/animal coevolution. Was this also a stimulus which enabled the straminipilous fungi, previously adapted to high protein/hydrocarbon/carbohydrate nutrition, to colonize roots?

*Hot by day, cool at night, under a clear sky, dew forms on leaves. With caducous sporangia, pro-*Phytophthora *taxa could reach the leaves. For advanced herbs,* Phytophthora *arrived.*

8. Angiosperm coevolution with animals and fungi

Animal nutrition forms an essential corollary to this hypothesis of DM coevolution with angiosperms. The coevolution of angiosperms with animals through seed dispersal, pollination and grazing (see Hughes, 1973, 1976) is intimately linked to the evolutionary development of secondary metabolites by angiosperms; species/specific relationships are well-known. Similarly, fungal/root symbioses should be considered.

8.1. FRUIT AND SEED DISPERSAL BY ANIMALS

Coevolution between animals and primary producers relating to fruits and seeds can perhaps be dated as early as the Permian with the evolution of the fleshy, butyraceous fruit of *Ginkgo*, the shape of which has remained unchanged throughout the fossil record. Although coevolved fruit and seed dispersal predates the origin of the angiosperms, the inherent high carbohydrate/protein factors apparently had no influence for any potential coevolution of DM precursors with 'pre-angiosperms'.

8.2. INSECT POLLINATION

Before the advent of the angiosperms most insects had biting mouthparts (Crepet and Friis, 1987). Adaptations to the use of fleshy plant parts (plant sap, and fruits) and exudates (nectar with pollen) were not common. However, angiospermy (the enclosure of the ovule, first occurring in the Bennettitales) is seen as a product of insect grazing in the Triassic. By the Barremian (late early-Cretaceous) differences in angiosperm pollen morphology provide evidence of provincialism in angiosperms and their different pollination mechanisms: pollen from north of the Tethys Ocean ('Europe') had a reticulate and columellate exine, whereas 'southern' pollen ('Africa & South America') was predominantly tectate and granular. It is from the South American and African plates that angiosperms with the potential for becoming parasitized by pre-DM fungi have had their origin. Pollen from the Cretaceous is known for the Caryophyllales (Amaranthaceae), Malpighiales (Euphorbiaceae) and Myrtales (Onagraceae) and early Tertiary pollen is known for the Malvales and Solanales (Convolvulaceae) (Muller, 1981, in Dick, 1988). These are all significant families for the DMs (see Figure 1).

8.3. GRAZING

The pasture grasses are the most prominent components of the floras of savannah and prairie. It is in these grasses (as compared with the earlier-evolved bamboos) that the graminicolous DMs are found. The loss of forest followed by grassland development in climates otherwise suitable for forest climax are considered to have arisen as the herds of herbivorous hoofed mammals coevolved with the pasture grasses (Clayton and Renvoize, 1986; Jacobs, Kingston and Jacobs, 1999). This phenomenon began in South America in the Tertiary (Eocene-Oligocene boundary) crossing to North America in the early Miocene as uplift eliminated the American seaway; later it was to sweep through Europe to Asia. Land bridges or island chains enabling immigrant grazing mammals to colonize Africa from Europe via southern Asia and western India (Coryndon & Savage, 1973) only began to appear in the Miocene. Animal *emigration* was not prevalent, except perhaps from east Africa to India. The Central Atlantic and Tethys Oceans (including the Mediterranean Sea) were also a barrier to north-south migration, at least for larger animals (Coryndon and Savage, 1973), but perhaps not for angiosperms between West Africa and Spain. Australasia was isolated, drifting northwards with a previously warm-temperate-adapted biota. It was only from the Oligocene that land bridges permitted tropical animals and plants to migrate southward, a migration which continued through the Quaternary. All groups of herbaceous plants, exposed to high UV and photosynthetic wavelengths, have developed protective mechanisms based on secondary metabolite production. It is this series of events that could also have determined the evolution of the graminicolous DMs in south east Asia, with species radiating into Australasia and Africa.

8.4. MYCORRHIZAL ASSOCIATIONS

Most, but not all, angiosperms have mycorrhizal associations which enable more efficient absorption of phosphorus and/or nitrogen. There is a concommitant redress to imbalance in C:N ratios, and possible reduction in secondary metabolite production. From fossil evidence, vascular plant axis-fungal associations are known to have existed at least from the Devonian (*Asteroxylon-Glomus*-type inclusions; Pirozynski, 1976a, b). Such associations would have gained importance under savannah climates because mycorrhizal associations are beneficial in climates producing increased water stress and depletion of nitrogen and phosphorus from the soil in storms and flash floods. Symbiotic associations other than with fungi also evolved, such as *Rhizobium*-root nodules, and the increased nitrogenous uptake may have contributed to the cyano-compounds characteristic of the Fabaceae, a family that showed increasing diversity during the drier savannah conditions of the Eocene. Host-root associations (mycorrhizas or weak root parasites) might so affect the well-being of the host as to make a significant change (either way) to its vulnerability to parasitism of *aerial* parts. In experimental studies of biological control (Elad and Chet, 1987), induced root communities have been explored within root systems. Communities on and in roots do not appear to have been evaluated in connection with the metabolism of the whole angiosperm plant in different growing conditions or with DM parasitism of the aerial shoot. Lack of data for whole-plant pathological communities is a serious gap in knowledge.

For present-day angiosperms, mycorrhizal associations are most researched for grassland and forest. Information on mycorrhizal associations in wetland species is mainly confined to ericoid mycorrhiza, which may not be relevant to coevolution of the

Peronosporomycetes, since few Ericales are hosts to these fungi. One interesting conjunction, which needs more substantiation, is that between ruderal herbs (e.g., Brassicaceae and Caryophyllaceae) and DMs. Ruderal herbs lack mycorrhizas (Gerdemann, 1968). It is interesting to speculate that perhaps the absence of mycorrhizal associations alters the flow of metabolites, and perhaps induces the manufacture of certain secondary metabolites, so that an accumulation of essential oils, alkaloids, and saponins is facilitated. For most organisms, including mycote fungi, these substances are antagonistic or repellent. For a few organisms (straminipilous fungi?) they may be attractants, or serve as unexpected intermediates to essential metabolism. The wetland Poaceae could be similar to ruderals in lacking mycorrhiza (but apparently this has not been researched - pers. comm., D. H. Lewis, University of Sheffield), and the pattern of habitat ranges (predominantly *not* mesic) of the panicoid grasses from marshes through saturation/drought-alternating savannah to semi-arid/rocky habitats could be consistent with such an hypothesis.

9. Evolution of the biotrophic habit: nutrition and biochemistry

Coevolution is often taken to mean co-phylogenetic evolution (e.g., Crute, 1981) but co-phylogenetic coevolution is seldom found. Reference to Figure 1 shows that *Peronospora* has been found on relatively few of the angiosperm families that are now grouped into orders on the basis of molecular data (APG, 1998). Even within the angiosperm orders there appear to be highly clustered centres of speciation, although it must be remembered that species-specific nomenclature has greatly exaggerated this skewness. The fact that the widest spectrum of angiosperm hosts is parasitized by the morphologically 'advanced' (i.e., lacking zoospores) genus *Peronospora* is not incongruous if the 'attraction' is by biochemical stimuli. For species-specific parasitism the more intensively studied phenomena of suppression and inhibition are likely to be less important than the elements of attraction and growth stimulation.

My hypothesis is that the propensity of Peronosporomycetes for the products of angiosperm secondary metabolism may be, simultaneously, the most important stimulus and constraint for their evolution. Secondary metabolite production may have been stimulated by high carbohydrate levels or C:N ratios that would otherwise be deleterious to the well-being of the angiosperm. Secondary biochemical pathways are thought by Ross and Sombrero (1991) to have evolved to remove excess photosynthate from the cell systems by producing osmotically inactive, secondary metabolites. Angiosperms may have evolved some of the secondary metabolites initially to provide a UV screening function, others subsequently acquired a protective function against animal predators, with the result that the web of angiosperm-insect-vertebrate coevolution became increasingly complex. It is immaterial whether high insolation, stress, or root system associations induced the developments of the secondary metabolic pathways. The interaction between the physiological activities of leaves, shoots, and roots upon secondary metabolite production and translocation could have a marked effect in creating conditions suitable for disease establishment. In particular, the distribution of metabolites in clones of a host that have been grown in different communities, especially root rhizosphere and hyphosphere communities, has rarely been examined (but see Paxton, 1983). Mycorrhizal associations apart, the nitrogen-fixing potential of the Fabaceae would have led to a different secondary metabolism involving proteins and cyano-compounds. Distinct sulphur-related metabolic pathways in *Allium* could similarly explain this otherwise anomalous host of *Peronospora destructor*. From the coevolutionary point of view, genotypic variation is less important than the community structure. The equilibrium between the composition and stability of the community with environmental pressures to produce secondary metabolites, allowing their accumulation in potential target organs, would be crucial. The exogenous organic compounds essential for the growth and reproduction of DMs (and other straminipilous fungi) and the complementary integration of these substances with the unique biochemical pathways of the straminipilous fungi have yet to be fully investigated.

Although vitamin requirements by Peronosporomycetidae are well-known (e.g., Hohl, 1983), auxotrophs can usually survive on the low concentrations normally leaked from autotrophs. Vitamin requirements are unlikely to be determining factors for obligate parasitism.

The study of the physiology of host-parasite relationships has not so far yielded hints as to the coevolution of the species-specific relationship. The problem with nutritional studies (cf. Hohl, 1983) is that they have to take place in artificial environments. Conditions conducive to maximum and prolonged vegetative growth may not be the most advantageous (i.e., fittest or most competitive) in the natural environment. Even for *Phytophthora* no sound case has yet been presented to suggest that particular nutrient requirements are *essential* prerequisites for successful attack.

Most of the DMs are leaf and herbaceous stem parasites (exceptions are two species of *Plasmopara*, *P. halstedii* and *P. lactucae-radicis*, which are root parasites; Stanghellini, Adaskaveg and Rasmussen, 1990). Photosynthesis may therefore be an important physiological influence. Diversifications from the C_3 metabolism (C_4 and Crassulacean Acid Metabolism, CAM) are common; they are assumed to have arisen frequently and independently in different angiosperm orders, but the charting of their systematic distributions (Moore, 1982; Harborne, 1988) is still in its infancy and does not appear to have been recently reviewed. It is probable that both carbon intermediates and ionic balances differ between C_4 and CAM. The availability of organic carbon to parasites may be different as well. If the angiosperm ordinal distributions of C_4 and CAM are compared, and also compared with phytoalexin distribution (Dick, 1988: fig. 5), the only host orders producing phytoalexins, but for which no C_4 or CAM has been reported, are in the more advanced orders of the Eurosids (Fabales, Rosales, including Urticaceae; Malvales) and Euasterids (Apiales). The Caryophyllales and Malpighiales (Euphorbiaceae), and probably the Asterales and Asparagales, are orders in which C_4, CAM and phytoalexin production are common. The Poales, and the panicoid grasses in particular (graminicolous DM hosts), are noted for C_4 metabolism and flavonoids. A C_4 metabolic relationship has coevolutionary potential for *Albugo* and *Peronospora* which, on molecular phylogeny, appear to be widely divergent although their host family profiles are very similar (Figure 1). However, the association of C_4 orders with *Plasmopara* is poor, and it is non-existent with *Pseudoperonospora*. For *Plasmopara* an association is possible with CAM in the Eurosids and, similarly, for *Pseudoperonospora* with the Cucurbitales and Rosales.   All the circumstantial evidence points to a biochemical, rather than a co-phylogenetic coevolution.

It is generally accepted that genetic lesions resulting in loss of biological function are unlikely to be restored in full, or in the same way.   This is particularly true for mutations, less so for transposons.   Genetic lesions that lead to absolute requirements for particular organic chemicals are known to occur among races of a parasite (e.g., *Phytophthora infestans*; Hohl, 1983).   Genotypic differences in the host production of secondary metabolites may contribute to the resistance or susceptibility of the individual, clone or cultivar.   An example of such intraspecific differences has been shown in relation to catechol production by *Allium* and its effect on *Colletotrichum* (Walker and Stahmann, 1955).   Perhaps a similar hypothesis could be tested for parasitism by *Peronospora destructor*.   Cantino (1955), Gleason (1976) and Hohl (1983) have reviewed mineral nutritional aspects of the Peronosporomycetidae.   Cantino's hypothesis that the loss of ability to utilize certain inorganic forms of nitrogen and sulphur is still accepted (Peronosporomycetidae: ability to use $SO_4^=$, ability to metabolize different inorganic N sources variable; Saprolegniomycetidae: inability to utilize $SO_4^=$, NO_3^-).   These physiological criteria correlate with the morphological criteria in classifications (compare Table 1).

The dichotomy of the DAP (α, ϵ-diaminopimelic acid) and AAA (α-aminoadipic acid) pathways of lysine synthesis (Vogel, 1964) is regarded as of phylogenetic importance. Unlike most fungal parasite-angiosperm host systems, both the angiosperm hosts and their straminipilous parasites have the same (DAP) lysine synthesis pathway.   Lysine is a precursor for many polyamines.   Galston (1983) linked polyamine metabolism with the metabolism of growth substances such as gibberellins and cytokinins.   The stunting (hypoplasia) of sunflowers caused by *Plasmopara* should be recalled (Sackston, 1981). While most attention has been given to secondary metabolites such as flavonoids and alkaloids, Galston regarded polyamines as neglected but probably important secondary metabolites.   He noted that the polyamine putrescine is probably to be found in all cells: could this be the cause of the characteristic odour (see Smith, 1884: 281) of rot caused by *Phytophthora*?   Other polyamines may help to control the nuclear cycle.   Galston also reported a link between CAM and polyamines mediating cellular pH stasis.   Perhaps the polyamines could be another component of the 'metabolic packages' for biotrophic parasitism?

FIGURE 4: Cretaceous drift map, 100 ± 10 m.y.B.P. with present-day continental outlines, from Dick (2001c), to show the probable limits of the tropical woody angiosperm flora (shaded black: note the island chains connecting Angaraland with southern Laurasia and North Africa respectively) and the proposed possible origins (open stars) of the dicotyledonicolous downy mildew genera *Peronospora* with *Paraperonospora* and *Bremia* (Miocene in Eurasia); *Plasmopara*, perhaps with *Pseudoperonospora* and *Bremiella*, (early Tertiary in South America); *Basidiophora* and *Benua* (Oligocene in Central America); the monocotyledonicolous (graminicolous) downy mildews, order Sclerosporales, in Asia and the family Verrucalvaceae in south-east Asia (Pliocene/Quaternary), together with the family Albuginaceae (Maastrichtian/Paleocene in western North Africa).   See text for further explanation.

From *Straminipilous Fungi: Systematics of the Peronosporomycetes Including Accounts of the Marine Straminipilous Protists, the Plasmodiophorids and Similar Organisms* / by Michael W. Dick. © 2001 Kluwer Academic Publishers, ISBN 0-7923-6780-4. Table III: 3; p. 149.

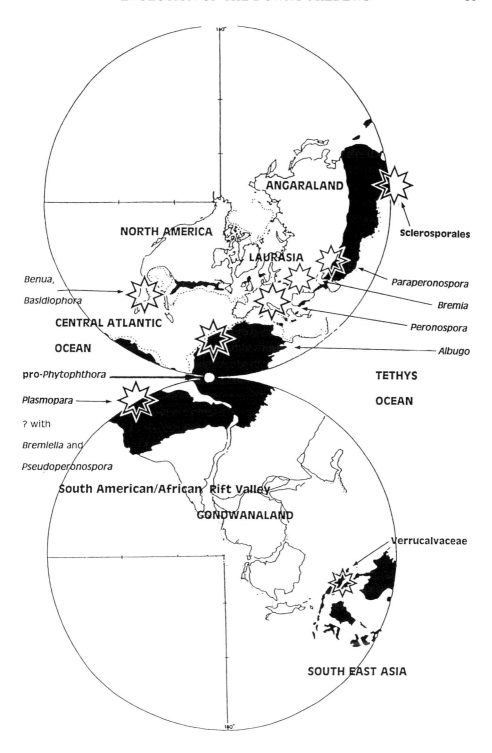

There are ample data to support the assumption that most of the Peronosporomycetidae have a dependence for the intermediates of sterol metabolism (Warner, Sovocool and Domnas, 1982; Elliott, 1983). High carbohydrate levels are conducive to the production of secondary metabolites (sterols, essential oils, flavonoids from phenylalanine; alkaloids from carotenoid precursors). Since the protoplasms of an haustorial obligate parasite and a host are never in contact, there must be a dynamic equilibrium between the components of the plasmamembranes to control the differential permeabilities. The metabolism of component phospholipids and sterols of the plasma membranes could obviously be involved (and could provide a role for haustoria; see Spencer-Phillips, 1997).

There seems to be a connection between phytoalexin phenomena, the sterol requirements of the parasitic species, and the secondary metabolites characteristic of the host order. The key could perhaps be an intense sterol-precursor requirement by the straminipilous fungus that can be satisfied by certain hosts which, for various reasons, have accumulations of a variety of secondary metabolites such as essential oils, saponins, and alkaloids. The kind of photosynthetic pathway (C_4, as in sugar cane and sweet corn), or high insolation, or absence of a mycorrhizal carbon 'sink' could all individually or collectively give rise to this accumulation. The phytoalexin theory originated from studies of resistance to *Ph. infestans* in potato. The parasite/host interaction, which results in the production of rishitin, involves an induced shift in the biochemical pathway normally leading to steroid glycoalkaloids in the potato (Friend and Threlfall, 1976). It could also involve diversion of sterols to the parasite, which probably requires an exogenous source of these substances. Investigations with *Phytophthora* (Hohl, 1983) have failed to reveal significant changes in sterol concentration or composition in relation to intra-specific variations in resistance or virulence. Stössel and Hohl (1981) found that high sterol ratios tend to correlate with susceptibility; this would be expected if host sterols were crucial to the evolution of the parasitic relationship. If this is so, the biochemical diversity of the phytoalexins elicited in Fabaceae (isoflavonoids) and Solanaceae (terpenoids) may be explained. The panicoid grasses, noted for C_4 metabolism, also produce sulphated flavonoids and phytoalexins. For recent phytoalexin reviews see Dixon and Piava (1995); Smith (1996) and Ebel (1998). The phytoalexin phenomenon could be regarded as a by-product of parasite-induced diversion of host-steroid metabolism.

Clarke (1983) has also suggested that phytoalexins are of secondary importance in hypersensitive reactions. Discussions of differences in phytoalexin elicitation reflect abnormalities and deviations in the coevolutionary relationship. In natural populations, phytoalexin elicitation may confer a biological advantage *to the parasite* by limiting the extent of parasitism so that perennation of heterogeneous gene pools of both parasite and host is maintained. Only in artificial systems of clonal culture will this mechanism assume undue prominence. Paradoxically, the production of secondary metabolite pathways, evolved to alleviate physiological stress in the host, may have been evolutionarily conserved because they conferred a biological advantage (defence systems) by deterring fungal parasites as well as animal predators. It is noteworthy that many biochemical defence systems, including phytoalexins, are not as efficient as might have been expected, were they *primarily* evolved, under strong selective pressure, for defence.

Sterol biosynthesis is linked through the terpenoids that relate to all these secondary metabolites.

A second steroid metabolism link is with the action of the polyene antibiotics (for early references, see Tsao, 1970). Polyene antibiotics act on Mycota but not on straminipilous fungi. Their selectivity suggests that there is a difference between these two groups of heterotrophs with respect to their membrane-bound sterols, as these antibiotics are thought to function by acting on membrane-bound sterols. The variety of sterols and their precursors which act as the messengers and recognition systems between the organism and its environment is not understood. In terms of response to the anti-straminipile fungicides Hymexazol and Metalaxyl (Kato *et al.*, 1990) the graminicolous DM relatives, the Verrucalvaceae, were relatively resistant to Metalaxyl but their response to Hymexazol was similar to that of *Pythium*, in contrast to the responses of the *Phytophthora* species tested. Ann and Ko (1992) surveyed antibiotic resistance and dependence in *Phytophthora*. Griffith, Davis and Grant (1992) have emphasised the importance of understanding the utilization of differences in metabolism and sterol production and requirements in assessing fungicide efficacy. The actions of sterol-inhibiting fungicides (Fletcher, 1987; Gisi, this volume) are also relevant to coevolution.

There must be, or have been, physical proximity between populations of the photosynthetic host and heterotrophic parasite taxa involved for coevolution to have taken place. Harlan (1976) has discussed, in general, the influence of parasitism on angiosperm evolution. Coevolution may embrace more than two taxa and the populations need not be large, stable, or of equivalent size. However, if the potential parasite is sexually reproducing, there is probably a minimum population size below which a coevolutionary relationship is unlikely to develop. This postulate could be argued on the out-breeding and colonization potentials of the parasite as well as the ability of the host population to develop resistance. Proximity may be of long standing, with a gradually established, balanced dependence. Alternatively, a conjunction may happen suddenly, either when a vulnerable alien host becomes introduced to an environment containing a previously stabilized host-parasite relationship, or when the spore population of a parasite is alien but able to exploit a susceptible but previously inaccessible host (Hijwegen, 1988). There must be physical or biochemical similarities or analogues that enable an appropriate degree of association between previously separated populations. Too great a vulnerability will lead to an unstable and ephemeral relationship. The essence of coevolution is adaptive change in balanced relationships. Chance associations leading to coevolution must be less likely for fungi that are essentially systemic, because there would be less likelihood that either host or parasite would survive long enough to reach reproductive maturity.

10. Coevolution of the downy mildews with angiosperms in the Tertiary

Interacting organisms must be able to come into contact, and there must be sufficient similarity for nutritional requirements to be satisfied. Sometimes, this will be because new hosts are phylogenetically close to former hosts. But phylogenetically distant or unrelated potential hosts may also provide the critical factors for the nutritional environment of the parasite. The factors involved are not necessarily the most evident and the major nutritional requirements, carbon and nitrogen sources, are unlikely to be critical.

The outstanding characteristic of the distribution of the DM parasites of dicotyledons is that it is not primarily the more primitive or ancient orders of angiosperms that are affected. Angiosperms parasitized by DMs are mostly in *highly evolved families*, OR in taxa that may have a propensity to produce *high levels of secondary metabolites*. If the ordinal rather than family level (*contra* Palti and Kenneth, 1981) of the angiosperms is reviewed, a possible pattern emerges. Despite the postulation of aquatic ancestral origins for both the angiosperms and the DMs, the host dicotyledonous angiosperm orders are not noted for their aquatic members. The habitats of these hosts are diverse and an enormous morphological and anatomical range is encompassed. Coevolution of the DMs is not, therefore, specialization from closely associated ancestral stocks.

10.1. A NOTE ON *PHYTOPHTHORA*

Phytophthora and the dicotyledonicolous DMs share a common ancestry but divergent groups of angiosperm hosts. *Phytophthora* possesses a range of species with different degrees of specialization (Erwin and Ribeiro, 1996). Any pattern of distribution with respect to the morphological (Roman numeral) groups of Stamps *et al.* (1990) either with host-range or with restriction to a geographic region is difficult to discern. The widespread species of *Phytophthora*, in terms of both host range and geographic distribution (lists in Erwin and Ribeiro, 1996), include: *Ph. cactorum* [I]; *Ph. nicotianae* [II]; *Ph. citricola* [III]; *Ph. cinnamomi* [VI] and *Ph. drechsleri* [VI]; no such widespread representatives of groups IV or V are recognized. However, *Ph. infestans* [IV] and *Ph. palmivora* [II] are thought to have a central American provenance, augmented by man. The angiosperm families Lauraceae and Fagaceae, prominent in the sub-tropical Eocene flora of Europe, include the genera *Cinnamomum* (much more widespread in the Tertiary) and *Quercus* respectively, with which one may bracket the parasites *Ph. cinnamomi* [VI] and *Ph. quercina* [I]. Several species are apparently more frequently recognized in particular regions of the world, such as India (*Ph. arecae* [II]; *Ph. cajani* [VI]; *Ph. cypari-bulbosi* [III]); south-east Asia (*Ph. colocasiae* [IV], *Ph. richardiae* [VI]); Australasia (*Ph. clandestina* [I]) and northern Europe (*Ph. fragariae* [V]; *Ph. idaei* [I]; *Ph. erythroseptica* [VI]).

Species with variously restricted host-ranges and geographic distributions can be recognized, perhaps usually confined to one angiosperm order (e.g., Solanales: *Ph. infestans*), or a genus, such as *Ph. inflata* [III] on *Ulmus* in North America; *Ph. quininea* [V] on *Chinchona* in north-western South America; *Ph. colocasiae* [IV] on *Colocasia esculenta*, or even a part of a plant, such as *Ph. megakarya* [II] on *Theobroma cacao* pods in west Africa. Other restricted species are apparently anomalous, such as *Ph. ilicis*, restricted to the European *Ilex aquifolium* [IV] introduced and growing in North America and *Ph. lateralis* [II] isolated from native *Chamaecyparis* in western North America, but which is thought to have been introduced from the Mediterranean.

Present-day *Phytophthora* species embrace a wide range of habits, from root parasites to leaf and fruit pathogens; the monotypic genera *Peronophythora* on *Litchi* (Sapindaceae) and *Trachysphaera* on *Theobroma* (Malvaceae [formerly Sterculiaceae]), *Citrus* (Rutaceae, Watson, 1971), *Coffea* (Rubiaceae), *Persea* (Lauraceae) (Spaulding, 1961) and *Musa* (Musaceae, Gingiberales - Kader, 1999) are genera which have been separated from *Phytophthora sensu stricto* on morphological criteria (branched sporangiophore and spiny sporangia respectively) as well as their habit, restricted to fruits, but may be encompassed by *Phytophthora sensu lato*.

Without human intervention, *Phytophthora sensu lato* would still be a mix of widespread-cosmopolitan taxa, with or without wide host ranges, and provincials adapted to much more restricted habitats. The DMs are almost certainly derived from the immediate antecedents of the gene pool for *Phytophthora sensu lato*; possibly even *ancestral* to some present-day *Phytophthora* species.

The position of the marine genus, *Halophytophthora*, associated with mangroves, is rather less certain but in some molecular phylogenies appears to be deeply rooted and doubtfully monophyletic with *Phytophthora sensu lato*. In view of this, it is worth noting that these estuarine and marine littoral vascular plants (*Avicennia* in the Avicenniaceae, Lamiales; and *Rhizophora* in the Rhizophoraceae, Malpighiales) can be traced back as far as the Paleocene (Plaziat, 1995). *Nypa*-like fossils (*affin.* Arecaceae ?) from similar habitats have earlier dates.

10.2. DICOTYLEDONICOLOUS DOWNY MILDEWS

Palti and Kenneth (1981) and Dick (1988, 2001*c*) have noted the restrictions in the ranges of hosts for the Albuginaceae and Peronosporaceae. It is unlikely that it can be a coincidence that *Albugo* and *Peronospora* have similarly restricted ranges of host angiosperm orders. Many of the hosts are mediterranean-climate ruderals from both the Old and New Worlds: it may be pertinent that much of the Mediterranean flora may have had a north-west African origin, and that plants of open disturbed ground often form large, genetically relatively homogeneous populations ideal for the spread of parasites. These highly evolved herbs are especially noted for the production of secondary metabolites, perhaps because they lack vesicular/mycorrhizal (VM) associations (Gerdemann, 1968). For most organisms, including eumycote fungi, these substances are antagonistic or repellent: for organisms such as the straminipilous fungi they may be attractants, or serve as fortuitous intermediates to essential metabolism, as discussed earlier. Many of these metabolites have biosynthetic links with steroid metabolism, and steroid dependence is a characteristic feature of the Peronosporales. Phytochemical research on the hosts of these parasites should provide the co-evolutionary link that is being sought. Attempts to fit a pattern of coevolution based on angiosperm relationships are unlikely to succeed.

Peronospora is predominantly a Northern Hemisphere genus (exceptions include: *Pe. andina* and *Pe. calindriniae* from Argentina and *Pe. mesembryanthemi* and *Pe. oxalidis* from South Africa) from the Old World (exceptions include: *Pe. floerkeae* and *Pe. oxybaphi* from North America) north of *ca* 35°N - roughly north of the Himalayas. (Exceptions include: *Pe. lycii* from China (Sechuan) and *Pe. satarensis* from India.) The taxonomy of *Peronospora* is complicated by the species/specific concepts championed by Gäumann (1923) and later authors from eastern Europe (e.g., Săvulescu, 1962; Săvulescu and Săvulescu, 1952), with the result that the species list is enormous and workers are sometimes resorting to *sensu lato* designations and sometimes to *sensu stricto* interpretations.   Apart from the host identity, simplistic morphological assessments are often all that are used. Gustavsson (1959*a*, 1959*b*) and Constantinescu (1979, 1989, 1991*a*, 1991*b*, 1992, 1996*a*, 1996*b*, 1998) have adopted a more considered, but still morphological, approach. See Hall (1996) for a more recent review of species concepts. *Peronospora* is most common on the Ranunculales-Caryophyllales, Eurosids I (Fabales, Rosales [including Urticaceae] and Malpighiales [Euphorbiaceae and Violaceae]); Eurosids II (Brassicales); Asterids (but not Asteraceae); all are late-evolving angiosperm orders, of which many members are herbaceous and found in conditions of high insolation (montane, mediterranean, or both).   Many of these angiosperms (Brassicaceae, Caryophyllaceae, and Amaranthaceae [Chenopodiaceae], Fabaceae and Euasterids I) have a centre of speciation in the Armeno-Iranian floristic sub-zone of Takhtajan (1969).   Species of the monocotyledonous genus *Allium*, family Alliaceae, order Asparagales, which also shares the same centre of origin and is noted for its sulphur-containing secondary metabolites, are hosts to the otherwise anomalous *Pe. destructor*.   An appreciable part of this modern mediterranean flora is thought to have been derived from North Africa at different times, through Asia Minor, the Italian peninsula or Iberia depending upon proximities in different epochs. A mid- to late-Tertiary, peri-montane origin for *Peronospora* is suggested.

The most plausible origin of *Plasmopara* (note hosts in Vitaceae and Euasterids II [Apiales, Asteraceae and Caprifoliaceae]) is from tropical South America in the mid- to late Palaeogene (Eocene - Oligocene): note that Wilson (1907) erected the genus *Rhysotheca* for all species *other* than the type species of *Plasmopara* (*P. pygmaea*) because these species produced zoospores (Table 3). Although there could have been early radiation through north-west Africa to Eurasia before the South Atlantic expanded, host origins are not entirely consistent with this suggestion. A tropical-host (not cold-temperature hardy?) element survives in *Plasmopara*, for example in *Mikania* (south American Asteraceae), *Plectranthus* (Lamiaceae), *Cissus* (west African Vitaceae) and the two anomalous parasites of *Oplismenus* and *Pennisetum* respectively (west African Poaceae).   However, as the two American tectonic plates drew together (*ca* 50 m.y.B.P.), a later radiation through Central America was possible. It may be speculated that, unlike the South American flora, that of North America, due to the early decline of the Laurentian tropical element, would have been particularly vulnerable to parasites evolving from South American taxa of pro-*Phytophthora* because these plants would have had little opportunity to develop genes for resistance to this genus. Alternatively, did *Ph. infestans* come from a precursor in south-east North America, becoming juxtaposed to *Solanum* from the northern Andean uplands?

The evolutionary pattern of *Basidiophora, Benua, Bremia* and *Paraperonospora* shows adaptive radiation in diverse Asteraceae. The species *Bremia graminicola* on the grass *Arthraxon* (Naoumoff, 1913; Patel, 1948) is anomalous. The relationship with the Asteraceae is not primarily cophylogenetic: it is probably relatively recent after the host family had become cosmopolitan, mostly developing during the late Tertiary, less than 10 million years ago, with *Basidiophora* and *Benua* evolving on the North and South

American cordilleras (the conspecificity of *Basidiophora montana* with *B. entospora* is not convincing, compare Barreto and Dick, 1991, with Constantinescu, 1998, and consider the differences between the cold-hardy *Erigeron* and tropical *Mikania* hosts), and *Bremia* and *Paraperonospora* evolving in the Old World, perhaps with *Bremia* in central Europe and *Paraperonospora* further to the east.

The origins of *Bremiella* (hosts in Violaceae, Balsaminaceae and Apiaceae) and *Pseudoperonospora* (hosts in Urticaceae and Cucurbitaceae) are less certain but may also be from the New World. The Peronosporales/angiosperm associations are similar in host ranges to those of *Aphanomyces* (Saprolegniomycetidae) indicating that there may be a fundamental physiological and biochemical requirement shared by the two sub-classes. (Likewise, compare plasmodiophorid host ranges - Dick, 2001*c*.)

Other possible biochemically-mediated organismal complexes have been envisaged, for example, the suggestion (Sansome and Sansome, 1974) that *Albugo* and *Peronospora* are synergistic on crucifers, and the susceptibility of beans to *Pseudoperonospora cubensis* only after prior infection by *Uromyces* (Yarwood, 1977).

10.3. MONOCOTYLEDONICOLOUS DOWNY MILDEWS

Almost all DMs in monocotyledons occur in grasses. The grass family, Poaceae, is very large (>9500 spp. world-wide) and Clayton and Renvoize (1986) suggest that although the origins of the grasses may have been in the Southern Hemisphere (Gondwanaland continents) with the subfamily Bambusoideae, the more advanced subfamilies and tribes have evolved in North Africa, South East Asia and Europe. The pasture grasses must have radiated from the tropical/warm temperate climate of what is now temperate North America into Eurasia in the early Tertiary, moving southwards and eastwards as the climate cooled and habitats were opened up with the burgeoning of the grazing mammals.

A totally different geographic centre of adaptive radiation (south-east Asia/Oceania/Australasia) has been proposed (Weltzien, 1981) for *Sclerospora* and related genera (Shaw, 1978) and accepted by Dick (1988). A 'pro-*Pythium*' ancestor might be presumed, but there are indications (e.g., Table 2: flagellar base of *Sclerospora*) that a saprolegniomycetidous origin is more likely. A link between *Verrucalvus* or *Pachymetra* with *Phytophthora* or *Pythium* is not supported by biochemical or mitochondrial DNA restriction patterns (Belkhiri and Dick, 1988; Dick *et al.*, 1989; pers. comm., G. R. Klassen, University of Manitoba); *Verrucalvus* has restriction patterns close to those of the Leptolegniaceae. *Pachymetra* is placed with *Aphanomyces* using 28S rDNA (Riethmüller *et al.*, 1999) and 18SrDNA (Spencer and Dick, unpublished). The Sclerosporales are primarily associated with the subfamily Panicoideae and its tribe Andropogoneae. This subfamily and tribe have tropical evolutionary origins; they are geographically centred in south-east Asia, having secondary developments in north East Africa with mediterranean Europe, and Australasia. The Himalayan uplift (5-23 m.y.B.P.), which drained the Tethys Ocean remnants and resulted in subsequent much colder montane and Siberian steppe climates to the north, and would have forced the early tropical grasses and their graminicolous downy mildews southwards into Africa, India, and south-east Asia. This pattern of change could also explain the occurrence of the north-eastern Australian genera *Verrucalvus* (Dick *et al.*, 1984) on the grass *Pennisetum clandestinum* (Kikuyu Grass, introduced from East Africa) and *Pachymetra* (Dick *et al.*, 1989) on sugar cane, which is thought to have originated in New Guinea. It is now known (pers. comm., R. C. Magarey, Bureau of Sugar Experiment Stations, Queensland) that *Pachymetra* occurs in the native Queensland panicoid grass, *Imperata cylindrica* var. *major* (Blady Grass). The absence of any associations with southern temperate grasses makes it unlikely that there was early evolution of these graminicolous fungi from Gondwanaland, which would have required isolated, three-pronged migration and evolution in Africa, India, and Australasia (but not southern South America). A centre of radiation from Africa (Williams, 1984) or India could only be supported if the Sclerosporales are considered to be of late Neogene (Miocene - Pliocene) or Quaternary origin. Such an origin is unlikely on morphological evidence from the fungi or biogeography of the grasses.

The subfamily Panicoideae contains species with leaf anatomy of both kranz and non-kranz kinds; kranz anatomy is associated with C_4 photosynthesis (see Sage and Monson, 1998). There are two kinds of kranz anatomy: PS (Parenchyma Sheath with chlorenchyma strongly radiating and both sheaths present, starch forming only in the PS) and MS (Mestome Sheath with chlorenchyma irregular, only the one sheath present, derived from the MS and containing starch). However, almost all of the known hosts of Sclerosporales are kranz MS species, irrespective of the tribe and subtribe to which they belong (e.g., tribe Paniceae: subtribe Setariinae; *Setaria*; subtribe Cenchrinae; *Pennisetum*, and all the genera of the subtribes of the tribe Andropogoneae). Kranz MS species have evolved with metabolic pathways that have accommodated to the high carbohydrate levels arising from their habitats, which although ranging from marshes, through ruderal habitats to stony deserts, are all habitats with high insolation and possibly unbalanced mineral nutrition. These habitats are not mesic. The water regime would appear to be of minor significance. Consider the sugar cane, *Saccharum*; cereal grasses such as *Sorghum* and *Pennisetum* (Pearl Millet); and grasses with essential oils such as *Cymbopogon* (Lemon Grass). The subfamily is also noted for its flavonoids which are, uniquely, sulphated. Thus, while the type of the Sclerosporales is *Sclerospora graminicola*, described from *Setaria* from Europe, this is still within the bounds of a coevolutionary development and ancestry located on the South East Asia plate. The occurrence of *Sclerophthora* in grasses of the subfamily Chloridoideae is not anomalous because this subfamily is regarded as a monophyletic line which has successfully proliferated throughout the tropics because of its consistent C_4 (kranz PS) metabolism. On the other hand, the occurrence of *Sclerospora* and *Sclerophthora* in the non-kranz genus *Oryza* (which shares a monophyletic clade with bambusoid species) is apparently anomalous on presently available information (but future studies might reveal host-metabolic similarities between *Oryza* and the Panicoideae because of their marsh/wetland habitats and submerged root systems). The wetland host habitat (often with periodic, prolonged seasonal droughts) is an obvious, but recent, coevolutionary factor for zoosporic fungal parasites.

10.4. THE NON-DOWNY MILDEW GENUS *ALBUGO*

Albugo may well be ancient, with a late Cretaceous origin in common with *Phytophthora*. *Albugo* and *Peronospora* have many host angiosperm orders in common (see Biga, 1955; Dick, 2001*c*: table III:3 and Figure 1), but this may reflect only similar metabolite preferences: *Peronospora* is predominantly Northern Hemisphere, but *Albugo* has hosts of evolutionarily early, occidental origin (Piperales, Amaranthaceae, Convolvulaceae (*Ipomoea*) in South America. *Albugo* shares with *Plasmopara* a warm-temperate/tropical element not found in *Peronospora*. The stimulus to parasitic diversification of *Albugo* may also have been the tectonic changes at the Cretaceous boundary.

10.5. ECONOMIC IMPORTANCE

It is not a coincidence nor is there a causal correlation that the occurrence of DM parasitism and the plants cultivated for food have depended on the production of secondary metabolites. The high levels of essential oils, mustard oils, and other secondary metabolites, that enhance the economic importance of crop plants in these taxa have rendered the plants vulnerable to the enzyme capacities and nutritional requirements of the DMs. Consider, for example, graminicolous grain crops (including flavour herbs such as Lemon Grass and sugar); pulses (plant proteins including enzymes); dicotyledonous grains (*Amaranthus*, *Fagopyrum*, *Helianthus*, *Linum*, *Raphanus*); green leaf (Brassicaceae, *Lactuca*, *Chenopodium*); tubers/roots (Solanaceae, Convolvulaceae, Apiaceae, Alliaceae); and fruits (Solanaceae, Cucurbitaceae, Rosaceae).

 Some secondary metabolites may be toxic to man, or otherwise unpalatable (Euphorbiaceae), but all represent a high carbon (and nitrogen or sulphur) source. Study of the occurrence of DMs in non-crop and non-horticultural hosts could be informative.

11. Conclusions

> "The family of parasites to which the potato fungus belongs existed in geological times, long prior to the potato plant or any of its relatives."
> - Worthington G. Smith, 1884: 275.

Comprehension of the interplay between the various facets of phylogenetic relationships, such as genome synteny and large genome repeats, genetic similarities, gene similarities and nucleotide sequences, simple sequence repeats, gene loss, gene transposition or suppression, and the progress of evolution through geologic time are aided by morphogenetic, systematic and molecular biological approaches. But molecular biology cannot yet explain the evolutionary pressures and advantages conferred by different physiological, biochemical or enzymic pathways. If a coevolutionary link based upon host production and parasite utilization of steroids or their precursors is confirmed, then the putative geographic origin of the host species and genus become more important than their evolutionary (phylogenetic) positions, especially since the angiosperm superorders are now cosmopolitan. Theories based on relict survival of obligately parasitic fungi become less attractive, particularly when the state of evolutionary advance of the hosts is also considered. Biodiversity depends on selective adaptation to closely circumscribed niches, based on enzyme capabilities for substrate exploitation and structures for function in these niches. Such biodiversity can be demonstrated using molecular markers at gene, species or ecosystem levels.

The size, age and genetic diversity of genera can be very different. If the various genera of the DMs arose from different 'pro-*Phytophthora*' ancestors, then the genetic distance between selected monophyletic DM genera and a particular group of species of *Phytophthora* may be less than the genetic distances within *Phytophthora sensu lato*. Comprehensive sequence data will be essential to assess whether the genetic distances between species of a more ancient genus (e.g., *Phytophthora*) may be greater than those between any recently evolved DM genera, even if these genera have arisen from different, though still relatively close, *Phytophthora* antecedents.

The affinity of *Phytophthora* for woody dicotyledonous hosts and the distinct, later origin of the pasture grasses makes it possible to suggest that, contrary to previously proposed phylogenies (Skalický, 1966; Barr, 1983; Shaw, 1981), some sections of *Pythium sensu lato* may have evolved later than some species of *Phytophthora*. On the other hand, molecular data may accumulate to suggest that *Halophytophthora*, despite its apparent anamorphy, may be ancestral; *Halophytophthora* should be included in any thorough phylogenetic molecular analyses.

Assuming *Phytophthora sensu lato* to be monophyletic, this genus must have evolved by the late Cretaceous, achieving a pan-tropical distribution in woody angiosperms (rarely gymnosperms), with appropriate speciation and adaptive radiation. The second stage in development would have been the parasitism of advanced herbaceous angiosperms acclimatized to cooler and more exposed environments, possibly through extinct precursors in the Armeno-Iranian floristic sub-region and the central American region. Geographic isolation could then have enabled the paraphyletic origins of the DM genera.

There would have been a comparable, parallel evolution of the graminicolous DMs. The graminicolous DMs are distinct, but in contrast to the dicotyledonicolous DMs, they appear to be cophylogenetically linked, and could have evolved when grazing became a selective force for the tropical, panicoid grasses of southern Asia. A 'pro-*Phytophthora*' or 'pro-*Pythium*' ancestor might be presumed, but there are indications that a saprolegniomycetid origin is more likely.

It is naïve to interpret studies of relatedness as leading to phylogenetic conclusions. Problems in determining *intra*-specific variability and *inter*-specific boundaries for variable and now-cosmopolitan DM species will need *independent* molecular checks to demonstrate the robustness of the model. Molecular data may not provide diagnostic traits for particular nodes, while Linnaean classifications require that the clade nodes be diagnosable and taxonomically valid. Taxonomically valid, hierarchical systematic ranks will not necessarily be co-ordinate because several branch steps may be encompassed within one hierarchical category. The type species must be characterized and placed in any systematic study. Given the estimates of percentages of undescribed species and percentages of known species studied by molecular techniques, morphologically-based taxonomy will coexist with molecular classifications for the foreseeable future. The absence of well-supported molecular biological sequences for the evolutionary origins of *Plasmopara, Peronospora*, or the Sclerosporales suggests that further revision of family and generic boundaries in the Peronosporales might be appropriate (Dick, 2001c).

Present-day geographical distributions are of little value in elucidating evolution. They merely confirm the rapid capabilities or opportunism of these parasites. The transcontinental 'marriage' of *Sclerospora*, from South East Asia, to *Zea*, from South America is one example: *Zea* was introduced to South East Asia and is a comparable host to South East Asian Panicoideae, having the same kranz MS (C_4) anatomy. Opportunism, inadvertently mediated by prehistoric and modern man, can and has been shown to enable rapid worldwide dissemination and subsequent evolution of diseased communities. The susceptibility of high latitude Gondwanaland species to the sub-tropical Northern Hemisphere Peronosporomycetidae is exemplified by the susceptibility of Southern Hemisphere Myrtaceae and Proteaceae to the spread of *Ph. cinnamomi* on *Eucalyptus* (Shearer and Tippett, 1989; Weste, 1994) and *Banksia* (Shearer and Hill, 1989), which is probably unstoppable yet less than a century since its introduction. Nevertheless, appreciable intraspecific coevolution can occur in relatively short time-scales. In this century accommodations between *Peronospora tabacina* on tobacco in North America and Europe and *Sclerospora graminicola* on agricultural grasses have occurred (Welzien, 1981). Ball and Pike (1984) showed that while fairly stable host-pathogen relationships have developed in Pearl Millet (*Pennisetum americanum*) since its introduction into India about 3000 years ago, the intercontinental variation in pathogenicity in *Sclerospora graminicola* suggests that further coevolutionary adjustment continues.

A complex web between angiosperm/animal coevolution and secondary metabolite-mediated straminipile/angiosperm coevolution has evolved. It is inconceivable to countenance evolutionary hypotheses that disregard palaeo-geographic or climatic factors, or the way in which these factors have determined the evolution of their primogenitors. For biotrophic parasites the evolution of the hosts is as important as the evolution of their pathogenicity. The importance of high insolation might be inferred for tropical uplands (Solanaceae), temporary clearings (climbing plants), and montane and mediterranean habitats (the last two habitats would also be more xeric). Relatively extreme habitats are potentially physiologically stressful for the hosts. In such habitats the selection for, or from, non-zoosporic biotrophic peronomycetes would be unrelated to their phylogeny, which should not therefore be based on zoosporic morphology and morphogenesis.

The physiological stresses induced in angiosperms in Tertiary ecosystems, and the resultant, genetically controlled metabolic responses, ensnared the most highly evolved angiosperms, so that they became susceptible to parasitism by the DMs. The link is not with the pattern of angiosperm phylogeny, as such, but with the highly evolved and greatly stressed angiosperms, characterized by their armoury of secondary metabolites functioning as deterents to animal predators and mycote fungal pathogens. But these same metabolites rendered such angiosperms uniquely vulnerable to the straminipile heterotrophs.

It is concluded that the phylogeny of the DMs is polyphyletic, with the dicotyledonicolous DMs arising from pro-*Phytophthora* Cretaceous ancestors, at different ages of the Tertiary or even the Quaternary, and in different continents: independent origins, both in geographic location and geologic time, are postulated for the genera *Albugo*, *Plasmopara* and *Peronospora*. The origin of the graminicolous DMs (the Sclerosporales) is uncertain: their antecedents may be from the Saprolegniomycetidae rather than the Peronosporomycetidae; the geologic age of origin may be the Quaternary; their centre of origin may have been south-east Asia.

The evolutionary origins of the downy mildews are recent compared with the origins of their hosts.

12. Acknowledgements

I am especially grateful to those authors who have kindly allowed me to refer to their work 'in press'. I thank Drs Julie Hawkins, Roland Goldring, Peter Goodenough, Deborah Hudspeth, Christopher Johnson and Frank Martin for their critical reading of sections of this chapter; I have tried to accommodate most of their suggestions. It is also a pleasure to acknowledge the help of Malcolm Beasely, the Botany Librarian of the Library and Information Services, The British Museum, and Denis Lamy, Museum National d'Histoire Naturelle, Laboratoire de Cryptogamie, Paris, in determining the correct citations for some of the early works.

13. References

Ainsworth, G.C. (1976) *Introduction to the History of Mycology*, Cambridge University Press, Cambridge, UK.

Angiosperm Phylogeny Group (APG) (1998) An ordinal classification for the families of flowering plants, *Annals of the Missouri Botanic Garden* **85**, 531-553.

Ann, P.J., and Ko, W.H. (1992) Survey of antibiotic resistance and dependence in *Phytophthora*, *Mycologia* **84**, 82-86.

Ball, S.L., and Pike, D.J. (1984) Intercontinental variation of *Sclerospora graminicola*, *Annals of Applied Biology* **104**, 41-51.

Barr, D.J.S. (1983) The zoosporic grouping of plant pathogens. Entity or non-entity? in S.T. Buczacki (ed.), *Zoosporic Plant Pathogens, a Modern Perspective*, Academic Press, London, UK, pp. 43-83.

Barr, D.J.S., and DéSaulniers, N.L. (1987) Ultrastructure of the *Lagena radicicola* zoospore, including a comparison with the primary and secondary *Saprolegnia* zoospores, *Canadian Journal of Botany* **65**, 2161-2176.

Barreto, R.W., and Dick, M.W. (1991) Monograph of *Basidiophora* (Oomycetes) with the description of a new species, *Botanical Journal of the Linnean Society* **107**, 313-332.

Baum, B.R., and Savile, D.B.O. (1985) Rusts (Uredinales) of Triticeae: evolution and extent of coevolution, a cladistic analysis, *Botanical Journal of the Linnean Society* **91**, 367-394.

Belkhiri, A., and Dick, M.W. (1988) Comparative studies on the DNA of *Pythium* species and some possibly related taxa, *Journal of General Microbiology* **134**, 2673-2683.

Belkhiri, A., Buchko, J., and Klassen, G.R. (1992) The 5S ribosomal RNA gene in *Pythium* species: two different genomic locations, *Molecular Biology and Evolution* **9**, 1089-1102.

Berbee, M.L., Carmean, D., Winka, K., and Eriksson, O. (1998) Phylogenetic resolution and the radiations of the Ascomycota, *Inoculum* **49**, 8.

Berkeley, M.J. (1846) Observations, botanical and physiological, on the potato murrain, Journal of the Horticultural Society of London, **1**, 9-34.

Berkeley, M.J. (1851) Observations of the vine mildew. By J. H. Léveillé. Translated from the French, with remarks by the Rev. M. J. Berkeley, M.A., F.L.S., Journal of the Horticultural Society of London, **6**, 284-295.

Berkeley, M.J., and Curtis, M.A. (1869) Fungi Cubenses (Hymenomycetes), Journal of the Linnean Society of London (Botany), **10**, 280-392.

Berlese, A.N. (1898) Ueber die Befruchtung und Entwickelung der Oosphäre bei den Peronosporeen, *Jahrbücher der Wissenschaftliche Botanik* **31**, 158-196.

Berlese, A.N. (1897-1902) Saggio di una monografia delle Peronosporaceae, *Revista di Patolologia vegetale, Padova* **6**, 78-101, 237-268 (1897); **7**, 19-37 (1898); **9**, 1-126 (1900); **10**, 185-298 (1902).

Berlin, J.D., and Bowen, C.C. (1964) The host-parasite interface of *Albugo candida* on *Raphanus sativus*, *American Journal of Botany* **51**, 445-452.

Bhattacharya, D., and Ehlting, J. (1995) Actin coding regions: gene family evolution and use as a phylogenetic marker, *Archiv für Protistenkunde* **145**, 155-164.

Biga, M.L.B. (1955) Riesaminazione delle specie del genere *Albugo* in base alla morfologia dei conidi, *Sydowia* **9**, 339-358.

Boussingault [forename initials not known] (1845) "... extraits suivants d'une lettre de M. le colonel Acosta sur la *maladie des pommes de terre* dans la Nouvelle-Granade", *Comptes Rendus, 2me Semestre* **21** (**No. 20**) [17 November 1845], 1114-1115.

Brasier, C.M. (1992) Evolutionary biology of *Phytophthora* Part I: genetic system, sexuality and the generation of variation, *Annual Review of Phytopathology* **30**, 153-171.

Brasier, C.M., and Griffin, M.J. (1979) Taxonomy of '*Phytophthora palmivora*' on cocoa, *Transactions of the British Mycological Society* **72**, 111-143.

Brasier, C.M., and Hansen, E.M. (1992) Evolutionary biology of *Phytophthora* Part II: phylogeny, speciation, and population structure, *Annual Review of Phytopathology* **30**, 173-200.

Briard, M., Dutertre, M., Rouxel, F., and Brygoo, Y. (1995) Ribosomal RNA sequence divergence within the Pythiaceae, *Mycological Research* **99**, 1119-1127.

Brummitt, R.K. (1996) In defence of paraphyletic taxa, in L.J.G. van der Maesen, X.M. Van Der Burgt and J.M. Van Medenbach De Rooy (eds.), *The Diversity of African Plants*, Kluwer Academic Publishers, Dordrecht, The Netherlands, pp. 371-384.

Cantino, E.C. (1955) Physiology and phylogeny in the water molds - a re-evaluation, *Quarterly Review of Biology* **30**, 138-149.

Cavalier-Smith, T. (1986) The Kingdom Chromista: origin and systematics, in F.E. Round, and D.J. Chapman (eds.), *Progress in Phycological Research*, Biopress, Bristol, UK, pp. 309-347.

Cavalier-Smith, T. (1989) The Kingdom Chromista, in J.C. Green, B.S.C. Leadbeater, W.L. Diver (eds.), *The Chromophyte Algae, Problems and Perspectives*, Clarendon Press, Oxford, UK, pp. 381-407.

Cavalier-Smith, T. (1998) A revised six-kingdom system of life, *Biological Reviews of the Cambridge Philosophical Society* **73**, 203-266.

Cavalier-Smith, T., Chao, E.E., and Allsopp, M.T.E.P. (1995) Ribosomal RNA evidence for chloroplast loss within Heterokonta: Pedinellid relationships and a revised classification of ochristan algae, *Archiv für Protistenkunde* **145**, 209-220.

Chilvers, G.A., Lapeyrie, F.F., and Douglass, P.A. (1985) A contrast between Oomycetes and other taxa of mycelial fungi in regard to metachromatic granule formation, *New Phytologist* **99**, 203-210.

Clark, J.S.C., and Spencer-Phillips, P.T.N. (1993) Accumulation of photoassimilate by *Peronospora viciae* (Berk.) Casp. and leaves of *Pisum sativum* L.: evidence for nutrient uptake via intercellular hyphae, *New Phytologist* **124**, 107-119.

Clarke, D.D. (1983) Potato late blight: a case study, in J.A. Callow (ed.), *Biochemical Plant Pathology*, Wiley, New York, USA, pp. 3-17.

Clayton, W.D., and Renvoize, S.A. (1986) *Genera Graminum, Grasses of the World*, HMSO Books, Norwich, UK.

Clinton, G.P. (1905) Downy mildew or blight of muskmelons and cucumbers, *Connecticut Agricultural Experiment Station, Annual Report 1904*, 329-362.

Coffey, M.D. (1975) Ultrastructural features of the haustorial apparatus of the white blister fungus *Albugo candida*, *Canadian Journal of Botany* **53**, 1285-1299.

Constantinescu, O. (1979) Revision of *Bremiella* (Peronosporales), *Transactions of the British Mycological Society* **72**, 510-515.

Constantinescu, O. (1989) *Peronospora* complex on Compositae, *Sydowia* **41**, 79-107.

Constantinescu, O. (1991*a*) An annotated list of *Peronospora* names, *Thunbergia* **15**, 1-110.

Constantinescu, O. (1991*b*) *Bremiella sphaerosperma* sp. nov. and *Plasmopara borreriae* comb. nov., *Mycologia* **83**, 473-479.

Constantinescu, O. (1992) The nomenclature of *Plasmopara* parasitic on Umbelliferae, *Mycotaxon* **43**, 471-477.

Constantinescu, O. (1996*a*) *Paraperonospora apiculata* sp. nov., *Sydowia* **48**, 105-110.

Constantinescu, O. (1996*b*) *Peronospora* on *Acaena* (Rosaceae), *Mycotaxon* **58**, 313-318.

Constantinescu, O. (1998) A revision of *Basidiophora* (Chromista, Peronosporales), *Nova Hedwigia* **66**, 251-265.

Cook, K.L., Hudspeth, D.S.S., and Hudspeth, M.E.S. (2001) A *cox2* phylogeny of representative marine Peronosporomycetes (Oomycetes), *Nova Hedwigia* **122**, 231-243.

Cooke, D.E.L., Kennedy, D.M., Guy, D.C., Russell, J., Unkles, S.E., and Duncan, J.M. (1996) Relatedness of group I species of *Phytophthora* as assessed by random amplified polymorphic DNA (RAPDs) and sequences of ribosomal DNA, *Mycological Research* **100**, 297-303.

Cooke, D.E.L., Jung, T., Williams, N.A., Schubert, R., Bahnweg, G., Oßwald, J.M., and Duncan, J.M. (1999) Molecular evidence supports *Phytophthora quercina* as a new species, *Mycological Research* **103**, 799-804.

Cooke, D.E.L., Williams, N.A., Williamson, B., and Duncan, J. M. (2001) Evolution and taxonomy of *Peronospora*: a molecular analysis, Kluwer Academic Publishers, Dordrecht, The Netherlands, **[this volume]**

Corda, A.J.K. (1837) *Icones Fungorum Hucusque Cognitorum, Vol. 1*, Praha, Czechslovakia.

Coryndon, S.C., and Savage, R.J.G. (1973) The origin and affinities of African mammal faunas, in N.F. Hughes (ed.), *Organisms and Continents through Time*, Special Papers in Palaeontology, No. 12, Palaeontological Association, London, UK, pp. 121-135.

Crepet, W.L., and Friis, E.M. (1987) The evolution of insect pollination in angiosperms, in E.M. Friis, W.G. Chaloner & P.R. Crane (eds.), *The Origins of Angiosperms and their Biological Consequences*, Cambridge Universty Press, Cambridge, UK, pp. 181-201.

Crute, I. (1981) The host specificity of peronosporaceous fungi and the genetics of the relationship between host and parasite, in D. M. Spencer (ed.), *The Downy Mildews*, Academic Press, London, UK, pp. 45-56.

Daggett, S.S., Knighton, J.E., and Therrien, C.D. (1995) Polyploidy among isolates of *Phytophthora infestans* from eastern Germany, *Journal of Phytopathology* **143**, 419-422.

Davison, E.M. (1968) Cytochemistry and ultrastructure of hyphae and haustoria of *Peronospora parasitica* (Pers. ex Fr.) Fr., *Annals of Botany* **32**, 613-621.

de Bary, A. (1863) Recherches sur le développement de quelques champignons parasites, *Annales des Sciences Naturelles, Paris, Série IV* **20**, 5-148.

de Bary, A. (1866) *Morphologie und Physiologie der Pilze, Flechten, und Myxomyceten*, Leipzig, Germany [or see de Bary, A. (1887)].

de Bary, A. (1876) Researches into the nature of the potato-fungus, *Phytophthora infestans*, *Journal of the Royal Agricultural Society, England. 2 Series* **12**, 239-269.

de Bary, A. (1887) *Comparative Morphology and Biology of the Fungi, Mycetozoa and Bacteria* (The authorised English translation by H. E. F. Garnsley revised by I.B. Balfour, of the German original of 1884), Clarendon Press, London, UK.

de Queiroz, K., and Gauthier, J. (1994) Toward a phylogenetic system of biological nomenclature, *TREE* **9**, 27-31.

Dick, M.W. (1972) Morphology and taxonomy of the Oomycetes, with special reference to Saprolegniaceae, Leptomitaceae and Pythiaceae. II. Cytogenetic systems, *New Phytologist* **71**, 1151-1159.

Dick, M.W. (1987) Sexual reproduction: nuclear cycles and life-histories with particular reference to lower-eukaryotes, *Biological Journal of the Linnean Society* **30**, 181-192.

Dick, M.W. (1988) Coevolution in the heterokont fungi (with emphasis on the downy mildews and their angiosperm hosts), in K.A. Pirozynski and D.L. Hawksworth (eds.), *Coevolution of Fungi with Plants and Animals*, Academic Press, London, UK, pp. 31-62.

Dick, M.W. (1989) *Phytophthora undulata* comb. nov., *Mycotaxon* **35**, 449-453.

Dick, M.W. (1990*a*) Phylum Oomycota, in L. Margulis, J. O. Corliss, M. Melkonian and D. Chapman (eds.), *Handbook of Protoctista*, Jones & Bartlett, Boston, USA, pp. 661-685.

Dick, M.W. (1990*b*) *Keys to Pythium*, Reading, UK, published by the author [ISBN 0-9516738-0-7].

Dick, M.W. (1992) Patterns of phenology in populations of zoosporic fungi, in G. C. Carroll and D. T. Wicklow (eds.), *The Fungal Community, Its Organization and Role in the Ecosystem*, Marcel Dekker Inc., New York, U.S.A., pp. 355-382.

Dick, M.W. (1995) Sexual reproduction in the Peronosporomycetes (chromistan fungi), *Canadian Journal of Botany, Supplement 1, Sections E-H* **73**, S712-S724.

Dick, M.W. (1997*a*) Fungi, flagella and phylogeny, *Mycological Research* **101**, 385-394.

Dick, M.W. (1997*b*) The Myzocytiopsidaceae, *Mycological Research* **101**, 878-882.

Dick, M.W. (1998) Impact of Molecular Methods on Fungal Systematics. A Joint (SGM Group) Symposium between the Systematics and Evolution Group of the SGM and the BMS, held at the University of Nottingham March 30-31, 1998, *The Mycologist* **12**, 116-117.

Dick, M.W. (2001*a*) The Peronosporomycetes, in K. Esser and P. A. Lemke (eds.), *The Mycota; Volume VII*: D. J. McLaughlin, E. McLaughlin & P. A. Lemke (eds.), *Volume VII, Systematics and Evolution, Part A*, Springer-Verlag, pp. 39-72.

Dick, M.W. (2001*b*) The Peronosporomycetes and other flagellate fungi, in D. H. Howard (ed.), *Fungi Pathogenic for Humans and Animals: Second Edition, Revised and Expanded*, American Society for Microbiology, Washington, USA, Marcel Dekker **[with press]**.

Dick, M.W. (2001*c*) *Straminipilous Fungi*, Kluwer Academic Publishers, Dordrecht, The Netherlands.

Dick, M.W. (2001*d*) Binomials in the Peronosporales, Sclerosporales and Pythiales, [this volume], Kluwer Academic Publishers, Dordrecht, The Netherlands, **[this volume]**.

Dick, M.W., Croft, B.J., Magarey, R.C., Cock, A.W.A.M. de, and Clark, G. (1989) A new genus of the Verrucalvaceae (Oomycetes), *Botanical Journal of the Linnean Society* **99**, 97-113.

Dick, M.W., Vick, M.C., Gibbings, J.G., Hedderson, T.A., and Lopez Lastra, C.C. (1999) 18S rDNA for species of *Leptolegnia* and other Peronosporomycetes: justification for the subclass taxa Saprolegniomycetidae and Peronosporomycetidae and division of the Saprolegniaceae *sensu lato* into the families Leptolegniaceae and Saprolegniaceae, *Mycological Research* **103**, 1119-1125.

Dick, M.W., and Win-Tin (1973) The development of cytological theory in the Oomycetes, *Biological Reviews of the Cambridge Philosophical Society* **48**, 133-158.

Dick, M.W., Wong, P.T.W., and Clark, G. (1984) The identity of the oomycete causing 'Kikuyu Yellows', with a reclassification of the downy mildews, *Botanical Journal of the Linnean Society* **89**, 171-197.

Dilcher, D.L. (1965) Epiphyllous fungi from Eocene deposits in western Tennessee, USA, *Palaeontographica* **116**, 1-54.

Dixon, R.A., and Paiva, N.L. (1995) Stress-induced phenylproanoid metabolism, *Plant Cell* **7**, 1085-1097.

Donoghue, M.J. (1985) A critique of the biological species concept and recommendations for a phylogenetic alternative, *The Bryologist* **88**, 172-181.

Douglas, J.G. (1973) The Mesozoic floras of Victoria. Part 3, *Memoirs of the Geological Survey of Victoria* **29**, 10-23.

Doyle, J.A., Jardine, S., and Doerenkamp, A. (1982) *Afropollis*, a new genus of early angiosperm pollen, with notes on the Cretaceous palynostratigraphy and paleoenvironments of northern Gondwana, *Bulletin, Centres Recherche Explor. Prod. Elf Aquitaine* **6**, 39-117.

Doyle, J.J. (1992) Gene trees and species trees: molecular systematics as one-character taxonomy, *Systematic Botany* **17**, 144-163.

Dudler, R. (1990) The single-copy actin gene of *Phytophthora megasperma* encodes a protein considerably diverged from any other known actin, *Plant Molecular Biology* **14**, 415-422.

Duncan, P.M. (1876) On some unicellular algae parasitic within Silurian and Tertiary coals, with a notice of their presence in *Calceola sandalina* and other fossils, *Quarterly Journal of the Geological Society of London* **32**, 205-210.

Ebel, J. (1998) Oligoglucoside elicitor-mediated activation of plant defense, *BioEssays* **20**, 569-576.

Edson, H.A. (1915) *Rheosporangium aphanidermatum*, a new genus and species of fungus parasitic on sugar beets and rhadishes, *Journal of Agricultural Research* **4**, 279-292.

Elad, Y., and Chet, I. (1987) Possible role of competition for nutrients in biocontrol of *Pythium* damping-off by bacteria, *Phytopathology* **77**, 190-195.

Elias, M.K. (1966) Living and fossil algae and fungi, formerly known as structural parts of marine bryozoans, *Palaeobotanist* **10**, 5-18.

Elliott, C.G. (1983) Physiology of sexual reproduction in *Phytophthora*, in D.C. Erwin, S. Bartnicki-Garcia & P.H. Tsao (eds.), Phytophthora, *Its Biology, Taxonomy, Ecology an Pathology*, American Phytopathological Society, St Paul, Minnesota, USA, pp. 71-80.

Emerson, R., and Held, A.A. (1969) *Aqualinderella fermentans* gen. et sp. nov., a phycomycete adapted to stagnant waters. II. Isolation, cultural characteristics and gas relations, *American Journal of Botany* **56**, 1103-1120.

Emerson, R., and Natvig, D.O. (1981) Adaptation of fungi to stagnant waters, in D.T. Wicklow & G.C. Carroll (eds.), *The Fungal Community, its Organization and Role in the Ecosystem*, Marcel Dekker Inc., New York, USA, pp. 109-128.

Erwin, D.C., and Ribeiro, O.K. (1996) Phytophthora *Diseases Worldwide*, American Phytopathological Society, St Paul, Minnesota, USA.

Farlow, W.G. (1876) On the American grape vine mildew, *Bulletin of the Bussey Institute* **1**, 415-428.

Feist, M., and Grambast-Fessard, N. (1991) The genus concept in Charophyta: evidence from the Palaeozoic to Recent, in R. Riding (ed.), *Calcareous Algae and Stromatolites*, Springer-Verlag, Berlin, Germany.

Fitzpatrick, H.M. (1930) *The Lower Fungi. Phycomycetes*, McGraw-Hill, New York, USA.

Fletcher, R.A. (1987) Plant growth regulating properties of sterol-inhibiting fungicides, in S.S. Purchit (ed.), *Hormonal Regulation of Plant Growth and Development*, Martinus Nijhoff, Dordrecht, pp. 103-114.

Francis, D.M., and Michelmore, R.W. (1993) Two classes of chromosome-sized molecules are present in *Bremia lactucae*, *Experimental Mycology* **17**, 284-300.

Fraymouth, J. (1956) Haustoria of the Peronosporales, *Transactions of the British Mycological Society* **39**, 79-107.

Friend, J., and Threlfall, D.R. (eds.) (1976) *Biochemical Aspects of Plant-Parasite Relationships*, Phytochemical Society Symposia Series No. 13, Academic Press, London, UK.

Galston, A.W. (1983) Polyamines as modulators of plant development, *BioScience* **33**, 382-388.

Gäumann, E. (1923) Beiträge zu einer Monographie der Gattung *Peronospora* Corda, *Beiträge zur Kryptogamenflora der Schweiz* **5** (**4**), 1-360.

Gerdemann, J.W. (1968) Vesicular-arbuscular mycorrhiza and plant growth, *Annual Review of Phytopathology* **6**, 397-418.

Gisi, U. (2001) Chemical control of downy mildews, [this volume], Kluwer Academic Publishers, Dordrecht, The Netherlands, **[this volume]**.

Gleason, F. (1976) The physiology of the lower freshwater fungi, in E.B.G. Jones (ed.), *Recent Advances in Aquatic Mycology*, Elek Press, London, UK, pp. 543-572.

Goosen, P.G., and Sackston, W.E. (1968) Transmission and biology of sunflower downy mildew, *Canadian Journal of Botany* **46**, 5-10.

Gray, S.F. (1821) *A Natural Arrangement of British Plants...*, **1**, p. 540, Baldwin, Crodock, and Joy, London, UK.

Griffith, J.M., Davis, A.J., and Grant, B.R. (1992) Target sites of fungicides to control Oomycetes, in W. Köllered (ed.), *Target Sites of Fungicide Action*, Chemical Rubber Company Press, Boca Raton, Florida USA, pp. 69-100.

Grosjean, M.C. (1992) Classification et identification des espèces du genre *Pythium*, champignons phytopathogènes du sol, par l'analyse de l'espace inerne transcrit de l'opéron ribosomique, Lyon I University.

Gunderson, J.H., Elwood, H., Ingold, A., Kindle, K., and Sogin, M.L. (1987) Phylogenetic relationships between chlorophytes, chrysophytes, and oomycetes, *Proceedings of the National Academy of Sciences, USA* **84**, 5823-5827.

Gustavsson, A. (1959a) Studies on Nordic Peronosporas. I. Taxonomic revision, *Opera Botanica* **3** (**1**), 1-271.

Gustavsson, A. (1959b) Studies on Nordic Peronosporas. II. General account, *Opera Botanica* **3** (**2**), 1-61.

Hall, G.[S.] (1996) Modern approaches to species concepts in downy mildews, *Plant Pathology* **45**, 1009-1026.

Harborne, J.B. (1988) *Introduction to Ecological Biochemistry*, 3rd edition, London, UK, Academic Press.

Harlan, J.R. (1976) Diseases as a factor in plant evolution, *Annual Review of Phytopathology* **14**, 31-51.

Held, A.A. (1970) Nutrition and fermentative energy metabolism of the water mold *Aqualinderella fermentans*, *Mycologia* **62**, 339-358.

Heller, A., Rozynek, B., and Spring, O. (1997) Cytological and physiological reasons for the latent type of infection in sunflower caused by *Plasmopara halstedii*, *Journal of Phytopathology* **145**, 441-445.

Herrado, M.L., and Klemsdal, S.S. (1998) Identification of *Pythium aphanidermatum* using the RAPD technique, *Mycological Research* **102**, 136-140.

Hibbett, D.S., and Donoghue, M.J. (1998) Integrating phylogenetic analysis and classification in fungi, *Mycologia* **90**, 347-356.

Hightower, R.C., and Meagher, R.B. (1986) The molecular evolution of actin, *Genetics* **114**, 315-332.

Hijwegen, T. (1988) Coevolution of flowering plants with pathogenic fungi, in K.A. Pirozynski and D.L. Hawksworth (eds.), *Coevolution of Fungi with Plants and Animals*, London, UK, Academic Press, pp. 63-77.

Hohl, H.R. (1983) Nutrition of *Phytophthora*, in D.C. Erwin, S. Bartnicky-Garcia & P.H. Tsao (eds.), Phytophthora, *Its Biology, Taxonomy Ecology and Pathology*, American Phytopathological Society, St Paul, Minnesota, USA, pp. 41-54.

Holwill, M.E.J. (1982) Dynamics of eukaryotic movement, in W.B. Amos & J.G. Duckett (eds.), *Prokaryotic and Eukaryotic Flagella*, Cambridge University Press, Cambridge, UK, pp. 289-312.

Hudspeth, D.S.S. , Nadler, S.A., and Hudspeth, M.E.S. (2000) A *COX2* molecular phylogeny of the Peronosporomycetes, *Mycologia* **92**: 674-684.

Hughes, N.F. (ed.) (1973) *Organisms and Continents through Time*, Special Papers in Palaeontology, No. 12, Palaeontological Association, London, UK.

Hughes, N.F. (ed.) (1976) *Palaeobiology of Angiosperm Origins*, Cambridge University Press, Cambridge, UK.

Idris, M.O., and Ball, S.L. (1984) Inter- and intracontinental sexual compatibility in *Sclerospora graminicola*, *Plant Pathology* **33**, 219-223.

Ingram, D.S. (1980) The establishment of dual cultures of downy mildew fungi and their hosts, in D.S. Ingram & J.P. Helgeson (eds.), *Tissue Culture Methods for Plant Pathologists*, Blackwell Scientific Publications, Oxford, UK, pp. 139-144.

Ito, S. (1913) Kleine Notizen über parasitischer Pilze Japans, *Botanical Magazine, Tokyo* **27**: 217-223.

Jacobson, D.J., Lefebvre, S.M., Ojerio, R.S., Berwald, N., and Heikkinen, E. (1998) Persistent, systemic, asymptomatic infections of *Albugo candida*, an oomycete parasite, detected in three wild crucifer species, *Canadian Journal of Botany* **76**, 739-750.

Jahn, T.L., Landman, M.D., and Fonseca, J.R. (1964) The mechanisms of locomotion of flagellates. II. Function of the mastigonems of *Ochromonas*, *Journal of Protozoology* **11**, 291-296.

Jacobs, B.F., Kingston, J.D., and Jacobs, L.L. (1999) The origin of grass-dominated ecosystems, *Annals of the Missouri Botanical Garden* **86**: 590-643.

Kader, A.A. (1999) Recommendations for maintaining post-harvest quality: cigar end rot [banana finger: *Verticillium theobromae* and or *Trachaesphaera fructigena*], hHp://postharvest.ucdavis.edu/Produce/ProduceFacts/Fruit/Banana.

Karling, J.S. (1981) *Predominantly Holocarpic and Eucarpic Simple Biflagellate Phycomycetes*, J. Cramer, Vaduz, Lichtenstein.

Kato, S., Coe, R., New, L., and Dick, M.W. (1990) Sensitivities of various Oomycetes to hymexazol and metalaxyl, *Journal of General Microbiology* **136**, 2127-2134.

Keen, N.T., and Yoshikawa, M. (1983) Physiology of disease and the nature of resistance to *Phytophthora*, in D.C. Erwin, S. Bartnicky-Garcia and P.H. Tsao (eds.), Phytophthora, *Its Biology, Taxonomy Ecology and Pathology*, American Phytopathological Society, St Paul, Minnesota, USA, pp. 279-287.

Kenneth, R.G. (1981) Downy mildews of graminaceous crops, in Spencer, D.M. (ed.), *The Downy Mildews*, Academic Press, London, UK, pp. 367-394.

Kenrick, P., and Crane, P.R. (1997) *The Origin and Early Diversification of Land Plants a Cladistic Study*, Smithsonian Institution Press, Washington, USA.

Klassen, G.R., McNabb, S.A., and Dick, M.W. (1987) Comparison of physical maps of ribosomal DNA repeating units in *Pythium*, *Phytophthora* and *Apodachlya*, *Journal of General Microbiology* **133**, 2953-2959.

Klebahn, H. (1905) Eine neue Pilzkrankheit der Syringen, *Centralblatt für Bakteriologie* **15**, 335-336.

Klenk, H.-P., Palm, P., and Zillig, W. (1994) DNA-dependent RNA polymerases as phylogenetic marker molecules, *Systematic and Applied Microbiology* **16**, 638-647.

Kochman, J., and Majewski, T. (1970) *Grzyby (Mycota). [Flora Polska] Tom IV. Glonowce (Phycomycetes) Wroślikowe (Peronosporales)*, Państwowe Wydawnictwo Naukowe, Warszawa, Poland.

Lafon, R., and Bulit, J. (1981) Downy mildew of the vine, in Spencer, D.M. (ed.), *The Downy Mildews*, Academic Press, London, UK, pp. 601-614.

Lange, L., Olson, L.W., and Safeeulla, K.M. (1984) Pearl millet downy mildew (*Sclerospora graminicola*): zoosporogenesis, *Protoplasma* **119**, 178-187.

Large, E.C. (1940) *The Advance of the Fungi*, Jonathan Cape, London, UK.

Leadbeater, B.S.C. (1989) The phylogenetic significance of flagellar hairs in the Chromophyta, in J.C. Green, B.S.C. Leadbeater and W.L. Diver (eds.), *The Chromophyte Algae, Problems and Perspectives*, Clarendon Press, Oxford, UK, pp. 145-165.

Lebeda, A., and Schwinn, F.J. (1994) The downy mildews - an overview of recent research progress (Falscher Mehitau - Übersicht über neuere Forschungsresultate), *Zeitschrift für Pflanzenkrankheiten und Pflanzenschutz* **101**, 225-254.

Leipe, D.D., Wainright, P.O., Gunderson, J.H., Porter, D., Patterson, D.J., Valois, F., Himmerich, S., and Sogin, M.L. (1994) The stramenopiles from a molecular perspective: 16S-like rRNA sequences from *Labyrinthuloides minuta* and *Cafeteria roenbergensis*, *Phycologia* **33**, 369-377.

Leipe, D.D., Tong, S.M., Goggin, C.L., Slemenda, S.B. Pieniazek, N.J., and Sogin, M.L. (1996) 16S-like rDNA sequences from *Developayella elegans*, *Labyrinthuloides haliotidis*, and *Proteromonas lacertae* confirm that stramenopiles are a primarily heterotrophic group, *European Journal of Protistology* **32**, 449-458.

Lévesque, C.A., Beckenbach, K., Baillie, D.L., and Rahe, J.E. (1993) Pathogenicity and DNA restriction fragment length polymorphisms of isolates of *Pythium* spp. from glyphosate-treated seedlings, *Mycological Research* **97**, 307-312.

Lévesque, C.A., Vrain, T.C., and De Boer, S.H. (1994) Development of a species-specific probe for *Pythium ultimum* using amplified ribosomal DNA, *Phytopathology* **84**, 474-478.

Lévesque, C.A., Harlton, C.E., and de Cock, A.W.A.M. (1998) Identification of some oomycetes by reverse dot blot hybridization, *Phytopathology* **88**, 213-222.

Liew, E.C.Y., MacLean, D.J., and Irwin, J.A.G. (1998) Specific PCR based detection of *Phytophthora medicaginis* using the intergenic spacer region of the ribosomal DNA, *Mycological Research* **102**, 73-80.

Lotsy, J.P. (1907) *Vorträge über Botanische Stammesgeschichte Gehalten an der Reichsuniversität, Erster Band: Algen und Pilze*, Gustav Fischer, Jena, Germany.

Lucas, J.A., Hayter, J.B.R., and Crute, I.R. (1995) The downy mildews: host specificity and pathogenesis, in K. Kohmoto, U.S. Singh and R.P. Singh (eds.), *Pathogenesis and Host Specificity in Plant Diseases*, Pergamon, UK, pp. 217-234.

Lucas, J.A., Shattock, R.C., Shaw, D.S., and Cooke, L.R. (eds.) (1991) Phytophthora, Cambridge University Press, Cambridge, U.K.

McNabb, S.A., Boyd, D.A., Belkhiri, A., Dick, M.W., and Klassen, G.R. (1987) An inverted repeat comprises more than three-quarters of the mitochondrial genome in two species of *Pythium*, *Current Genetics* **12**, 205-208.

McNabb, S.A., and Klassen, G.R. (1988) Uniformity of mitochondrial DNA complexity in Oomycetes and the evolution of the inverted repeat, *Experimental Mycology* **12**, 233-242.

Martin, F.N. (1995*a*) Electrophoretic karyotype polymorphisms in the genus *Pythium*, *Mycologia* **87**, 333-353.

Martin, F.N. (1995*b*) Linear mitochondrial genome organization in vivo in the genus *Pythium*, *Current Genetics* **28**, 225-234.

Millardet, P.M.A. (1885) Traitment du mildou et du rot, *Journal d'Agriculture Practique* **2**, 513-516 [See Schneiderhan, F. J. (1933) below.]

Miyabe, K., and Kawakami, T. (1903) *Kawakamia* Miyabe. A new genus belonging to Peronosporaceae, *Botanical Magazine, Tokyo*, **17** 306.

Money, N.P. (1998) Why oomycetes have not stopped being fungi, *Mycological Research* **102**, 767-768.

Montagne, J.F.C. (1845) [Untitled], *L'Institut, Journal Universel des Sciences et des Sociétés Savantes en France et à l'Étranger. 10 Section Sciences mathimatiques, physiques et naturelles, 1845,* **13 (no. 609)**, 312-314.
  [Notes to clarify confusion in recent citations: published: 03 Septembre, 1845; diagnosis p. 313; the *Contents* page at the end of the volume cites: "Observations sur la maladie des pommes de terre. Montagne..."; reprinted as: *Société Philomatique de Paris. Extraits des procès-verbaux des Séances pendant l'année 1845, Botanique [Séance du 30 Août 1845],* pp. 98-101 (probably published during 1846); see also *L'Institut, Journal Universel des Sciences et des Sociétés Savantes en France et à l'Étranger. 10 Section Sciences mathimatiques, physiques et naturelles, 1845,* **13 (no. 616)**: (published 22 Octobre 1845): a report of a note presented by M. Montagne at the Séance on 20 October 1845 of the Academy of Sciences (*Compte-rendu hebdomadàire des Séances de l'Académie des Sciences*)].
Moore, G., Devos, K.M., Wang, Z., and Gale, M.D. (1995) Grasses, line up and form a circle. The genomes of six major grass species can be aligned by dissecting the individual chromosomes into segments and rearranging these linkage blocks into highly similar structures, *Current Biology* **5**, 737-739.
Moore, P.D. (1982) Evolution of photosynthetic pathways in flowering plants, *Nature, London* **295**, 647-648.
Mugnier, J., and Grosjean, M.C. (1995) *PCR Catologue in Plant Pathology:* Pythium, Rhone-Poulenc Agro, Lyon, France.
Muller, J. (1981) Fossil pollen records of extant angiosperms, *Botanical Review* **47**, 1-142.
Naoumoff, N. (1913) Matériaux pour la Flore mycologique de la Russie, *Bulletin de la Société mycologique de France* **29**, 273-290.
Nes, W.D. (1990) Stereochemistry, sterol metamorphosis and evolution, in P.J. Quinn and J.L. Harwood (eds.), *The Biochemistry, Structure and Utilization of Plant Lipids,* Portland Press, Portland, Oregon, USA, pp. 308-319.
Palti, J., and Kenneth, R. (1981) The distribution of downy mildew genera over the families and genera of higher plants, in D.M. Spencer (ed.), *The Downy Mildews,* Academic Press, London, UK, pp. 45-56.
Pampaloni, I. (1902) I resti organici nel disodile di Melilli in Sicilia, *Palaeontographica Italiana* **8**, 121-130.
Panabières, F., Marais, A., Trentin, F., Bonnet, P., and Ricci, P. (1989) Repetitive DNA polymorphism analysis as a tool for identifying *Phytophthora* species, *Phytopathology* **79**, 1105-1109.
Panabières, F., Ponchet, M., Allasia, V., Cardin, L., and Ricci, P. (1997) Characterization of border species among Pythiaceae: several *Pythium* isolates produce elicitins, typical proteins from *Phytophthora* spp., *Mycological Research* **101**, 1459-1468.
Patel, M.K. (1948) *Bremia* species on *Arthraxon lancifolius* Hoch in India, *Indian Phytopathology* **1**: 104-106.
Parrish, J.T. (1987) Global palaeogeography and palaeoclimate of the late Cretaceous and early Tertiary, in E.M. Friis, W.G. Chaloner, and P.R. Crane (eds.) *The Origins of Angiosperms and their Biological Consequences,* Cambridge Universty Press, Cambridge, UK, pp. 51-73.
Paxton, J.D. (1983) *Phytophthora* root rot and stem rot of soybean: a case study, in J.A. Callow (ed.) *Biochemical Plant Pathology,* Wiley, New York, USA, pp. 19-30.
Pemberton, C.M., Davey, R.A., Webster, J., Dick, M.W., and Clark, G. (1990) Infection of *Pythium* and *Phytophthora* species by *Olpidiopsis gracilis* (Oomycetes), *Mycological Research* **94**, 1081-1085.
Petersen, A.B., and Rosendahl, S. (2000) Phylogeny of the Peronosporomycetes (Oomycota) based on partial sequences of the large ribosomal subunit (LSU rDNA), *Mycological Research,* **104**, 1295-1303.
Petersen, H.E. (1909) Studier over Ferskvands-Phycomyceter. Bidrag til Kundskaben om de Submerse, Phycomyceters Biologi og Systematik, samt om deres Udbredelse i Danmark, *Botanisk Tidsskrift* **29**, 345-440.
Pethybridge, G.H. (1913) On the rotting of potato tubers by a new species of *Phytophthora* having a method of sexual reproduction hitherto undescribed, *Scientific Proceedings of the Royal Dublin Society* **13**, 529-565.
Peyton, G.A., and Bowen, C.C. (1963) The host-parasite interface of *Peronospora manshurica* on *Glycine max, American Journal of Botany* **50**, 787-797.
Pirozynski, K.A. (1976a) Fossil Fungi, *Annual Review of Phytopathology* **14**, 237-246.
Pirozynski, K.A. (1976b) Fungal spores in the fossil record, *Biological Memoirs* **1**, 104-120.
Pirozynski, K.A., and Malloch, D.W. (1975) The origin of land plants: a matter of mycotrophism, *BioSystems* **6**, 153-164.
Plaats-Niterink, A.J. Van der (1981) Monograph of the genus *Pythium, Studies in Mycology, Centraalbureau voor Schimmelcultures, Baarn* **21**, 1-242.

Plaziat, J.-C. (1995) Modern and fossil mangroves and mangals: their climatic and biogeographic variability, in D.W.J. Bosence & P.A. Allison (eds.), *Marine Palaeoenvironmental Analysis from Fossils*, Geological Society Special Publication No. 83, Geological Society of London, London, UK, pp. 73-96.

Podger, F.D. (1972) *Phytophthora cinnamomi*, a cause of lethal disease in indigenous plant communities in Western Australia, *Phytopathology* **62**, 972-981.

Potter, D., Saunders, G.W., and Andersen, R.A. (1997) Phylogenetic relationships of the Raphidophyceae and Xanthophyceae as inferred from nucleotide sequences of the 18S ribosomal RNA gene, *American Journal of Botany* **84**, 966-972.

Preisig, H.R. (1999) Systematics and evolution of the algae: phylogenetic relationships of taxa within the different groups of algae, *Progress in Botany* **60**, 369-412.

Pringsheim, N. (1858) Beitrag für Morphologie und Systematik, der Algen. Die Saprolegnieen, *Jahrbuch für Wissenschaftlichen Botanik*, **1**, 284-306.

Regel, E. (1843) Beiträge zur Kenntnis einiger Blattpilze, *Botanische Zeitung* **1**, 665-667.

Retallack, G.J., and Dilcher, D.L. (1981*a*) Arguments for a Glossopterid ancestry of angiosperms, *Paleobiology* **7**, 54-67.

Retallack, G.J., and Dilcher, D.L. (1981*b*) A coastal hypothesis for the dispersal and rise to dominance of flowering plants, in K. J. Niklas (ed.), *Paleoecology and Evolution*, Praeger, New York, USA, pp. 27-77.

Riethmüller, A., Weiss, M., and Oberwinkler, F. (1999 [publ. 2000]) Phylogenetic studies of Saprolegniomycetidae and related groups based on nuclear large subunit ribosomal DNA sequences, *Canadian Journal of Botany* **77**, 1790-1800.

Ristaino, J.B., Madritch, M., Trout, C.L., and Parra, G. (1998) PCR amplification of ribosomal DNA for species identification in the plant pathogen genus *Phytophthora*, *Applied and Environmental Microbiology* **64**, 948-954.

Ross, J.D., and Sombrero, C. (1991) Environmental control of essential oil production in Mediterranean plants, in J.B. Harborne and F.A. Barberan (eds.) *Ecological Chemistry and Biochemistry of Plant Terpenoids*, Proceedings of the Phytochemical Society of Europe, Volume 31, Clarendon Press, Oxford, UK, pp. 83-94.

Rostowzew, S.J. (1903) Beiträge zur Kenntnis der Peronosporeen, *Flora, Jena* **92**, 405-430.

Roze, E., and Cornu, M. (1869) Sur deux nouveaux types génériques pour les familles des Saprolégniées et Péronosporées, *Annales des Sciences Naturelles, Botanique, Série V* **11**, 72-91.

Saccardo, P.A. (1876) Fungi veneti novi vel critici. Series V, *Nuovo Giornale Botanico Italiano* **8**, 161-211.

Sackston, W.E. (1981) Downy mildew of sunflower, in D. M. Spencer (ed.), *The Downy Mildews*, Academic Press, London, UK, pp. 545-575.

Sage, R.F., and Monson, R.K. (eds.) (1998) *C₄ Plant Biology*, Academic Press, San Diago, California, USA.

Sansome, E., and Sansome, F.W. (1974) Cytology and life-history of *Peronospora parasitica* on *Capsella bursa-pastoris* and of *Albugo candida* on *C. bursa-pastoris* and on *Lunularia annua*, *Transactions of the British Mycological Society* **62**, 323-332.

Savile, D.B.O. (1968) Possible interrelationships between fungal groups, in G.C. Ainsworth and A.L Sussman (eds.), *The Fungi, An Advanced Treatise, Volume III, The Fungal Population*, Academic Press, New York, USA, pp. 649-675.

Savile, D.B.O. (1976) Evolution of the rust fungi (Uredinales) as reflected by their ecological problems, *Evolutionary Biology* **9**, 137-207.

Săvulescu, O. (1962) A systematic study of the genera *Bremia* Regel and *Bremiella* Wilson, *Revue de Biologie. Academia Republicii Populare Romîne, Bucarest* **7**, 43-62.

Săvulescu, T., and Săvulescu, O. (1952) Studiul *Sclerospora*, *Basidiophora* si *Peronoplasmopara*, *Buletin ştiinţific. Academia Republicii Populare Romîne* pp. 327-457.

Sawada, K. (1922) Materials for the mycological study of Formosa, *Transactions of the Natural History Society of Formosa*, **62**, 77-84.

Schneiderhan, F.J. (1933) *The Discovery of Bordeaux Mixture*, Phytopathological Classics Number 3, American Phytopathological Society, Ithaca, N.Y., USA.

Schroeter, J., (1886-1887 in fascicles; 1889) Die Pilze Schlesiens. Oomycetes, in *Kryptogamen-Flora von Schlesien* **3** (1): 225-257.

Schröter, J. (1879) *Protomyces graminicola* Saccardo, *Hedwigia* **18**, 83-87.

Schröter, J. (1893) Peronosporineae, in A. Engler (ed.) *Die Natürlichen Pflanzenfamilien nebst ihren Gattungen und wichigeren Arten insbesondere den Nutzplanzen, unter Mitwirkung zahlreicher hervorragender Fachgelehrten begründet von A. Engler und K. Prantl. I. Teil, Abteilung 4*: 108-119.

Seward, A.C. (1898) *Fossil Plants, Volume I*, Cambridge University Press, Cambridge, UK, pp. 205-222.

Shaw, C.G. (1978) *Peronosclerospora* species and other downy mildews of Gramineae, *Mycologia* **70**, 594-604.

Shaw, C.G. (1981) Taxonomy and evolution, in D.M. Spencer (ed.) *The Downy Mildews*, Academic Press, London, UK, pp. 17-29.

Shearer, B.L., and Hill, T.C. (1989) Diseases of *Banksia* woodlands on the Bassendean and Spearwood dune systems, *Journal of the Royal Society of Western Australia* **71**, 113-114.

Shearer, B.L., and Tippett, J.T. (1989) Jarrah dieback: the dynamics and management of *Phytophthora cinnamomi* in the Jarrah (*Eucayptus marginata*) forest of south-western Australia, *Research Bulletin No. 3, Department of Conservation and Land Management, Como, Western Australia*, i-xvii, 1-76.

Shirai, M., and Hara, K. (1927) *A List of Japanese Fungi Hitherto Known*. Third Ed., Shikuoka, Japan.

Silberman, J.D., Sogin, M.L., Leipe, D.D., and Clark, C.G. (1996) Human parasite finds taxonomic home, *Nature (London)* **380**, 398.

Skalický, V. (1964) Beitrag zur infraspezifischen Taxonomie der obligat parasitischen Pilze (Prispevek k vnitrodruhove taxonomii nekterych obligatne parasitickych plisni a hub), *Acta Universitatis Carolinae (Praha) 1964*, 25-89.

Skalický, V. (1966) Taxonomie der Gattungen der Familie Peronosporaceae (Taxonomie rodu celedi Peronosporaceae), *Preslia (Praha)* **38**, 117-129.

Skidmore, D.I., and Ingram, D.S. (1985) Conidial morphology and the specialization of *Bremia lactucae* Regel (Peronosporaceae) on hosts in the family Compositae, *Botanical Journal of the Linnean Society* **91**, 503-522.

Smith, C.J. (1996) Tansley Review No. 86. Accumulation of phytoalexins: defence mechanism and stimulus response system, *New Phytologist* **132**, 1-45.

Smith, W.G. (1884) *Diseases of Field and Garden Crops Chiefly such as are caused by Fungi*, Macmillan, London, UK.

Soltis, D.E., Soltis, P.S., Chase, M.W., Albach, D., Mort, M.E., Savolainen, V., and Zanis, M. (1998*b*) Molecular phylogenetics of angiosperms: congruent patterns inferred from three genes. Part II, *American Journal of Botany* **85**, 157 [abstract].

Soltis, P.S., Soltis, D.E., Chase, M.W., Albach, D., Mort, M.E., Savolainen, V., and Zanis, M. (1998*a*) Molecular phylogenetics of angiosperms: congruent patterns inferred from three genes. Part I, *American Journal of Botany* **85**, 157-158 [abstract].

Soltis, P.S., Soltis, D.E., and Chase, M.W. (1999) Angiosperm phylogeny inferred from multiple genes as a tool for comparative biology, *Nature (London)* 402: 402-404.

Sparrow, F.K. (1960) *Aquatic Phycomycetes, Second Revised Edition*, University of Michigan Press, Ann Arbor, Michigan, USA.

Spaulding, P. (1961) Foreign diseases of forest trees of the world, *U.S. Agricultural Research Service, Agriculture Handbook no. 197*: 1-361.

Spencer, D.M. (ed.) (1981) *The Downy Mildews*, Academic Press, London, UK.

Spencer, M.A., and Dick, M.W. (2002, in press) Aspects of downy mildew biology: perspectives for plant pathology and peronosporomycete phylogeny, in R. Watling (ed.), *Tropical Mycology, British Mycological Society Meeting, Liverpool, April 2000*, CABI, Wallingford, UK.

Spencer, M.A., Vick, M.C., and Dick, M.W. (submitted) *Aplanopsis* and *Pythiopsis*, the systematics of the common terrestrial Saprolegniaceae, with a revision of the 'subcentric' grouping of *Achlya* species, *Mycological Research*.

Spencer-Phillips, P.T.N. (1997) Function of fungal haustoria in epiphytic and endophytic infections, *Advances in Botanical Research* **24**, 309-333.

Stamps, J., Waterhouse, G.M., Newhook, F.J., and Hall, G.S. (1990) Revised tabular key to the species of *Phytophthora*, *Mycological Papers* **162**, 1-28.

Stanghellini, M.E., Adaskaveg, J.E., and Rasmussen, S.L. (1990) Pathogenesis of *Plasmopara lactucae-radicis*, a systemic root pathogen of cultivated lettuce, *Plant Disease* **74**, 173-178.

Stidd, B.M., and Consentino, K. (1975) *Albugo*-like oogonia from the North American Carboniferous, *Science* **190**, 1092-1093.

Stössel, P., and Hohl, H.R. (1981) Sterols in *Phytophthora infestans* and their role in the parasitic interactions with *Solanum tuberosum*, *Bericht der Schweitzerischen Botanischen Gesellschaft* **90**, 118-128.

Takhtajan, A. (1969) *Flowering Plants Origin and Dispersal*, [translation by C. Jeffrey], Oliver & Boyd, Edinburgh, UK.

Tarling, D.H. (1980) *Continental Drift & Biological Evolution*, Carolina Biology Readers, Carolina Biological Supply Company, Burlington, North Carolina, USA.

Taylor, G. (1952) Analysis of the swimming of long and narrow animals, *Proceedings of the Royal Society of London* **214A**, 158-183.

Thirumalachar, M.J., Shaw, C.G., and Narasimhan, M.J. (1953) The sporangial phase of the downy mildew on *Eleusine coracana* with a discussion of the identity of *Sclerospora macrospora* Sacc, *Bulletin of the Torrey Botanical Club* **80**, 299-307.

Tsao, P. (1970) Selective media for isolation of pathogenic fungi, *Annual Review of Phytopathology* **8**, 157-186.

Unger, F. (1833) *Die Exantheme der Pflanzen und einige mit diesen verwandte krankheiten der Gewächte pathogenetisch und notographisch dergstellt*, Carl Gerold, Vienna.

Unger, F. (1847) Beitrag zur Kenntniss der in der Kartoffelkrankheit vorkommenden Pilze und der Ursache ihres Entstehens, *Botanische Zeitung, 5 Jahrgang* **18**, 305-317, plate 6 [N.B.: column numbers not page numbers].

Unkles, S.E., Moon, R.P., Hawkins, A.R., Duncan, J.M., and Kinghorn, J.R. (1991) Actin in the oomycetous fungus *Phytophthora infestans* is the product of several genes, *Gene* **100**, 105-112.

Upchurch, G.R., and Wolfe, J.A. (1987) Mid-Cretaceous to Early Tertiary vegetation and climate: evidence from fossil leaves and woods, in E.M. Friis, W.G. Chaloner, and P.R. Crane (eds.) *The Origin of Angiosperms and their Biological Consequences*, Cambridge University Press, Cambridge, UK, pp. 75-105.

Van Der Auwera, G., and De Wachter, R. (1998) Structure of the large subunit rDNA from a diatom, and comparison between small and large subunit ribosomal RNA for studying stramenopile evolution, *Journal of Eukaryote Microbiology* **45**, 521-527.

Van Der Auwera, G., De Baere, R., Van Der Peer, Y., De Rijk, P., Van Den Broek, I., and De Wachter, R. (1995) The phylogeny of the Hyphochytriomycota as deduced from ribosomal DNA sequences of *Hyphochytrium catenoides*, *Molecular Biology and Evolution* **12**, 671-678.

Vaughan, H.E., Taylor, S., and Parker, J.S. (1996) The ten cytological races of the *Scilla autumnalis* species complex, *Heredity* **79**, 371-379.

Vaughan, J.G., and Geissler, C.A. (1997) *The New Oxford Book of Food Plants*, Oxford University Press, Oxford, UK.

Vercesi, A., Tornaghi, R., Sant, S., Burruano, S., and Faoro, F. (1999) A cytological and ultrastructural study on the maturation and germination of oospores of *Plasmopara viticola* on overwintering vine leaves, *Mycological Research* **103**, 193-202.

Vogel, H.J. (1964) Distribution of lysine pathways among fungi: evolutionary implication, *American Naturalist* **98**, 435-446.

Voglmayr, H., Bonner, L., and Dick, M.W. (1999) Taxonomy and oogonial ultrastructure of a new aero-aquatic peronosporomycete, *Medusoides* gen. nov. (Pythiogetonaceae fam. nov.), *Mycological Research* **103**, 591-606.

Voglmayr, H., and Greilhuber, J. (1998) Genome size determination in Peronosporales (Oomycota), *Fungal Genetics and Biology* **25**, 181-195.

Walker, J.C., and Stahmann, M.A. (1955) Chemical nature of disease resistance in plants, *Annual Review of Plant Physiology* **6**, 351-366.

Warner, S.A., Sovocool, G.W., and Domnas, A.J. (1982) Fungal transformations of triparanol, *Applied and Environmental Microbiology* **44**, 1471-1475.

Waterhouse, G.M. (1964) The genus *Sclerospora*. Diagnoses (or descriptions) from the original papers and a key, *Commonwealth Mycological Institute, Commonwealth Agricultural Bureaux, Miscellaneous Publication, Number 17*, 1-30.

Waterhouse, G.M. (1968) The genus *Pythium* Pringsheim. Diagnoses (or descriptions) and figures from the original papers, *Mycological Papers* **110**, 1-71.

Waterhouse, G.M., and Brothers, M.P. (1981) The taxonomy of *Pseudoperonospora*, *Mycological Papers* **148**, 1-28.

Watson, A.J. (1971) Foreign bacterial and fungal diseases of food, forage and fiber crops, *U.S. Agricultural Research Service, Agriculture Handbook no. 418*: 1-111.

Weltzien, H.C. (1981) Geographical distribution of downy mildews, in D.M. Spencer (ed.), *The Downy Mildews*, Academic Press, London, UK, pp. 45-56.

Weste, G. (1994) Impact of *Phytophthora* species on native vegetation of Australia and Papua New Guinea, *Australian Journal of Plant Pathology* 23, 190-209.

Whitfield, P.R., and Bottomley, W. (1983) Organization and structure of chloroplast genes, *Annual Review of Plant Physiology* 34, 279-310.

Williams, R.J. (1984) Downy mildews of tropical cereals, in *Advances in Plant Pathology* 2, 1-103.

Wilson, G.W. (1907) Studies in North American Peronosporales - II. Phytophthoreae and Rhysotheceae, *Bulletin of the Torrey Botanical Club* 34, 387-416.

Wilson, G.W. (1914) Studies in North American Peronosporales - VI. Notes on miscellaneous species, *Mycologia* 6, 192-210.

Win-Tin, and Dick, M.W. (1975) Cytology of Oomycetes: evidence for meiosis and multiple chromosome associations in Saprolegniaceae and Pythiaceae with an introduction to the cytotaxonomy of *Achlya* and *Pythium*, *Archives of Microbiology* 105, 283-293.

Wolfe, J.A., and Upchurch, G.R. (1987) North American nonmarine climates and vegetation during the Late Cretaceous, *Palaeogeography, Palaeoclimatology, Palaeoecology* 61, 33-77.

Woodham-Smith, C. (1962) *The Great Hunger, Ireland 1845-9*, Hamish Hamilton, London, UK.

Yarwood, C.E. (1977) *Pseudoperonospora cubensis* in rust-infected bean, *Phytopathology* 67, 1021-1022.

HOST RESISTANCE TO DOWNY MILDEW DISEASES

B. MAUCH-MANI

University of Neuchâtel, Institute of Botany, Rue Emile-Argand 13, CH-2007 Neuchâtel, Switzerland

1. Introduction

Plants are exposed during their whole life to a wide range of potential pathogens and pests, and have developed a number of resistance mechanisms to protect themselves. These defence mechanisms have been classified broadly as avoidance, tolerance and resistance (Parlevliet, 1992). In the case of avoidance, contact between the plant and the potential parasite is strongly reduced. This type of resistance is mainly found against animal pests and herbivores where repelling smells, tastes or structures such as thorns lead to a reduced interaction between the two partners of the interaction. Tolerance of a plant against a given pathogen does not interfere with the development of the pathogen itself but enables an overall reduction of the damage resulting from the pathogen's presence in the host. This review will mainly deal with resistance defined as the ability of a given host plant to hinder the development of a pathogen (Robinson, 1969). For agriculture and horticulture, host resistance is still the most important way of controlling diseases because it leads to the most effective cost-benefit ratio for the grower. Breeders try to produce plants with the highest possible level of resistance while maintaining or even improving the agronomic qualities of the crop.

Resistance operates at different levels and can accordingly be subdivided into different classes. This does not imply that at the biochemical level some of the actual defence mechanisms cannot be the same. The diversity of cellular responses induced in host and non-host plants suggests that there are many different stages during an attempted infection where an interaction between the parasite and the host can take place. Each of these stages might represent a "switching point" and the specific response at each stage could then determine the further development of the interaction (Heath, 1974).

The broadest type of resistance is non-host resistance. Plants are constantly in contact with many different microbial organisms and from these only a very restricted number is able to successfully infect a given plant species. Thus, non-host resistance is the most common and the most effective form of resistance in higher plants. General resistance (conferring a partial, quantitative protection), gene-for-gene resistance (based on the specific interaction between the products of avirulence genes in the pathogen and

P.T.N. Spencer-Phillips et al. (eds.), Advances in Downy Mildew Research, 59–83.

resistance genes in the host) and acquired, also called induced resistance will be discussed in the following chapters in relation to downy mildew diseases.

The downy mildews occupy a somewhat special niche due to their taxonomic position. Although the downy mildews, as well as the other oomycetes *Phytophthora* and *Pythium*, have been included within the kingdom Fungi, recent findings concerning their evolutionary phylogeny have lead to a re-classification into the new kingdom Chromista (Cavalier-Smith, 1986) or Straminipila (Dick, 1995 and this volume). From a phytopathological point of view, the behavior of oomycetes is doubtlessly fungus-like. Certain features such as cell wall composition (Bartnicki-Garcia, 1970) or the pathway of sterol metabolism (Nes, 1987; Griffith *et al.*, 1992) limit the chemical control of these organisms with conventional fungicides (see Gisi, this volume). All the genera known as downy mildews, *Basidiophora, Bremia, Bremiella, Paraperonospora, Peronospora, Peronosclerospora, Plasmopara, Pseudosperonospora, Sclerospora, Sclerophthora* and have an obligate parasitic mode of life and are for this reason more difficult to investigate and handle than their facultative parasitic relatives. However, progress has been made in recent years in the molecular genetics of oomycetes due to the availability of a *Peronospora*/Arabidopsis pathosystem (Koch and Slusarenko, 1990; Holub *et al.*, 1993; Mauch-Mani *et al.*, 1993), and new insights have been be gained into the complex relationship between a downy mildew and its host plant. The lettuce-*Bremia lactucae* model system is explained in detail by Lebeda, Pink and Ashley (this volume).

2. General resistance

General resistance, also called field resistance, partial resistance, quantitative, or horizontal resistance is usually not race-specific and commonly assumed to be polygenic. Plants showing this type of resistance can be infected by pathogens but the rate of disease progress and corresponding symptom expression is significantly reduced compared to a susceptible plant. Although general resistance is more durable than monogenic race-specific resistance, it is the latter that is mainly selected for by breeders due to its easier manipulation.

To measure general resistance, rating parameters have to be defined so that a reduction in disease intensity can easily be evaluated under field conditions. In the case of the downy mildews, this will essentially be the assessment of surface area of the plant showing asexual sporulation and/or macroscopic disease symptoms such as discoloration. Field resistance cannot be measured in absolute terms and has to be compared to a defined standard under the same conditions. Environmental conditions, pathogen pressure, age and growth conditions of the plants are among other factors which can influence the expression of this type of resistance. The desired level of resistance will vary from crop to crop depending on the parts of the plant that will be exploited commercially. For example lettuce should show a very high degree of field resistance since the whole above-ground parts of the plant are destined for consumption. In contrast, in pea lower levels of resistance could be tolerated on the leaves as long as the pods and seeds themselves are not affected.

In lettuce, certain cultivars rated as susceptible to *B. lactucae* based on artificial inoculations display reduced disease susceptibility when tested in the field under natural epidemic conditions (Dixon *et al.*, 1973; Crute and Dixon, 1981). Based on the investigation of a large number of lettuce cultivars, Crute and Norwood (1981) found that the characteristics of this field resistance were a lower disease incidence, restriction of infection to the outer leaves, less infected leaves per plant, fewer and smaller lesions, and a reduced rate of disease progress compared to a susceptible control. Comparing a butterhead lettuce cultivar (Hilde) and a crisphead lettuce (Iceberg) they deduced that the observed resistance might be due to morphological or physiological differences between the two types of lettuce (Crute and Norwood, 1981). In a large-scale experiment in the field, 756 lettuce genotypes tested showed significant differences in partial resistance against *Bremia lactucae* (Eenink, 1981). No linkage was found between presence or absence of known *R*-genes and the partial resistance level expressed in these plants. Other characters of the plants such as lettuce type, size or erectness could not be correlated to the degree of observed field resistance, although results for chosen genotypes which had been tested twice and in different years and/or environments showed a good correlation. Yuen and Lorbeer (1984) studied crisphead lettuce cultivars for the expression of field resistance in New York state using a graphic digitizer to determine leaf and lesion areas, thus circumventing subjective rating differences. The field resistance they observed was characterized by a smaller total lesion area in resistant plants. Interestingly, one of the crisphead lettuce cultivars (Mesa 659) tested in their study was the same cultivar tested earlier by Dixon *et al.* (1973) in the UK. Where Yuen and Lorbeer (1984) reported 26% of infected leaf area for Mesa 659, Dixon *et al.* (1973) found 25.3% of basal infected area for the same cultivar under field conditions.

In the above mentioned studies the resistance level of the plants was assessed based on macroscopic symptoms observed in the field. For selection in breeding programmes though, it would be desirable to be able to assess field resistance under laboratory conditions. To this end Lebeda and Reinink (1991) analysed variations in the early development of *B. lactucae* on lettuce cultivars with different levels of field resistance at the microscopic level on cotyledons and leaf discs. Differences were observed between the resistant cultivar Iceberg and susceptible cultivars in the percentage of spores germinating and the incidence and speed of development of infection structures. On cultivar Iceberg, the germ tubes were significantly longer compared to the ones on susceptible cultivars. In the same cultivar, secondary vesicles were first observed 24 h after inoculation compared to 6 h in susceptible cultivars. No intercellular hyphae and haustoria were found in Iceberg within the first 24 h after inoculation. The results show that incidence and speed of germination of a single isolate of *B. lactucae* can differ on host genotypes although the hosts lack the corresponding *R*-gene. The observed differences in the early infection stages are most probably based on differences in the ability and/or the speed of a given cultivar to react to an infection at the biochemical level. In this context the correlation between a high level of peroxidase activity prior to infection and the degree of field resistance observed in lettuce might play a crucial role (Reuveni *et al.*, 1991). The same observation holds true in the *Cucumis sativus/Pseudoperonospora cubensis* interaction where zymograms for peroxidase isoenzyme patterns offer a quick and reliable method for the discrimination of specific cucumber genotypes with high levels of field resistance against cucumber downy mildew (Lebeda and Dolezal, 1995). A high peroxidase activity could enable the rapid polymerization of phenolics at the site of an attempted penetration and thus contribute to resistance. A primary selection for high peroxidase activity prior to field selection for resistance could help to shorten and focus specific breeding programmes.

Partial resistance of vining pea lines has been investigated in detail by Stegmark (1990, 1991, 1992, 1994). Peas grown for canning and freezing are an important crop in southern Sweden and are unfortunately susceptible to downy mildew. Improved resistance is therefore a desired goal in breeding. While investigating variation for virulence among Scandinavian isolates of *Peronospora viciae* f. sp. *pisi* on different *Pisum sativum* genotypes, Stegmark (1990) found one pea breeding line showing a stable partial resistance to different isolates of *P. viciae* and this line was also highly resistant to 'race 8' from the Netherlands which is considered to be virulent on most pea genotypes carrying known *R*-genes. To gain insight into the inheritance of partial resistance, and more specifically to see if the partial resistance of three sister-lines was determined by identical genes, these lines and the susceptible parent line were crossed with each other and the downy mildew resistance of the different combinations was analyzed. Resistance turned out to be caused by different genes in the three sister-lines. In one moderately resistant line (X309), resistance was monogenic recessive. Line X282, which was more resistant than X309, carried at least one gene conferring resistance that showed dominance in crosses with X309. Line X311 carried, in addition to the resistance in X309, some gene or genes with an intermediate effect (Stegmark, 1992). In order to assess how breeding lines for partial resistance selected in the greenhouse behave under field conditions, a pea line partially resistant to downy mildew was back-crossed to a susceptible cultivar with more pods per node and a lower seed weight. Lines showing different degrees of infection in greenhouse tests were selected and investigated in field trials and in the greenhouse for four generations. Artificial infection of seedlings in the greenhouse correlated with natural infection of pods in the field (Stegmark, 1991).

Recently, molecular techniques have permitted a better analysis of the genetic factors (quantitative trait loci, QTL) determining quantitative resistance (Michelmore, 1995). The identification of genomic regions contributing to pathogen resistance will allow assesment of the effect and the number of major genetic factors responsible for quantitative resistance. Although several QTLs conferring resistance to the oomycete *Phytophthora infestans* (Leonards-Schippers, 1994) as well as those for resistance against downy mildew of pearl millet caused by *Sclerospora graminicola* (Jones *et al.*, 1995) turned out to be race-specific, further research in this area is likely to provide isolate- or race-non-specific QTLs, as is the case for other host-pathogen systems (reviewed in Michelmore, 1995). Once identified, such QTLs for downy mildew resistance could be introduced into susceptible host plants and build the foundation for (QTL-)marker assisted breeding of these crops.

3. Gene-for-gene type of resistance

Race-specific resistance, also called gene-for-gene resistance or vertical resistance is based on the specific interaction between the products of avirulence (*avr*) genes in the pathogen and resistance (*R*) genes in the host (Flor, 1971). Genetic studies of many different plant pathosystems suggest that resistance results when an appropriate *avr-R* gene pair is present; if either member of the gene pair is absent or inactivated the host plant becomes susceptible to the pathogen. The molecular basis for such a gene-for-gene system is currently explained by an elicitor-receptor model (Gabriel and Rolfe, 1990), where *avr* genes code for products that are recognized by the corresponding *R* gene-encoded receptors. Recognition of the pathogen triggers subsequent signal transduction events which finally lead to the activation of the defence responses that will eventually limit or stop the pathogen's ingress. Most R gene proteins share common motifs, such as leucine-rich repeats (LRR), with or without nucleotide binding site (NBS), a leucine zipper (LZ), a signalling domain resembling the Toll and interleukin1 transmembrane receptors (TIR) or a kinase domain (reviewed in Baker *et al.*, 1997). The structure of *R* genes against downy mildews is known for some *R* gene candidates conferring resistance to *B. lactucae*, the *Dm* genes (Shen *et al.*, 1998) and for some of the *RPP* (recognition of *P. parasitica*) genes against *Peronospora parasitica* (Parker *et al.*, 1997; McDowell *et al.*, 1998; Botella *et al.*, 1998). The putative *Dm* genes are characterized by the presence of LRR regions in C-terminal position relative to the NBS and thus resemble the *RPM1* (Grant *et al.*, 1995) and *RPS2* (Bent *et al.*, 1994; Mindrinos *et al.*, 1994) genes in *A. thaliana* conferring resistance to *Pseudomonas syringae* pv. *maculicola* and *Pseudomonas syringae* pv. *syringae*, respectively. Genes from the *RPP1* complex resistance locus (Botella *et al.*, 1998) and *RPP5* (Parker *et al.*, 1997) encode functional products from the TIR-LRR-NBS R protein class with a strong structural similarity to the plant resistance gene products N from tobacco (Whitham *et al.*, 1994) and L6 from flax (Lawrence *et al.*, 1995). *RPP8* encodes a LRR-NBS protein with a putative N-terminal leucine zipper and therefore seems to be more closely related to *RPS2* and *RPM1* than to the other known *RPP* genes (McDowell *et al.*, 1998).

The high similarity found between *R* genes in different plant species suggests that they might also share a common signal transduction pathway or components of it. LRRs play a role in protein-protein interactions (Kobe and Deisenhofer, 1995) and this has given rise to the speculation that LRRs could be the binding domain of R gene proteins for pathogen-derived elicitors (Bent, 1996; Baker *et al.*, 1997). Warren *et al.* (1998) isolated a mutation within the LRR domain of the Arabidopsis *RPS5* gene (which confers resistance to the bacterial pathogen *Pseudomonas syringae* pv *tomato* DC3000) that partially suppresses several *R* genes mediating resistance to other isolates of *P.s. tomato* and, interestingly, to several isolates of *P. parasitica*. The sporulation of 5 different *P. parasitica* isolates was significantly enhanced in *rps5-1* mutants compared to wild type plants. However, resistance was not fully compromised in any of the *rps5-1/P. parasitica* interactions since sporulation never reached wild type levels, and necrotic flecks indicating resistance reactions were present even in the most susceptible interactions. The authors suggest that the *rps5-1* mutation acts either by delaying pathogen recognition itself or by interfering with a subset of events that occur after pathogen recognition. Other mutants affecting resistance mediated by multiple *R* genes have been described (Hammond-Kossack and Jones, 1996; Baker *et al.*, 1997). The *ndr1* mutation in Arabidopsis leads to a suppression of resistance conferred by *RPS2, RPM1, RPS5*, and several *RPP* loci (Century *et al.*, 1995, 1997), and the *eds1* mutation (enhanced disease susceptibility) affects several *RPP* specificities (Parker *et al.*, 1996). Resistance genes often occur in clusters (Crute and Pink, 1996; Michelmore and Meyers, 1998) and recently much progress was reported concerning complex loci for resistance against *P. parasitica* or *B. lactucae* and the possible mechanisms determining their structure (Meyers *et al.*, 1998; Botella *et al.*, 1998; McDowell *et al.*, 1998).

4. Acquired resistance

In addition to constitutively present barriers against invaders and the defence reactions activated only upon contact with a pathogen, plants possess a further means of defending themselves: systemic or acquired resistance. Acquired resistance is the phenomenon by which a genetically susceptible plant can be induced to defend itself against a broad range of virulent pathogens, thus displaying phenotypic resistance. Conceptually, this form of resistance is reminiscent of the immunization system found in animals. Acquired resistance as such has been known for a long time (Chester, 1933; Gäumann, 1946) but it was only with the studies by Ross (1966) and Kuc and coworkers (Madamanchi and Kuc, 1991; Hammerschmidt and Kuc, 1995) that the resistance could be linked to an activation of defence mechanisms such as the accumulation of pathogenesis-related (PR) proteins, peroxidases or glycine-rich proteins (reviewed in Hammerschmidt and Kuc, 1995). Induction of a resistant state can be achieved by a pre-inoculation of a small part of the plant with necrotizing pathogens which then leads to the expression of resistance in other, systemic parts of the plant.

A striking feature of acquired resistance is the strong, early increase of endogenously produced salicylic acid (SA) in the induced plants (Malamy *et al.*, 1990; Métraux *et al.*, 1990; Rasmussen *et al*, 1991). Interestingly, it was shown that exogenous application of salicylic acid alone, in absence of necrotising pathogens, also leads to an induction of resistance (White, 1979; Ward *et al.*, 1991). The same effect was achieved by treating plants with synthetic resistance-inducing chemicals such as INA (2,6-dichloroisonicotinic acid) (Métraux *et al.*, 1991) or BTH (benzo-(1,2,3)-thiadiazole-7-carbothioic acid S-methyl ester) (Görlach *et al.*, 1996) which are thought to mimic the action of salicylic acid. Induction with necrotizing pathogens and with chemicals is now referred to as systemic acquired resistance (SAR) (Van Loon, 1997). A similar type of resistance, called induced systemic resistance (ISR) has been observed when plants are grown in soil colonized by certain resistance-inducing bacteria known as PGPRs (plant growth promoting bacteria) belonging to the group of fluorescent Pseudomonads (Van Loon, 1997). Essential elements of the phenomenology of acquired resistance have been comprehensively reviewed recently (Hunt and Ryals, 1996; Ryals *et al.*, 1996; Schneider *et al.*, 1996; Sticher *et al.*, 1997; Mauch-Mani and Métraux, 1998).

Induced resistance against downy mildews has been described for both monocots and dicots. Pearl millet plants were shown to exhibit biologically induced resistance against *Sclerospora graminicola* after an induction treatment with the same organism (Kumar *et al.*, 1993). Morris *et al.* (1998) reported chemical (BTH) induction of resistance in maize against *Peronosclerospora sorghi*. Used at a concentration of 1 or 2 g of BTH per kg of seed, they obtained a level of protection similar to the one achieved by treatment with the oomycete-specific fungicide metalaxyl. BTH also activated the expression of the maize *PR-1* and *PR-5* genes.

Until recently, research concerning biologically induced acquired resistance in dicots against downy mildews dealt mainly with the tobacco/*Peronospora tabacina* system where the inducer was either a virus or *Peronospora* itself (Stolle *et al.*, 1988; Ye *et al.*, 1989, 1991). Protection of the plants by natural compounds has also been reported. Kamoun *et al.* (1993) induced SAR in tobacco plants against *Peronospora parasitica* var *nicotinianae* by treating them with cryptogein, a small cytotoxic peptide secreted by *Phytophthora* species (Ricci and Pernollet, 1989; Yu, 1995). Another natural substance, the non-protein amino acid DL-3-aminobutyric acid (BABA) has a good potential for protecting plants systemically against downy mildews. It was shown to be effective in tobacco against blue-mold (Cohen, 1994), downy mildew in grapes (Cohen *et al.*, 1999 and this volume), and *P. parasitica* in Arabidopsis (Zimmerli et al., 2000; Jakab et al, 2001).

Since the Arabidopsis/*Peronospora* system is now amenable to molecular genetics, progress in this field has been accelerated. SAR against *P. parasitica* in Arabidopsis has been shown to be induced biologically (Uknes *et al.*, 1993; Mauch-Mani and Slusarenko, 1994) as well as chemically (Uknes *et al.*, 1992; Lawton *et al.*, 1996; Zimmerli *et al.*, 2000). ISR also operates in the Arabidopsis/*Peronospora* system (L. Schellenbaum and B. Mauch-Mani, unpublished results; Ton *et al.*, 2001). Treatment of Arabidopsis accession Columbia with PGPRs lead to a 30-50% reduction in infection by *P. parasitica.* The restriction in fungal growth seems to be independent of SAR gene induction: in Arabidopsis a SA-producing PGPR strain induced *PR-1*, while no *PR-1* accumulation took place in plants treated with a non-SA-producing PGPR strain. Together with the fact that both bacterial strains were able to induce resistance in Arabidopsis NahG plants, this points to a SA-independent resistance induction pathway by PGPRs (L. Schellenbaum and B. Mauch-Mani, unpublished results). Testing different inducers of SAR and ISR, Ton *et al.* (2001) showed, that although resistance could be induced in all cases, SAR-mediated resistance was more effective against *P. parasitica* than the ISR-mediated one.

Parts of plants distant from local induction by pathogens or chemicals, marker genes for SAR (SAR genes) show an increased expression tightly correlated to the expression of SAR itself (Ward *et al.*, 1991; Uknes *et al.*, 1992). The analysis of Arabidopsis mutants either compromised in SAR or showing constitutive SAR expression has allowed a dissection of the signal transduction pathway leading from induction to expression of SAR. The Arabidopsis mutants *npr* (nonexpresser of PR genes), *nim* (non immunity) and *sai* (salicylic acid insensitive) react with a normal HR when inoculated with avirulent pathogens, but no longer express acquired resistance (Cao *et al.*, 1994, 1997; Delaney *et al.*, 1995; Ryals *et al.*, 1996; Shah *et al.*, 1997). However, the SA level produced by these mutants is comparable to wild type levels but there is a block in the pathway downstream of SA perception. Chemicals such as INA or BTH cannot induce resistance in these mutants indicating that they must use a similar signal transduction pathway as SA. The NahG Arabidopsis plants (Lawton *et al.*, 1995) and the *ndr1* mutant (Century *et al.*, 1995) are unable to build up SAR but they both can be rescued by INA treatment. *Sid* (salicylic acid induction deficient) mutants impaired in the induction of salicylic acid after pathogen inoculation show pronounced susceptibility to avirulent isolates of *P. parasitica* and a decrease in SAR. The *sid* mutant phenotype can be rescued by SA, INA, and BTH treatment, respectively (Nawrath and Métraux, 1999). A number of mutants show constitutive expression of SAR. The *lsd* (lesion simulating disease) mutant series (Dietrich *et al.*, 1994; Weyman *et al.*, 1995) as well as the *cpr1* (constitutive PR) mutant (Bowling *et al.*, 1994) are resistant against virulent isolates of *P. parasitica* and all of them have constitutively elevated SA and SAR gene levels. The phenotypes of different Arabidopsis mutants described above show clearly the importance of the SAR signal transduction pathway in resistance against downy mildews. A more detailed understanding of this pathway might lead to the engineering of more resistant crop plants in the future.

5. Mechanisms of resistance

5.1. HYPERSENSITIVE RESPONSE

The first macroscopic event in gene-for-gene resistance is normally the hypersensitive response (HR). According to Agrios (1988) HR is defined as the rapid death of host cells, usually within a few hours after contact with the pathogen. The extent of cell death depends on the specific interaction, ranging from single cell HR to necrotic areas comprising a large number of dead cells (Holub *et al.*, 1994). The extent of cell death usually reflects at what stage of infectional growth a pathogen was stopped and depends on the *R*-gene/*Avr*-gene combination present. In interactions between plants and obligately biotrophic pathogens, such as the downy mildews, where the partners typically form a very close association between haustoria and living plant cells, it might seem quite obvious that an infection can be stopped by depriving the parasite of its nutritional basis. Whether HR alone is sufficient to stop the pathogen remains unclear. The haustoria of the downy mildews range from simple filamentous branches to more elaborate vesicle-like structures which can even be lobed. Fraymouth (1956) observed a correlation between the extent of haustorial development and food supply in the plant. She suggested that in regions of the host where food supply is abundant, nutrients diffusing from the host cells might be taken up by hyphae in the intercellular space, the haustoria therefore remaining simple, whereas more elaborate forms of haustoria with a more extensive surface would be needed in regions of restricted nutritional supply. The apoplast of plants contains concentrations of amino acids (2-3mM), sugars (up to 10mM) and other nutrients which are sufficient to allow growth of micro-organisms (Hancock and Huisman, 1981). Clark and Spencer-Phillips (1993) clearly demonstrated that in the interaction between *Peronospora viciae* and *Pisum sativum* [^{14}C] from labelled sucrose introduced into the apoplast was taken up by the hyphae. Such an uptake of nutrients from the intercellular space though demonstrates that the plant must have other defence responses in addition to cell death to prevent invasion by the pathogen.

A further problem correlating HR and pathogen death is encountered when trying to define the exact timing of cell death. The observed cellular collapse and browning is merely the endpoint of a series of reactions taking place in the cell, the actual death though might have happened long before. Mansfield and coworkers (Woods *et al.*, 1988) investigated cell death in the gene-for-gene system between lettuce and *B. lactucae*. They assumed that cells having undergone an irreversible membrane damage are not able to recover, but that many biochemical changes, specially those involving oxidative processes, can still take place in cells after irreversible membrane damage and subsequent decompartimentalization and organelle disruption. Thus they used the failure of cells to plasmolyse as a marker to identify cell death in lettuce cotyledons. In this system, the timing of irreversible membrane damage is clearly determined by the specific resistance gene (*Dm*) involved. In lettuce cultivars carrying the resistance genes *Dm1*, *Dm2*, *Dm3* or *Dm5/8* a rapid hypersensitive cell death was observed. Based on the timing of the appearance of irreversible membrane damage, these cultivars could be divided into two groups. Plants carrying *Dm2* or *Dm 5/8* showed irreversible membrane damage during the expansion of the primary vesicle 5 hours after inoculation. *Dm1* and *Dm3* in contrast lead to a delay in irreversible membrane damage of about one hour that allowed formation and establishment of the secondary vesicle. The *Dm6* and *Dm7* genes, which confer intermediate resistance, did not lead to irreversible membrane damage in epidermal cells within the first 12 hours after inoculation. Shortly after *Bremia* formed haustoria in mesophyll cells, these cells died.

Despite differences in the time-course of the plant response, the activity of *Dm* genes none-the-less results in a common theme: only penetrated cells show an HR and cell death always precedes restriction of hyphal growth. Nothing is known about the initial triggering of signalling events in this system. Attempts to find specific elicitors of the HR in intercellular washing fluids of infected tissues and from infection structures of *Bremia* has not proven successful (Crucefix *et al.*, 1987).

Another well investigated gene-for-gene system is the interaction between *P. parasitica* and *A. thaliana*. Here too, resistance is expressed in the form of a hypersensitive reaction ranging from one dead cell to whole clusters of dead cells (Holub *et al.*, 1994). Holub and colleagues used a different method of assessment of the interaction as described above for *Bremia*. Scoring large numbers of plants and classifying them according to macroscopically defined parameters, they created a grid of recognition specificities between numerous accessions of Arabidopsis and isolates of *P. parasitica*. A nomenclature system was developed using single letter abbreviations that allows researchers of different groups to exactly define an interaction between Arabidopsis and *Peronospora* and to make comparisons with the results obtained by other groups. The responses they observed range from small flecks (F), medium sized cavities (C) to large pits (P). Translated to the cellular level, these descriptions correspond to single to few-cell HR, gradually increasing in dead cell number until covering large areas. They refined their scoring system by also taking into account the degree of asexual sporulation (heavy, H; medium, M; low, L; rare, R; none, N) and the time of the onset of sporulation (early, E; delayed, D). Based on this system they were able to postulate the presence of *RPP* resistance loci in Arabidopsis. Many of these genes have now been mapped or even cloned (Holub *et al.*, 1993; Parker *et al.*, 1993; Tör *et al.*, 1994; Joos *et al.*, 1996; Parker *et al.*, 1997; McDowell *et al.*, 1998; Botella *et al.*, 1998; Bittner-Eddy *et al*, 2000).

A major difference between the extent of HR in lettuce infected by *Bremia* as opposed to Arabidopsis infected by *Peronospora* is that in the latter case, HR cell death is not restricted to cells containing a haustorium but can spread to several cell layers surrounding the penetrating hyphae.

HR cell death is also a common feature of many other incompatible interactions of downy mildews and their hosts (Spencer, 1981). Resistant pearl millet cultivars react with a typical HR when attacked by *Sclerospora graminicola* (Mauch-Mani *et al.*, 1989). For this host-pathogen system it has been proposed to use arachidonic acid-induced hypersensitive cell death (or rather a phenocopy thereof) as an assay of downy mildew resistance (Geetha *et al.*, 1996). Pearl millet seedlings were treated with arachidonic acid and the time between treatment and appearance of the phenocopy HR was positively correlated to the degree of resistance a given cultivar expressed when inoculated with *S. graminicola*.

Despite being clearly genetically defined, expression of HR can depend on other genes or environmental factors. Parker and co-workers (1996) described *EDS1*, a *RPP*-non-specific locus in Arabidopsis, that is required for the expression of resistance mediated by *RPP* resistance genes. The *eds1* mutation is recessive and is not a defective allele of any known *RPP* gene. Mutants in the *eds1* locus are highly susceptible to isolates of *P. parasitica* to which the wild type is resistant. In addition, the mutants were also shown to become susceptible to isolates of *P. parasitica* from *Brassica oleracea* for which Arabidopsis is a non-host. These results indicate that following the recognition event by the different *R* genes there is a common step mediating the upstream transduction of the signal.

In many host-pathogen systems with fungi (Dickson *et al.*, 1959; Dyck and Johnson, 1983; Gousseau *et al.*, 1985; Islam *et al.*, 1989), viruses (McKinney and Clayton, 1945; Schroeder *et al.*, 1965; Pfannenstiel and Niblett, 1978) and bacteria (Holliday *et al.*, 1981) temperature has shown to influence the outcome of an infection. This phenomenon is also encountered in downy mildew systems. Judelson and Michelmore (1992) studied the influence of temperature, over a range from 5 to 22°C, and genotype on infection types by 13 gene-for-gene interactions between lettuce and *Bremia.* Interestingly, some of the interactions could be influenced by temperature while others seemed to be non-responsive. Plants carrying the resistance genes *Dm6*, *Dm7*, *Dm11*, *Dm15*, and *Dm16* showed the tendency to become less resistant at lower temperatures, while plants with *Dm1*, *Dm2*, *Dm3*, *Dm4*, *Dm5*, *Dm10*, *Dm13* and, *Dm14* did not show changes in resistance. In temperature shift experiments they showed that a continued exposure to permissive temperatures was required for an effective expression of resistance. A similar phenomenon was described by Balass *et al.* (1993) for the *Cucumis melo /Pseudoperonospora cubensis interaction.* The inbred line PI 124111F of muskmelon is resistant at 21-25 °C by virtue of the resistance genes *Pc-1* and *Pc-2*, but becomes susceptible at 12-15 °C. The described loss of resistance is specific for *P. cubensis* since the same muskmelon line did not lose its resistance against *Sphaerotheca fuliginea* (powdery mildew) at 12 °C.

Recently, evidence was presented for functioning gene-for-gene resistance against *P. parasitica* without HR (Yu *et al.*, 1998). Arabidopsis plants mutated at the *DND1* locus are not able to form a HR but are nonetheless capable of defending themselves against *Peronospora.* In *dnd1* plants infection with the virulent isolate NOCO lead to a three-fold lower sporulation intensity than on the wild type. Mycelial growth in the tissue was not accompanied by cell necrosis or autofluorescence. The *dnd1* mutant has elevated levels of SA and *PR* gene transcripts that probably condition the constitutive systemic acquired resistance phenotype observed in these plants.

5.2. ELICITORS AND AVR GENE PRODUCTS

Knowledge of the genetics governing a plant/pathogen interaction is the first step in understanding how a plant might defend itself against its aggressor. However, it does not in itself explain the recognition event, the signal transduction cascade following this event and finally, the actual biochemical mechanisms based on defence gene activation that lead to a successful confinement and death of the pathogen.

Recognition of small protein elicitors secreted from the invading fungus, as elegantly described for *Cf*-gene mediated resistance of tomato to *Cladosporium fulvum* (DeWit, 1992), is not known in downy mildew systems. Attempts to find race-specific elicitors in *Bremia* using a similar approach as for *C. fulvum* were not successful (Crucefix *et al.*, 1987). A first step towards the identification of race-specific factors of a downy mildew might be found in the report on the occurrence of race-specific proteins at the surface of *S. graminicola* conidia (Kumar *et al.*, 1993). The proteins were shown to bind only to the cell wall of susceptible pearl millet genotypes. By developing restriction fragment length polymorphisms (RFLPs) as genetic markers for *B. lactucae*, and using them for co-segregation analysis with, among others, avirulence loci, linkage was detected between a RFLP locus and an avirulence gene, thus providing a potential starting point for chromosome walking to clone an avirulence gene (Hulbert *et al.*, 1988).

Knowledge on the possible structure of race-specific elicitors from downy mildews or their receptors in the plant might be deduced from studies with other oomycete pathogen/plant systems. Kamoun *et al.* (1994) described a gene product (elicitin) which might be considered as a species-specific elicitor in the interaction between *Phytophthora* species and tobacco. Specific binding sites on plasma membranes from soybean and parsley, respectively, have been described for a 13 amino acid fragment of a glycoprotein elicitor as well as for a heptoglucan elicitor from *Phytophthora sojae* (Cosio *et al.*, 1992; Nürnberger *et al.*, 1994, 1995).

5.3. STRUCTURAL BARRIERS

The strengthening of the cell wall upon pathogen attack can have several consequences for plant resistance. It might either directly hinder penetration or, by slowing down the infection process, allow the plant to activate additional defence mechanisms. Since downy mildews do not enter the plant tissue via wounds, the first major barrier to penetration is the epidermal cell wall or, when entering through stomata, the wall of the cells lining the substomatal cavity. By toughening the cell wall and therefore rendering it more resistant to pathogen-secreted hydrolases, the establishment of the first haustoria might be delayed or even made impossible. Since the pathogen possesses only a limited amount of nutrients from the spore the timely establishment of haustoria could be vital for its survival. In most interactions between plants and downy mildews, no differences between resistant and susceptible plants have been observed concerning the very initial penetration events. Cell wall strengthening seems to be more important in later phases of pathogen growth restriction.

Plants have different means of strengthening their cell walls. Among those, lignification (or accumulation of phenolics) and callose deposition have been described for many downy mildew systems, and usually, a good correlation between deposition of these substances and expression of resistance can be made. In muskmelon infected by *Pseudoperonospora cubensis,* major differences were observed between a susceptible and a resistant cultivar (Cohen *et al.*, 1989). In both cultivars the fungus was able to proliferate intercellularly and to form digitate haustoria in the host cells. Whereas no ultrastructural changes were observed in the susceptible cultivar within the first 144 h, major changes occurred in cells of resistant plants as early as 20 h after inoculation. These changes included a heavy deposition of paramural, layered, callose-like material along the inner surfaces of the host cell walls, enrichment of host cell walls with lignin-like material, and encasement of the haustoria with heavy deposits of callose-like materials. Interestingly, all these structural responses were not induced in the genotypically resistant cultivar at 12°C, the temperature where, as mentioned before, the plants express phenotypical susceptibility (Balass *et al.*, 1993).

Dai and co-workers (Dai *et al.*, 1995) investigated the reactions of three different species of grapevine, ranging from highly resistant, intermediate resistant to susceptible, to inoculation with *Plasmopara viticola*. In this case lignification only played a role in intermediate resistance. No lignification was observed in the resistant or the susceptible plants, respectively. Their data suggest that in this system it is mainly the rapidity of flavonoid formation that is determining a highly or intermediate resistance. Lettuce infected by *B. lactucae* also shows differential deposition of phenolics, measured as autofluorescence, depending on the genotype of the interaction (Bennett *et al.*, 1996). In a study involving 3 different isolates of *Bremia*, one causing a rapid HR (V0/11), a second one causing a later HR (CL9W), and finally, a virulent isolate (TV), two phases of phenolic deposition were identified. The primary deposition was localised around the penetration point and was found in incompatible as well as in compatible interactions prior to detection of irreversible membrane damage. The second more prominent phase of phenolics deposition started only after the occurrence of irreversible membrane damage, first in interactions leading to an early HR and later in those leading to a delayed HR. Significantly, inhibitors of mRNA synthesis (actinomycin D and cordycepin) reduced the autofluorescence response of the second phase deposition but the localised deposition around the penetration points was not affected.

Using Arabidopsis plants transgenic for a phenylalanine ammonia-lyase 1 promoter (*PAL1*)-β-glucuronidase (GUS) reporter construct and *P. parasitica* as the pathogen, Mauch-Mani and Slusarenko (1996) demonstrated that lignification played an important role in the establishment of resistance but was not the deciding factor between resistance and susceptibility. Treatment of plants with 2-hydroxyphenylaminosulphinyl acetic acid (1,1-dimethyl ester) (OH-PAS), a specific inhibitor of cinnamyl alcohol dehydrogenase in the lignification pathway, lead only to a shift towards susceptibility. However, when 2-aminoindan-2-phosphonic acid (AIP), a specific inhibitor of PAL, was used to treat the plants the incompatible interaction was changed into a compatible one. In this case not only lignification but also SA synthesis had been suppressed. Phloroglucinol-HCl staining of the incompatible interaction revealed that not only plant cell walls became lignified but *Peronospora* cell walls as well. Lignification of hyphae was also reported by Bennett *et al.* (1996) for *B. lactucae*. Plant cells usually have a pool of preformed phenylpropanoids in their vacuole (Friend, 1985) which, upon release, could activate peroxidases present in the apoplast leading to the oxidative linkage of the phenolic substances to the plant cell wall. In such a case, the fungal as well as the plant cell wall could serve as matrix for the polymeriation of these phenolics. The lignification of the fungal cell wall might not even be dependent on plant peroxidase activity, since studies have shown that the cell wall of *Bremia* itself is rich in peroxidase activity (Zinkernagel, 1986).

Callose deposition constitutes a further difference between the *Bremia*/lettuce and the *Peronospora*/Arabidopsis pathosystems. While no major depositions of callose in cells undergoing HR have been observed in the first system (Bennett, 1996), Arabidopsis cells react with a strong deposition of callose in the penetrated and surrounding cells (Parker *et al.*, 1993) and directly around the haustoria (B. Mauch-Mani, unpublished results). Interestingly, treatment of lettuce seedlings with an inhibitor of callose formation (2-deoxy-D-glucose) prior to infection with an avirulent isolate of *Plasmopara lactucae-radicis* lead to a switch to susceptibility. In this system, the haustoria usually get encased in a callose sheath which seems to play a prominent role in the defence ability of the plant (Stanghellini *et al.*, 1993). Further support for a role of callose in the *Peronospora*/Arabidopsis system is provided by observations of *lsd1* (lesion simulating disease resistance) mutants. These plants develop expanding lesions after inoculation with various pathogens and display resistance against further infections. Strong callose deposits were observed in and around cells penetrated by *Peronospora* isolates virulent on the wild type plants (Dietrich *et al.*, 1994). In this context it is interresting to note that in BABA-induced resistance against *P. parasitica* in Arabidopsis, callose deposition at the attempted points of penetration is sufficient to change a compatible intertaction into an incompatible one (Zimmerli et al., 2000).

5.4. PATHOGENESIS AND DEFENCE RELATED PROTEINS

Pathogenesis-related (PR) proteins were first described in 1970 independently by two groups. Both reported the synthesis of several proteins induced in tobacco plants exhibiting a hypersensitive response to tobacco mosaic virus (Gianinazzi *et al.*, 1970; van Loon and van Kammen, 1970). Since then, *PR* gene induction has been described for many host-pathogen systems and today the term PR protein comprises the intra and extra-cellular proteins that accumulate in plant tissues or cell cultures after pathogen attack or treatment with an elicitor (Bowles, 1990). Despite their ubiquitous presence and, for some of them, their known *in vitro* function, it is still not clear what exact role they play in plant defence.

The best correlation between PR expression and resistance where downy mildews are concerned has been presented by Alexander *et al.* (1993). They showed that by over-expressing *PR-1a* in tobacco plants, a significant increase in resistance against *Peronospora tabacina* and another oomycete, *Phytophthora parasitica* var *nicotianae* was achieved. Unfortunately, the biochemical function of *PR-1* is not known. A role for PR-proteins in defence against downy mildews can also be deduced from studies with Arabidopsis mutants. The constitutive lesion formation in absence of pathogens displayed in the *lsd* mutants is correlated with a constitutively increased expression of PRs (Dietrich *et al.*, 1994). In the constitutive immunity mutant *cim3*, PRs are constitutively expressed although necrotic lesions are absent, and the plants are resistant towards a virulent isolate of *P. parasitica* (Ryals *et al.*, 1996). An increase in resistance against *P. parasitica* was also described for the *cpr1* mutant, a constitutive PR expresser (Bowling *et al.*, 1994). The opposite is the case for the three following allelic mutants. The *nim* (Delaney *et al.*, 1995), the *npr* (Cao *et al.*, 1994) and the *sai1* mutant (Shah *et al.*, 1997) show a reduced accumulation of *PR* gene transcripts and a loss of resistance to avirulent isolates of *P. parasitica* compared to wild type plants. The gene was recently cloned. It codes for a IKB-like protein containing an ankyrin repeat (Cao *et al.*, 1997; Ryals *et al.*, 1997). This protein seems to function in the nucleus in response to salicylic acid (SA) treatment (cited in Zhou *et al.*, 1998). In most investigated systems, evidence was found that *PR-1*, *PR-2* and *PR-5* gene activation was mediated by a SA-dependent signal transduction pathway.

Another class of defence-related genes induced upon pathogen attack, the thionins and defensins, that encode small antifungal peptides, are controlled by a jasmonic acid (JA) mediated pathway (Bohlmann, 1994; Penninckx, 1996). Using the Arabidopsis *coi1* mutant which is affected in the JA-response pathway (Xie *et al.*, 1998) Thomma *et al.* (1998) showed that susceptibility to infection by *P. parasitica* was not altered in this mutant, whereas the plants were clearly more susceptible to *Alternaria brassicicola* and *Botrytis cinerea*. The results were confirmed by external application of methyl jasmonate to wild type plants: methyl jasmonate induced resistance to *Alternaria* but not to *P. parasitica*. However, a independence from the jasmonate pathway concerning induction of resistance to other downy mildews cannot be deduced since some oomycetes, as *Phytophthora infestans* on potatoes and tomatoes (Cohen *et al.*, 1993), *Pythium mastophorum* on Arabidopsis (Vijayan *et al.*, 1998) or *Pythium ultimum* on Norway spruce (Kozlowski *et al.*, 1998) can be controlled, at least to some extent, by jasmonate treatment. Indirect evidence pointing to a possible involvement of a jasmonate mediated pathway in plant defence is the observed rapid increase in lipoxygenase (LOX) activity in infected plants (reviewed in Siedow, 1991; Slusarenko,

1996). LOX activity is believed to generate, among others, signal molecules such as jasmonate or methyl jasmonate and to play a role in the initiation of lipid peroxidation leading to irreversible membrane damage, and thus, cell death.

Different genotypes of pearl millet with different susceptibility to *S. graminicola* were tested for LOX activity in seeds and seedlings (Nagarathna *et al.*, 1992). A good correlation between enzyme activity of the seeds and their reaction to downy mildew in the field was recorded. For example, maximum activity was found in seeds of highly resistant genotypes and minimum activity in the highly susceptible ones. The authors proposed therefore to use LOX activity as a biochemical marker for screening different genotypes of pearl millet for downy mildew resistance. When testing LOX activity in seedlings, they noted that in susceptible plants the activity decreased after inoculation with *S. graminicola* zoospores when compared with uninoculated controls, whereas in resistant plants a significant increase in the enzyme activity was observed on the second and third days after inoculation.

5.5. SALICYLIC ACID

In incompatible interactions between plants and pathogens an increase in free and conjugated SA has frequently been observed and is usually associated with the HR (Raskin, 1992). Despite intensive studies, the exact role(s) of SA in plant defence has not yet been established and it is also not clear whether SA synthesis is a cause or a consequence of a hypersensitive reaction. SA, applied in high concentrations has been shown to inhibit catalase activity (Chen and Klessig, 1991) and it was proposed that H_2O_2 may act as a second messenger for SA in the induction of defence reactions (Chen *et al.*, 1993) although these results have been questioned later (Schneider *et al.*, 1996; Sticher *et al.*, 1997; Mauch-Mani and Métraux, 1998). SA might also exert a direct antimicrobial effect (Raskin, 1992) and has been shown to play an important role in the induction of certain *PR* genes (Yalpani *et al.*, 1991; Enyedi *et al.*, 1992; Ryals *et al.*, 1996). The importance of SA in gene-for-gene resistance concerning downy mildews is best demonstrated in the following two examples. Arabidopsis plants transgenic for the *NahG* gene, encoding a bacterial salicylate hydroxylase catalyzing the conversion of SA into catechol, have constitutively very low levels of SA and have lost the ability to defend themselves against an avirulent isolate of *P. parasitica* (Delaney *et al.*, 1994). Treatment of genotypically resistant Arabidopsis plants with the PAL inhibitor AIP to block the phenylpropanoid biosynthetic pathway leading to the formation of SA results in a phenoptypically susceptible reaction when the plants are inoculated with an avirulent isolate of *P. parasitica* (Mauch-Mani and Slusarenko, 1996). In both cases, the absence of SA leads to a switch from resistance to susceptibility.

5.6. PHYTOALEXINS

Many plants react to pathogen and non-pathogen attack as well as to treatment with biotic and abiotic elicitors by synthesizing low molecular weight, generally lipophilic compounds which inhibit the growth of fungi and bacteria *in vitro* and usually accumulate at or around the site of infection (Kuc, 1995; Mansfield, 2000). These phytoalexins have a broad antimicrobial activity. They are produced from primary metabolic precursors feeding into secondary metabolic pathways and their synthesis depends on the coordinate activity of many enzymes.

Phytoalexin synthesis has also been described for plant/downy mildew systems. The major phytoalexin found in lettuce tissue after infection by *B. lactucae* is lettucenin A, a terpenoid. Its accumulation is closely associated with cell necrosis and the amount of phytoalexin produced is proportional to the number of cells having undergone an HR (Bennett *et al.*, 1994). In grapevine plants the restriction of hyphal development has been correlated with an increase in blue autofluorescence, mainly due to the presence of trans-resveratrol, a stilbene derivative, appearing early in the necrotic tissue after infection by *P. viticola* (Langcake and Pryce, 1976). Based on fungitoxicity studies against *P. viticola* sprorangia, Dercks and Creasy (1989) then proposed that trans-resveratrol should be considered as a phytoalexin of grapevine. Sunflowers infected by *Plasmopara halstedii* accumulate scopoletin, a coumarin, as a major phytoalexin (Tal and Robeson, 1986; Spring *et al.*, 1991). Interestingly, scopoletin seems to play a dual role in the resistance response of sunflower. Besides its antifungal activity, it is an inhibitor of IAA oxidase which is induced upon *P. halstedii* infection in sunflowers (Benz and Spring, 1995). Wildtype Arabidopsis plants produce the phytoalexin camalexin (an indole ring substituted at the 3 position with a sulphur containing moiety) (Slusarenko and Mauch-Mani, 1991; Tsuji *et al.*, 1992). Several phytoalexin deficient mutants (*pad*) which accumulate different, but in every case strongly reduced, amounts of camalexin have been isolated (Glazebrook and Ausubel, 1994). Inoculation of these mutants with 6 different avirulent isolates of *P. parasitica* showed, that all four *PAD* genes are required for the expression of resistance. The degree to which resistance was lost depended on which *PAD* gene was mutated. The most dramatic change, expressed as full susceptibility to four of six isolates, was observed in the *pad4* mutant. Double mutants between *pad1, pad2* and *pad3* exhibited additive shifts to moderate or full susceptibility to most of the *Peronospora* isolates tested (Glazebrook *et al.*, 1997). The *PAD4* gene was recently shown to play a role in the regulation of the defence response of the plants and not in the biosynthesis of camalexin itself (Zhou *et al.*, 1998).

It still remains to be shown whether phytoalexins play a role in restricting growth of an invading primary pathogen, are formed to prevent secondary infections at the initial penetration site, or are involved in both defence strategies.

6. Concluding remarks

A large, rapidly growing amount of information exists on the mechanisms plants utilise to defend themselves against pathogens. Despite the severe economical impact downy mildew diseases have worldwide in agriculture, horticulture and viticulture, unfortunately, plant/downy mildew interactions have not been the first choice research systems in the past, probably due to the difficulties encountered when working with obligate biotrophic organisms. Despite this fact, and thanks to colleagues such as Richard Michelmore and Ian Crute, who kept "faithful" to their downy mildew over the years, the volume of information about the mechanisms, and the underlying genetics, governing the interactions between plants and downy mildews has been constantly increasing. Through the availability of Arabidopsis mutants and *Peronospora* as a pathogen of Arabidopsis, research on resistance against downy mildews is now leading the field. We have powerful tools and the technology to dissect and analyze the components and the role of specific genes in host resistance. Hopefully, the newly gained knowledge will help to improve crop plants.

7. Acknowledgements

Support from the Swiss National Foundation (grant 31-42979) is gratefully acknowledged.

8. References

Agrios, G. N. (1988) *Plant Pathology* (London: Academic Press).

Alexander, D., Goodman, R.M., Gut-Rella, M., Glascock, C., Weyman, K., Friedrich, L., Maddox, D., Ahl Goy, P., Luntz, T., Ward, E. and Ryals, J.A. (1993) Increased tolerance to two oomycete pathogens in transgenic tobacco expressing pathogenesis-related protein 1a, *Proceedings of the National Academy of Science USA* **90**, 7327-7331.

Baker, B., Zambryski, P., Staskawicz, B. and Dinesh-Kumar, S.P. (1997) Signalling in plant-microbe interactions, *Science* **276**, 726-733.

Balass, M., Cohen, Y. and Bar-Joseph, M. (1993) Temperature-dependent resistance to downy mildew in muskmelon: Structural responses, *Physiological and Molecular Plant Pathology* **43**, 11-20.

Bartnicki-Garcia, S. (1970) Cell wall composition and other biochemical markers in fungus phylogeny, in J. B. Harborne (ed.) *Phytochemical Phylogeny*, Academic Press London, pp. 81-102.

Bennett, M., Gallagher, M., Fagg, J. Bestwick, C.S., Paul, T., Beale, M. and Mansfield, J.W. (1996) The hypersensitive reaction, membrane damage, and accumulation of autofluorescent phenolics in lettuce cells challenged by *Bremia lactucae*, *The Plant Journal* **9**, 851-865.

Bennett, M. H., Gallagher, M.D.S., Bestwick, C.S., Rossiter, J.T. and Mansfield, J.W. (1994) The phytoalexin response of lettuce to challenge by *Botrytis cinerea*, *Bremia lactucae* and *Pseudomonas syringae* pv. *Phaseolicola*, *Physiological and Molecular Plant Pathology* **44**, 321-333.

Bent, A. F. (1996) Plant disease resistance genes: Function meets structure, *The Plant Cell* **8**, 1757-1771.

Bent, A. F., Kunkel, B.N., Dahlbeck, D., Brown, K.L., Schmidt, R.L., Giraudat, J., Leung, J.L. and Staskawicz, B.J. (1994) RPS2 of *Arabidopsis thaliana*: A leucine-rich repeat class of plant disease resistance genes, *Science* **265**, 1856-1860.

Benz, A. and Spring, O. (1995) Identification and characterization of an auxin-degrading enzyme in downy mildew infected sunflower, *Physiological and Molecular Plant Pathology* **46**, 163-175.

Bittner-Eddy, D., Crute, I.R., Holub, E.B. and Beynon, J.L. (2000) RPP13 is a simple locus in *Arabidopsis thaliana* for alleles that specify downy mildew resistance to different avirulence determinants in *Peronospora parasitica*, *Plant Journal* **21**, 177-188.

Bohlmann, H. (1994) The role of thionins in plant protection, *Critical Reviews in Plant Science* **13**, 1-16.

Botella, M. A., Parker, J. E., Frost, L. N., Bittner-Eddy, P. D., Beynon, J. L.,Daniels, M. J. , Holub, E. B., Jones, J. D. G. (1998) Three genes of the Arabidopsis *RPP1* complex resistance locus recognize distinct *Peronospora parasitica* avirulence determinants, *The Plant Cell* **10**, 1847-1860.

Bowles, D. J. (1990) Defense-related proteins in higher plants, *Annual Review of Biochemistry* **59**, 873-907.

Bowling, S. A., Guo, A., Cao, H., Gordon, A.S., Klessig, D.F. and Dong, X. (1994) A mutation in Arabidopsis that leads to constitutive expression of systemic acquired resistance, *The Plant Cell* **6**, 1845-1857.

Cao, H., Glazebrook, J., Clarke, J.D., Volko, S. and Dong, X. (1997) The Arabidopsis *NPR1* gene that controls systemic acquired resistance encodes a novel protein containing ankyrin repeats, *Cell* **88**, 57-63.

Cao, H., Bowling, S.A., Gordon, A.S. and Dong, X. (1994) Characterization of an Arabidopsis mutant that is nonresponsive to inducers of systemic acquired resistance *The Plant Cell* **6**, 1583-1592.

Cavalier-Smith, T. (1986) The kingdom Chromista: Origins and systematics, in I. Round, Chapman, D.J. (ed) Progress in Phycological Research, Biopress Bristol, pp. 309-347.

Century, K. S., Holub, E.B. and Staskawicz, B.J. (1995) *NDR1*, a locus of *Arabidopsis thaliana* that is required for disease resistance to both a bacterial and a fungal pathogen, *Proceedings of the National Academy of Science USA* **92**, 6597-6601.

Century, K. S., Shapiro, A.D., Repetti, P.P., Dahlbeck, D., Holub, E.B. and Staskawicz, B.J. (1997) *NDR1*, a pathogen-induced component required for Arabidopsis disease resistance, *Science* **278** , 1963-1965.

Chen, Z. X. and Klessig, D.F. (1991) Identification of a soluble salicylic acid-binding protein that may function in signal transduction in the plant disease-resistance response, *Proceedings of the National Academy of Science USA* **88**, 8170-8183.

Chen, Z. X., Silva, H. and Klessig, D.F. (1993) Active oxygen species in the induction of plant systemic acquired resistance by salicylic acid, *Science* **262**, 1883-1885.

Chester, K. S. (1933). The problem of acquired physiological immunity in plants, *Quarterly Review of Biology* **8**, 275-324.

Clark, J. S. C. and Spencer-Phillips, P.T.N. (1993) Accumulation of photoassimilate by *Peronospora viciae* (Berk.) Casp. and leaves of *Pisum sativum*L.: evidence for nutrient uptake via intercellular hyphae, *New Phytologist* **124**, 107-119.

Cohen, Y. (1994) 3-Aminobutyric acid induces resistance against *Peronospora tabacina*, *Physiological and Molecular Plant Pathology* **44**, 273-288.

Cohen, Y., Gisi, U. and Niederman, T. (1993) Local and systemic protection against *Phytophthora infestans* induced in potato and tomato plants by jasmonic acid and jasmonic methyl ester, *Phytopathology* **83**, 1054-1062.

Cohen, Y., Eyal, H., Hanania, J. and Malik, Z. (1989) Ultrastructure of *Pseudoperonospora cubensis* in muskmelon genotypes susceptible and resistant to downy mildew, *Physiological and Molecular Plant Pathology* **34**, 27-40.

Cohen, Y., Reuveni, M. and Baider, A. (1999) local and systemic activity of BABA (DL-3-aminobutyric acid) against *Plasmopara viticola* in grapevines, European Journal of Plant Pathology **105**, 351-361.

Cosio, E. G., Frey, T. and Ebel, J. (1992) Identification of a high-affinity binding protein for a hepta-b-glucoside phytoalexin elicitor in soybean *European Journal of Biochemistry* **204**, 1115-1123.

Crucefix, D. N., Rowell, P.M., Street, P.F.S. and Mansfield, J.W. (1987) A search for elicitors of the hypersensitive reaction in lettuce downy mildew disease, *Physiological and Molecular Plant Pathology* **30**, 39-54.

Crute, I. R. and Dixon, G.R. (1981) Diseases caused by *Bremia* Regel, in D. M. Spencer (ed.) *The Downy Mildews*, London, Academic Press New York, San Francisco, pp. 421-460.

Crute, I. R. and Pink, A.C. (1996) Genetics and utilization of pathogen resistance in plants *The Plant Cell* **8**, 1747-1755.

Crute, I. R. and Norwood, J.M. (1981) The identification and characteristics of field resistance to lettuce downy mildew (*Bremia lactucae* Regel), *Euphytica* **30**, 707-717.

Dai, G. H., Andary C., Mondolot-Cosson L. and Boubals D. (1995) Histochemical studies on the interaction between three species of grapevine, *Vitis vinifera, V. rupestris* and *V. rotundifolia* and the downy mildew fungus, *Plasmopara viticola, Physiological and Molecular Plant Pathology* **46**, 177-188.

Delaney, T., Friedrich, L. and Ryals, J. (1995) Arabidopsis signal transduction mutant defective in chemically and biologically induced disease resistance, *Proceedings of the National Academy of Science USA* **92**, 6602-6606.

Delaney, T. P., Uknes, S., Vernoij, B., Friedrich, L., Weymann, K., Negrotto, D., Gaffney, T., Gut-Rella, M., Kessmann, H., Ward, E. and Ryals, J. (1994) A central role of salicylic acid in plant disease resistance, *Science* **266**, 1247-1250.

Derks, W. and Creasy, L.L. (1989) The significance of stilbene phytoalexins in the *Plasmopara viticola*-grapevine interaction, *Physiological and Molecular Plant Pathology* **34**, 189-202.

DeWit, P. J. G. M. (1992) Molecular characterization of gene-for-gene systems in plant-fungus interactions and the application of avirulence genes in control of plant pathogens, *Annual Review of Phytopathology* **30**, 391-418.

Dick, M. W. (1995) The straminipilous fungi. A new classification for the biflagellate fungi and their uniflagellate relatives with particular reference to Lagenidiaceous fungi., CAB International Mycological Paper No. 168.

Dickson, J. G., Syamananda, R. and Flangas, A.C. (1959) The genetic approach to the physiology of parasitism of the corn rust pathogens, *American Journal of Botany* **46**, 614-620.

Dietrich, R. A., Delaney, T.P., Uknes, S.J., Ward, E.J., Ryals, J.A. and Dangl, J.L. (1994) Arabidopsis mutants simulating disease resistance response, *Cell* **77**, 565-578.

Dixon, G. R., Tonkin, M.H. and Doodson, J.K. (1973) Colonization of adult lettuce plants by *Bremia lactucae*, *Annals of Applied Biology* **74**, 307-313.

Dyck, P. L. and Johnson, R. (1983) Temperature sensitivity of genes for resistance in wheat to *Puccinia recondita*, *Canadian Journal of Plant Pathology* **5**, 229-234.

Eenink, A.H. (1981) Partial resistance in lettuce to downy mildew (*Bremia lactucae*). I. Search for partially resistant genotypes and the influence of certain plant characters and environments on the resistance level, *Euphytica* **30**, 619-628.

Enyedi, A. J., Yalpani, N., Silverman, P. and Raskin, I. (1992) Localization, conjugation, and function of salicylic acid in tobacco during the hypersensitive reaction to tobacco mosaic virus, *Proceedings of the National Academy of Science USA* **89**, 2480-2484.

Flor, H. (1971) Current status of the gene-for-gene concept, *Annual Review of Phytopathology* **9**, 275-296.

Fraymouth, J. (1956) Haustoria of the Peronosporales, *Transactions of the British Mycological Society* **39**, 79-107.

Friend, J. (1985) Phenolic substances and plant disease, in C. F. van Sumere, Lea, P.J. (ed.) *The biochemistry of plant phenolics*, Clarendon Oxford, pp. 367-392.

Gabriel, D. W. and Rolfe, B.G. (1990) Working models of specific recognition in plant-microbe interactions, *Annual Review of Phytopathology* **28**, 365-391.

Gäumann, E. (1946) *Pflanzliche Infektionslehre* (Basel: Birkhäuser Verlag).

Geetha, S., Shetty, S. A., Shetty, H. S. and and Prakash, H. S. (1996) Arachidonic acid-induced hypersensitive cell death as an assay of downy mildew resistance in pearl millet *Annals of Applied Biology* **129**, 91-96.

Gianinazzi, S., Martin, C. and Vallée, J.C. (1970) Hypersensibilité aux virus, température et protéines solubles chez le *Nicotiana xanthi* n.c. Apparition de nouvelles macromolécules lors de la répression de la synthèse virale, *Comptes Rendus de l'Académie des Sciences Paris* **270**, 2383-2386.

Glazebrook, J. and Ausubel, F.M. (1994) Isolation of phytoalexin defficient mutants of *Arabidopsis thaliana* and characterization of their interactions with bacterial pathogens, *Proceedings of the National Academy of Science* **143**, 8955-8959.

Glazebrook, J., Zook, M., Mert, F., Kagan, I., Rogers, E. E., Crute, I. R., Holub, E. B., Hammerschmidt, R. and Ausubel, F. M. (1997) Phytoalexin-deficient mutants of Arabidopsis reveal that PAD4 encodes a regulatory factor and that four PAD genes contribute to downy mildew resistance, *Genetics* **146**, 381-92.

Görlach, J., Volrath, S., Knauf-Beiter, G., Hengy, G., Oostendorp, M., Staub, T., Ward, E., Kessmann, H. and Ryals, J. (1996) Benzothiadiazole, a novel class of inducers of systemic acquired resistance, activates gene expression and disease resistance in wheat, *The Plant Cell* **8**, 629-643.

Gousseau, H. D. M., Deverall, B.J. and McIntosh, R.A. (1985) Temperature-sensitivity of the expression of resistance to *Puccinia graminis* conferred by *Sr15,Sr9b*, and *Sr14* genes in wheat, *Physiological Plant Pathology* **27**, 335-343.

Grant, M. R., Godiard, L., Straube, E., Ashfield, T., Lewald, J., Sattler, A., Innes, R.W. and Dangl, J.L. (1995) Structure of the Arabidopsis RPM1 gene enabling dual specificity disease resistance, *Science* **269**, 843-846.

Griffith, J. M., Davis, A.J. and Grant, B.R. (1992) Target sites of fungicides to control oomycetes, in, W. Köller (ed.) *Target Sites of Fungicide Action*, CRC Press Boca Raton, pp. 69-100.

Hammerschmidt , R. and Kuc, J. (1995) *Induced resistance to disease in plants*, Kluwer Dordrecht.

Hammond-Kosack, K. E. and Jones, J.D.G. (1996) Resistance gene-dependent plant defense responses, *The Plant Cell* **8**, 1773-1791.

Hancock, J. G. and Huisman, O.C. (1981) Nutrient Movement in host-pathogen systems, *Annual Review of Phytopathology* **19**, 309-331.

Heath, M. C. (1974) Light and electron microscope studies of the interactions of host and non-host plants with cowpea rust - *Uromyces phaseoli* var. *vignae*, *Physiological Plant Pathology* **4**, 403-414.

Holliday, M. J., Long, M. and Keen, N.T. (1981) Manipulation of the temperature-sensitive interaction between soybean leaves and Pseudomonas syringae pv. glycinea - implications on the nature of determinative events modulating hypersensitive resistance *Physiological Plant Pathology* **19**, 209-216.

Holub, E., Crute, I., Brose, E. and Beynon, J. (1993) Identification and mapping of loci in Arabidopsis for resistance to downy mildew and white blister, in K. Davis, Hammerschmidt, R. (ed.) *Arabidopsis as a model for plant-pathogen interactions,*: American Phytopathological Society Press St Paul, MN, pp. 21-35.

Holub, E. B., Beynon, J. L. and Crute, I. R. (1994) Phenotypic and genotypic characterization of interactions between isolates of *Peronospora parasitica* and accessions of *Arabidopsis thaliana*, *Molecular Plant-Microbe Interactions* **7**, 223-239.

Hulbert, S. H., Ilott T. W., Legg E. J,. Lincoln S. E., Lander E. S. and Michelmore R. W. (1988) Genetic analysis of the fungus *Bremia lactucae* using restriction fragment length polymorphisms, *Genetics* **120**, 947-958.

Hunt, M. and Ryals, J. (1996) Systemic acquired resistance signal transduction, *Critical Reviews in Plant Science* **15**, 583-606.

Islam, M. R., Shepherd, K.W. and Mayo, G,M.E. (1989) Effect of genotype and temperature on the expression L genes in flax conferring resistance to rust, *Physiological and Molecular Plant Pathology* **35**, 141-150.

Jakab G., Cottier V., Toquin V., Rigoli G., Zimmerli L., Métraux J.-P., Mauch-Mani, B. (2001) beta-aminobutyric acid-induced resistance in plants, *European Journal of Plant Pathology*, **107**: 29-37.

Jones, E. S., Liu, C.J., Gale, M.D., Hash, C.T. and Witcombe, J.R. (1995) Mapping quantitative trait loci for downy mildew resistance in pearl millet, *Theoretical and Applied Genetics* **91**, 448-456.

Joos, H. J., Mauch-Mani, B. and and Slusarenko, A. J. (1996) Molecular mapping of the Arabidopsis locus RPP11 which conditions isolate-specific hypersensitive against downy mildew in ecotype RLD, *Theoretical and Applied Genetics* **92**, 281-284.

Judelson, H. S. and Michelmore, R. W. (1992) Temperature and genotype interactions in the expression of host resistance in lettuce downy mildew, *Physiological and Molecular Plant Pathology* **40**, 233-245.

Kamoun, S., Young, M., Glascock, C.B. and Tyler, B. (1993) Extracellular protein elicitors from *Phytophthora*: Host-specificity and induction of resistance to bacterial and fungal phytopathogens, *Molecular Plant-Microbe Interactions* **6**, 15-25.

Kamoun, S., Young, M., Förster, H., Coffey, M.D. and Tyler, B.T. (1994) Potential role of elicitins in the interaction between *Phytophthora* species and tobacco, *Applied and Environmental Biology* **60**, 1593-1598.

Kobe, B. and Deisenhofer, J. (1995) A structural basis of the interactions between leucine-rich repeats and protein ligands, *Nature* **374**, 183-186.

Koch, E. and Slusarenko, A. (1990) Arabidopsis is susceptible to infection by a downy mildew fungus *Plant Cell* **2**, 437-45.

Kozlowski, G., Buchala, A. and Métraux, J.-P. (1998) Methyl jasmonate protects Norway spruce (*Picea abies* L. Karst.) seedlings against *Pythium ultimum* Trow, *Physiological and Molecular Plant Pathology* **55**, 53-58.

Kuc, J. (1995) Phytoalexins, stress metabolism, and disease resistance in plants, *Annual Review of Phytopathology* **33**, 275-297.

Kumar, V. U., Meera, M.S., Hindumathy, C.K. and Shetty, H.S. (1993) Induced systemic resistance protects pearl millet plants against downy mildew disease due to *Sclerospora graminicola*, *Crop Protection* **12**, 458-462.

Kumar, V. U., Shishupala, S., Shetty, H. S. and Umesh-Kumar, S. (1993) Serological evidence for the occurrence of races in *Sclerospora graminicola* and identification of a race-specific surface protein involved in host recognition, *Canadian Journal of Botany* **71**, 1467-1471.

Langcake, P. and Pryce, R.J. (1976) The production of trans-resveratrol by *Vitis vinifera* and other members of the *Vitaceae* as a response to infection or injury, *Physiological Plant Pathology* **9**, 77-86

Lawrence, G. J., Finnegan, E.J., Ayliffe, M.A. and Ellis, J.G. (1995) The *L6* gene for flax rust resistance is related to the Arabidopsis bacterial resistance gene *RPS2* and the tobacco viral resistance gene *N*, *The Plant Cell* **7**, 1195-1206.

Lawton, K., Friedrich, L., Hunt, M., Weymann, K., Staub, T., Kessmann, H. and Ryals, J. (1996) Benzothiadiazole induces disease resistance in Arabidopsis by activation of the systemic acquired resistance signal transduction pathway, *Plant Journal* **10**, 71-82.

Lebeda, A. and Dolezal, K. (1995) Peroxidase isozyme polymorphism as a potential marker for detection of field resistance in *Cucumis sativus* to cucumber downy mildew (Pseudoperonospora cubensis (Berk. et Curt.) Rostov.), *Zeitschrift Fuer Pflanzenkrankheiten und Pflanzenschutz* **102**, 467-471.

Lebeda, A. R. K. (1991) Variation in the early development of *Bremia lactucae* on lettuce cultivars with different levels of field resistance, *Plant Pathology* **40**, 232-237.

Leonards-Schippers, C., Gieffers, W., Schafer-Pregel, R., Ritter, E., Knapp, S.J., Salamini,F. and Gebhardt, C. (1994) Quantitative resistance to *Phytophthora infestans* in potato: a case study for mapping in an allogamous plant species, *Genetics* **137**, 67-77.

Madamanchi, N. R. and Kuc, J. (1991) Induced systemic resistance in plants, in G. T. Cole, Hoch, H.C. (ed.) *The fungal spore and disease initiation in plants and animals*, Plenum Press New York, pp. 347-362.

Malamy, J., Carr, J.P., Klessig, D.F. and Raskin, I. (1990) Salicylic acid, a likely endogenous signal in the resistance response of tobacco to viral infection, *Science* **250**, 1002-1004.

Mansfield, J.W. (2000) Antimicrobial compounds and resistance. The role of phyoalexins and phytoanticipins, in A. Slusarenko, R.S.S. Fraser, and L.C. van Loon (eds) *Mechanisms of resistance to plant diseases*, Kluwer Academic Publishers, pp. 325-370.

Mauch-Mani, B., Schwinn, F.J. and Guggenheim, R. (1989) Early infection stages of the downy mildew fungi *Sclerospora graminicola* and *Peronosclerospora sorghi* in plants and cell cultures, *Mycological Research* **92**, 445-452.

Mauch-Mani, B., Croft, K.P.C. and Slusarenko. A.J. (1993) The genetic basis of resistance of *Arabidopsis thaliana* (L.) Heynh. to *Peronospora parasitica*, in K. Davis, Hammerschmidt, R. (ed.) *Arabidopsis as a model for plant-pathogen interactions*, American Phytopathological Society Press St Paul, MN, pp. 5-20.

Mauch-Mani, B. and Métraux, J.-P. (1998) Salicylic acid and systemic acquired resistance to pathogen attack, *Annals of Botany* **82**, 535-540.

Mauch-Mani, B. and Slusarenko, A. J. (1996) Production of salicylic acid precursors is a major function of phenylalanine ammonia-lyase in the resistance of Arabidopsis to *Peronospora parasitica*, *Plant Cell* **8**, 203-212.

Mauch-Mani, B. and and Slusarenko, A. J. (1994) Systemic acquired resistance in *Arabidopsis thaliana* induced by a predisposing infection with a pathogenic isolate of *Fusarium oxysporum*, *Molecular Plant Microbe Interactions* **7**, 378-383.

McDowell, J. M., Dhandaydham, M., Long, T. A., Aarts, M. G. M., Goff, S., Holub, E. B. and Dangl, J. L. (1998) Intragenic recombination and diversifying selection contribute to the evolution of downy mildew resistance at the *RPP8* locus of Arabidopsis, *The Plant Cell* **10**, 1861-1874.

McKinney, H. H. and Clayton, E.E. (1945) Genotype and temperature in relation to symptoms caused in Nicotiana by the mosaic virus, *Journal of Heredity* **36**, 323-331.

Métraux, J.-P., Signer, H., Ryals, J., Ward, E., Wyss-Benz, M., Gaudin, J., Raschdorf, K., Schmid, E., Blum, W. and Inverardi, B. (1990) Increase in salicylic acid at the onset of systemic acquired resistance in cucumber, *Science* **250**, 1004-1006.

Métraux, J.-P., Ahl Goy, P., Staub, T., Speich, J., Steinemann, A., Ryals, J., Ward, E. (1991) Induced systemic resistance in cucumber in response to 2,6-dichloro-isonicotinic acid and pathogens, in H. Hennecke, Verma, D.P.S. (eds.) *Advances in molecular genetics of plant microbe interactions*, Kluwer Academic Publishers Dordrecht, pp. 432-439.

Meyers, B. C., Chin, D.B., Shen, K.A., Sivaramakrishnan, S., Lavelle, D.O., Zhang, Z. and Michelmore, R.W. (1998) The major resistance gene cluster in lettuce is highly duplicated and spans several megabases, *The Plant Cell* **10**, 1817-1832.

Meyers, B. C., Shen, K. A., Rohani, P., Gaut, B. S. and Michelmore , R. W. (1998) Receptor-like genes in the major resistance locus of lettuce are subject to divergent selection, *The Plant Cell* , 1833-1846.

Michelmore, R. W. and Meyers, B.C. (1998) Clusters of resistance genes in plants evolve by divergent selection and a birth-and-death process, *Genome Research* **8**, 1113-11130.

Michelmore, R. W. (1995). Molecular approaches to manipulation of disease resistance, *Annual Review of Phytopathology* **33**, 393-427.

Mindrinos, M., Katagiri, F., Yu, G.L. and Ausubel, F.M. (1994) The Arabidopsis thaliana disease resistance gene RPS2 encodes a protein containing a nucleotide binding site and leucine-rich repeats, *Cell* **78**, 1089-1099.

Morris, S. W., Vernooij, B., Titatarn, S., Starrett, M., Thomas, S., Wiltse, C. C., Frederiksen, R. A., Bhandhufalck, A., Hulbert, S. and and Uknes, S. (1998) Induced resistance responses in maize, *Molecular Plant-Microbe Interactions* **11**, 643-658.

Nagarathna, K. C., Shetty, S. A., Bhat, S. G. and Shetty, H. S. (1992) The possible involvement of lipoxygenase in downy mildew resistance in pearl millet, *Journal of Experimental Botany* **43**, 1283-1287.

Nawrath , C. and Métraux, J.-P. (1999) Salicylic acid induction-deficient mutants of Arabidopsis express PR-2 and PR-5 and accumulate high levels of camalexin after pathogen inoculation, *Plant Cell*, **11**, 1393-1404.

Nes, W. D. (1987) Biosynthesis and requirement for sterols in the growth and reproduction of oomycetes, in G. Fuller, Nes, W.D. (ed.) *Ecology and Metabolism of Plant Lipids*, American Chemical Society Washington, pp. 304-328.

Nürnberger, T., Nennstiel, D., Hahlbrock, K. and Scheel, D. (1995) Covalent cross-linking of the *Phytophthora megasperma* oligopeptide elicitor to its receptor in parsley membranes, *Proceedings of the National Academy of Sciences USA* **92**, 2338-2342.

Nürnberger, T., Nennstiel, D., Jabs, T., Sacks, W.R., Hahlbrock, K. and Scheel, D. (1994) High affinity binding of a fungal oligopeptide elicitor to parsley plasma membranes triggers multiple defense responses, *Cell* **78**, 449-460.

Parker, J. E., Szabo, V., Staskawicz, B.J., Lister, C., Dean, C., Daniels, M.J. and Jones, J.D.G. (1993) Phenotypic characterization and molecular mapping of the *Arabidopsis thaliana* locus, *RPP5*, determining disease resistance to *Peronospora parasitica*, *The Plant Journal* **4**, 821-831.

Parker, J. E., Coleman, M. J., Szabo, V., Frost, L. N., Schmidt, R., van der Biezen, E. A., Moores, T., Dean, C., Daniels, M. J. and Jones, J. D. (1997) The Arabidopsis downy mildew resistance gene *RPP5* shares similarity to the toll and interleukin-1 receptors with *N* and *L6*, *Plant Cell* **9**, 879-94.

Parker, J. E., Holub, E. B., Frost, L. N., Falk, A., Gunn, N. D. and Daniels, M. J. (1996) Characterization of eds1, a mutation in Arabidopsis suppressing resistance to *Peronospora parasitica* specified by several different *RPP* genes, *Plant Cell* **8**, 2033-46.

Parlevliet, J. E. (1992) Selecting components of partial resistance, in H. T. Stalker, Murphy, J.P. (ed.) *Plant Breeding in the 1990s, Proceedings of the symposium on plant breeding in the 1990s*, CAB International Wallingford.

Penninckx, I. A. M. A., Eggermont, K., Terras, F.R.G., Thomma, B.P.H.J., De Samblanx, G.W., Buchala, A., Métraux, J.-P., Manners, J.M. and Broekaert, W.F. (1996) Pathogen-induced systemic activation of a plant defensin gene in Arabidopsis follows a salicylic acid-independent pathway involving components of the ethylene and jasmonic acid responses, *The Plant Cell* **8**, 2309-2323.

Pfannenstiel, M. A. and Niblett, C.L. (1978) The nature of resistance of Agroticums to wheat streak mosaic virus, *Phytopathology* **68**, 1204-1209.

Raskin, I. (1992) Role of salicylic acid in plants, *Annual Review of Plant Physiology and Plant Molecular Biology* **43**, 439-463.

Raskin, I. (1992) Salicylate, a new plant hormone, *Plant Physiology* **99**, 799-803.

Rasmussen, J. B., Hammerschmidt, R. and Zook, M.N. (1991) Systemic induction of salicylic acid accumulation in cucumber after inoculation with *Pseudomonas syringae* pv *syringae*, *Plant Physiology* **97**, 1342-1347.

Reuveni, R., Shimoni M. and Crute I. R. (1991) An association between high peroxidase activity in lettuce *Lactuca sativa* and field resistance to downy mildew *Bremia lactucae*, *Journal of Phytopathology* **132**, 312-318.

Ricci, P., Bonnet, P., Huet, J.C., Sallantin, M., Beauvais-Cante, F., Bruneteau, M., Billard, V., Michel, G. and Pernollet, J.C. (1989) Structure and activity of proteins from pathogenic *Phytophthora* eliciting necrosis and acquired resistance in tobacco. *European Journal of Biochemistry* **183**, 555-563.

Robinson, R. A. (1969) Disease resistance terminology, *Review of Applied Mycology* **48**, 593-606.

Ross, A. F. (1966) Systemic effects of local lesion formation, in A. B. R. Beemster, Dijkstra, J. (ed.) *Viruses of Plants*, North-Holland Publishing Amsterdam, pp. 127-150.

Ryals, J. A., Weymann, K., Lawton, K., Friedrich, L., Ellis, D., Steiner, H.-Y., Johnson, J., Delaney, T.P., Jesse, T., Vos, P. and Uknes, S. (1997) The Arabidopsis NIM1 protein shows homology to the mammalian transcription factor inhibitor IκB, *The Plant Cell* **9**, 425-439.

Ryals, J. A., Neuenschwander, U.H., Willits, M.G., Molina, A., Steiner, H.-Y. and Hunt, M.D. (1996) Systemic acquired resistance, *The Plant Cell* **8**, 1809-1819.

Schneider, M., Schweizer, P., Meuwly, P. and Métraux, J.-P. (1996) Systemic acquired resistance in plants, *International Review of Cytology* **168**, 303-340.

Schroeder, W. T., Provvidenti, R., Barton, D.W. and Mishanec, W. (1965) Temperature differentiation of genotypes for BV2 resistance in *Pisum sativum*, *Phytopathology* **56**, 113-117.

Shah, J., Tsui, F. and Klessig, D.F. (1997) Charaterization of a salicylic acid insensitive mutant (*sai1*) of Arabidopsis thaliana, identified in a selective screen utilizing the SA-inducible expression of the *tms2* gene, *Molecular Plant-Microbe Interactions* **10**, 69-78.

Shen, K. K, Meyers, B.C., Islam-Faridi, M.N., Chin, D.B., Stelly. D.M. and Michelmore, R.W. (1998) Resistance gene candidates identified by PCR with degenerate oligonucleotide primers map to clusters of resistance genes in lettuce, *Molecular Plant-Microbe Interactions* **11**, 815-823.

Siedow, J. N. (1991) Plant lipoxygenase: structure and function, *Annual Review of Plant Physiology and Plant Molecular Biology* **42**, 145-188.

Slusarenko, A. J. and Mauch-Mani, B. (1991) Downy mildew of *Arabidopsis thaliana* caused by *Peronospora parasitica*: a model system for the investigation of the molecular biology of host-pathogen interactions, in H. Hennecke, Verma, D.P.S.i (eds.) *Advances in molecular genetics of plant microbe interactions*, Kluwer Academic Publishers Dordrecht, pp. 280-283.

Slusarenko, A. J. (1996) The role of lipoxygenase in resistance of plants to infection, in G. J. Piazza, (ed.) *Lipoxygenase and lipoxygenase pathway enzymes*, AOCS Press Champaign, I L, pp. 176-197.

Spencer, D. M. (1981) in D. M. Spencer (ed.) *The Downy Mildews*, London, Academic Press New York, San Francisco.

Spring, O., Benz, A. and Faust, V. (1991) Impact of downy mildew (*Plasmopara halstedii*) infection on the development and metabolism of sunflower, *Journal of Plant Disease and Protection* **98**, 597-604.

Stanghellini, M. E., Rasmussen, S.L. and Vandermark, G.J. (1993) Relationship of callose deposition to resistance of lettuce to *Plasmopara lactucae-radicis*, *Phytopathology* **83**, 1498-1501.

Stegmark, R. (1991) Comparison of different inoculation techniques to screen resistance of pea lines to downy mildew, *Journal of Phytopathology* **133**, 209-215.

Stegmark, R. (1992) Diallel analysis of the inheritance of partial resistance to downy mildew in peas, *Plant Breeding* **108**, 111-117.

Stegmark, R. (1994) Downy mildew on peas (*Peronospora* viciae f. sp. pisi). *Agronomie* **14**, 641-647.

Stegmark, R. (1991) Selection for partial resistance to downy mildew in peas by means of greenhouse tests, *Euphytica* **53**, 87-96.

Stegmark, R. (1990) Variation of virulence among Scandinavian isolates of *Peronospora viciae* fsp *pisi* pea downy mildew and responses of pea genotypes, *Plant Pathology* **39**, 118-124.

Sticher, L., Mauch-Mani, B. and Métraux, J.-P. (1997) Systemic acquired resistance, *Annual Review of Phytopathology* **35**, 235-270.

Stolle, K., Zook, M., Shain, L., Hebard, F. and Kuc, J. (1988) Restricted colonization of *Peronospora tabacina* and phytoalexin accumulation in immunized tobacco leaves, *Phytopathology* **78**, 1193-1197.

Tal, B. and Robeson, D.J. (1986) The induction, by fungal inoculation, of ayapin and scopoletin biosynthesis in *Helianthus annuus*, *Phytochemistry* **25**, 77-79.

Thomma, B. P. H. J., Eggermont, K., Penninckx, I.A.M.A., Mauch-Mani, B., Vogelsang, R., Cammue, B.P.A. and Broekaert, W.F. (1998) Separate jasmonate-dependent and salicylate-dependent defense-response pathways in Arabidopsis are essential for resistance to distinct microbial pathogens, *Proceedings of the National Academy of Science USA* **95**, 15107-15111.

Ton, J., Van Pelt, J.A., Van Loon, L.C. and Pieterse, C.M.J. (2001) Differential effectiveness of salicylate-dependent, and jasmonate- and ethylene-dependent induced resistance in Arabidopsis. *Molecular Plant Microbe Interactions*, in press.

Tör, M., Holub, E. B., Brose, E. , Musker, R. , Gunn , N., Can, C., Crute, I. R. and Beynon ,J. L. (1994) Map positions of three loci in *Arabidopsis thaliana* associated with isolate-specific recognition of *Peronospora parasitica* (Downy Mildew), *Molecular Plant-Microbe Interactions* **7**, 214-222.

Tsuji, J., Jackson, E.P., Gage, D.A., Hammerschmidt, R. and Somerville, S.C. (1992) Phytoalexin accumulation in *Arabidopsis thaliana* during the hypersensitive response to *Pseudomonas syringae* pv *syringae*, *Plant Physiology* **98**, 1304-1309.

Uknes, S., Mauch-Mani, B., Moyer, M., Potter, S., Williams, S., Dincher, S., Chandler, D., Slusarenko, A., Ward, E. and Ryals, J. (1992) Acquired resistance in Arabidopsis, *The Plant Cell* **4**, 645-656.

Uknes, S., Winter, A., Delaney, T., Vernoij, B., Friedrich, L., Morse, A., Potter, S., Williams, S., Ward, E. and Ryals, J. (1993) Biological induction of systemic acquired resistance in Arabidopsis, *Molecular Plant-Microbe Interactions* **6**, 692-698.

Van Loon, L. C. (1997) Induced resistance in plants and the role of pathogenesis-related proteins, *European Journal of Plant Pathology* **103**, 753-765.

Van Loon, L. C. and van Kammen, A. (1970) Polyacrylamide disc electrophoresis of the soluble leaf proteins from *Nicotiana tabacum* var. "Samsun" and "Samsun NN" II. Changes in protein constitution after infection with tobacco mosaic virus, *Virology* **40**, 199-211.

Vijayan, P., Shockey, J., Lévesque, C.A., Cook, R.J. and Browse, J. (1998). A role for jasmonate in pathogen defense of Arabidopsis, *Proceedings of the National Academy of Science USA* **95**, 7209-7214.

Ward, E. R., Uknes, S.J., Williams, S.C., Dincher, S.S., Wiederhold, D.L., Alexander, D.C., Ahl-Goy, P., Métraux, J.-P. and Ryals, J.A. (1991) Coordinate gene activity in response to agents that induce systemic acquired resistance, *The Plant Cell* **3**, 1085-1094.

Warren, R. F., Henk, A., Mowery, P., Holub, E. and Innes, R.W. (1998) A mutation within the leucine-rich repeat domain of the Arabidopsis disease resistance gene *RPS5* partially suppresses multiple bacterial and downy mildew resistance genes, *The Plant Cell* **10**, 1439-1452.

Weymann, K., Hunt, M., Uknes, S., Neuenschwander, U., Lawton, K., Steiner, H. Y. and Ryals, J. (1995) Suppression and restoration of lesion formation in Arabidopsis lsd mutants, *Plant Cell* **7**, 2013-2022.

White, R. (1979) Acetyl salicylic acid (aspirin) induces resistance to tobacco mosaic virus in tobacco, *Virology* **99**, 410-412.

Whitham, S., Dinesh-Kumar, S.P., Choi, D., Hehl, R., Corr, C. and Baker, B. (1994) The product of the tobacco mosaic resistance gene N: similarity to Toll and the interleukin-1 receptor, *Cell* **78**, 1101-1115.

Woods, A. M., Faggs, J. and Mansfield, J.W. (1988) Fungal development and irreversible membrane damage in cells of *Lactuca sativa* undergoing the hypersensitive reaction to the downy mildew fungus *Bremia lactucae*, *Physiological and Molecular Plant Pathology* **32**, 483-498.

Xie, D.-X., Feys, B.F., James, S., Nieto-rostro, M. and Turner, J.G. (1998) COI1: An Arabidopsis gene required for jasmonate-regulated defense and fertility, *Science* **280**, 1091-1094.

Yalpani, N., Silverman, P., Wilson, T.M.A., Kleier, D.A. and Raskin, I. (1991) Salicylic acid is a systemic signal and an inducer of pathogenesis-related proteins in virus-infected tobacco, *The Plant Cell* **3**, 809-818.

Ye, X. S., Jarlfors, U., Tuzun, S., Pan, S.Q. and Kuc, J. (1991) Biochemical changes in cell walls and cellular responses of tobacco leaves related in systemic resistance to blue mold inducted by tobacco mosaic virus, *Canadian Journal of Botany* **70**, 49-57.

Ye, X. S., Pan, S.Q. and Kuc, J. (1989) Pathogenesis-related proteins and systemic resistance to blue mold and tobacco mosaic virus induced by tobacco mosaic virus, *Physiological and Molecular Plant Pathology* **35**, 161-175.

Yu, I. C., Perker, J. and Bent, A.F. (1998) Gene-for-gene disease resistance without the hypersensitive response in Arabidopsis *dnd1* mutant, *Proceedings of the National Academy of Science USA* **95**, 7819-7824.

Yu, L. M. (1995) Elicitins from *Phytophthora* and basic resistance in tobacco, *Proceedings of the National Academy of Science USA* **92**, 4088-4094.

Yuen, J.E. and Lorbeer, J.W. (1984) Field resistance of crisphead lettuce to *Bremia lactucae*, *Phytopathology* **74**, 149-152.

Zhou, N., Tootle, T.L., Tsui, F., Klessig, D.F. and Glazebrook, J. (1998) *PAD4* functions upstream from salicylic acid to control defense responses in Arabidopsis, *The Plant Cell* **10**, 1021-1030.

Zimmerli, L., Jakab, G., Métraux, J.P. and Mauch-Mani, B. (2000) Potentiation of pathogen-specific defense mechanisms in Arabidopsis by beta-aminobutyric acid, *Proceedings of the National Academy of Sciences USA* **97**:12920-12925.

Zinkernagel, V. (1986) Untersuchungen zur Anfälligkeit und Resistenz von Kopfsalat (*Lactuca sativa*) gegen falschen Mehltau (*Bremia lactucae*) III. Peroxidase-, peroxidatische Katalase- und Polyphenoloxidase-Aktivitäten, *Journal of Phytopathology* **115**, 257-266.

ASPECTS OF THE INTERACTIONS BETWEEN WILD *LACTUCA* SPP. AND RELATED GENERA AND LETTUCE DOWNY MILDEW (*BREMIA LACTUCAE*)

A. LEBEDA[1], D.A.C. PINK[2] and D. ASTLEY[2]
[1]*Department of Botany, Faculty of Science, Palacký University, Šlechtitelů 11, CZ-783 71 Olomouc-Holice, Czech Republic*
[2]*Department of Plant Genetics and Biotechnology, Horticulture Research International, Wellesbourne, Warwick CV35 9EF, UK*

1. Introduction

Biological diversity (biodiversity) is defined as the total variability within all living organisms and the ecological complexes they inhabit. Genetic diversity at the level of the species, i.e. genetic variability or differences between populations of a single species and among individuals within a population (Glowka *et al.*, 1994), is the basic component of ecosystem biodiversity and is of crucial importance. Recent estimates consider that 85-90% (i.e. 250,000 out of 270,000-300,000) of species of higher plants are known (IPGRI, 1993; Glowka *et al.*, 1994). Only about 5,000 species are used as human food, of which only three crops (maize, rice and wheat) supply almost 60% of the calories and protein that humans derive from plants (IPGRI, 1993). The genetic base and variability of the main agricultural crops are generally very narrow and the contribution of genes from wild relatives has often been limited (Zohary and Hopf, 1994; Smartt and Simmonds, 1995). This is the main reason for the collection, maintenance, research and utilization of plant genetic resources (Guarino *et al.*, 1995). In this decade, the discussion on biological variation has broadened to include the biodiversity of fungi (Hawksworth, 1991). Currently at least 70,000 (to 120,000) species of fungi are known. However, the total number of species is probably about 1.5 million (Hawksworth, 1991). Some of these are parasites of land plants (Ellis and Ellis, 1997), and on current estimates there are about 8,000 species of parasitic fungi (Allen *et al.*, 1999). Downy mildews, one of the most important groups of plant parasitic fungi (Spencer, 1981), belong taxonomically to the oomycetes. Downy mildews form a very large and diverse group of highly specialized fungi within the orders Peronosporales and Sclerosporales (Dick, 2000, this volume). All downy mildews are obligate biotrophic parasites with growth and sporulation on undamaged host tissue, mostly leaves (Lebeda and Schwinn, 1994; Lucas *et al.*, 1995). Parasites such as these represent an important part of plant pathosystems.

Pathosystems can be classified as natural (wild pathosystem) or artificial (crop pathosystem). A wild plant pathosystem is autonomous, i.e. control is primarily due to interactions between the three basic components: the host, the pathogen and the environment (disease triangle). Current knowledge of this type of pathosystem is still

P.T.N. Spencer-Phillips et al. (eds.), Advances in Downy Mildew Research, 85–117.

limited. Recently there has been increasing interest in the variability and coevolutionary consequences in wild plant pathosystems (Burdon *et al.*, 1989; Heath, 1991; Burdon, 1993; Dinoor and Eshed, 1997) as a basic key for better potential management of plant-pathogen interactions. There is only limited information on the variability of the interactions between wild *Lactuca* species and *Bremia lactucae* (Lebeda, this volume; Lebeda and Pink, 1995, 1998; Lebeda *et al.*, 2001c). A crop pathosystem is more or less deterministic and can be considered to have fourth component represented by humans (disease square). Methods of "human-guided" integrated control of downy mildew on lettuce (*L. sativa*) have been developed (Crute, 1989). However, the effectiveness of these control measures is not complete or durable (Lebeda, 1998a). Genetic resources of wild *Lactuca* spp. are considered as very important sources of many characters, including disease resistance, and play an unreplaceable role in lettuce breeding (Pink and Keane, 1993; Ryder, 1998).

1.1. PLANT-PARASITE INTERACTIONS AND THEIR VARIATION IN WILD PLANT PATHOSYSTEMS

The interactions between plants and fungal parasites vary enormously at different levels. In general, the basis of this variation (biodiversity) can be seen at the level of individuals. Wild and crop plants interacting with microorganisms are characterized by two main types of expression, basic incompatibility (nonhost plant resistance or nonspecific basic resistance) and basic compatibility (host plant resistance or parasite-specific resistance) (Lebeda, 1984a; Heath, 1991; Mauch-Mani, this volume). Basic or nonhost resistance is maintained by a mixture of passive and active defences that exhibit inter- and probably intra-specific differences and is characteristic for most of the plant-parasite interactions in nature. There are two types of parasite-specific host plant resistance: race-specific and race-nonspecific. The first is very often controlled by single dominant genes in the plant and is the result of a gene-for-gene interaction between host and pathogen (Crute, 1998). The second is often, but not exclusively, under the control of many genes (Crute, 1985). In crop plants the first type of resistance is most common and is sometimes considered as "an artifact of agriculture" (Barrett, 1985). However, intensive research on the variation in wild plant pathosystems has confirmed that this type of resistance is also very common in natural plant-parasite associations (Burdon *et al.*, 1996). In contrast, race-nonspecific resistance has not been found to be a common phenomenon in nature. The coevolutionary consequences, involving reciprocal adaptations of interacting plant and fungus lineages, of the above mentioned types of resistance have been discussed in detail by Heath (1997). Apart from a general overview given by Lebeda and Pink (1995, 1997) and Lebeda *et al.* (2001c) the variation for resistance in wild species of *Lactuca* and related genera has not been summarized.

Variation in host plant resistance is mostly mirrored by the diversity of fungal pathogens. Genetic variation in pathogen populations is generated by the processes of spontaneous mutation, sexual recombination and somatic hybridization. Variation in population structure can also occur through migration or a range of cytological and molecular changes (Burdon, 1993). The study of the structure and variation of pathogen populations in natural plant communities is still limited in comparison with those in crop pathosystems (Burdon, 1993). Genetic variation for virulence in *B. lactucae* in natural plant communities is poorly understood (Lebeda, this volume).

2. Characterization of the host plant (*Lactuca* spp.)

2. 1. TAXONOMY, VARIABILITY AND ECOLOGY OF *LACTUCA* SPP. AND RELATED GENERA

Currently the genus *Lactuca*, in the broad sense, is considered as part of a very large and heterogeneous group included in the family Compositae (Asteraceae). The Compositae is one of the largest families of flowering plants, with about 1,100 currently accepted genera and about 25,000 species. Most of its members are evergreen shrubs, subshrubs or perennial rhizomatous herbs, but tap-rooted or tuberous-rooted perennials, and biennial and annual herbs are also frequent (Heywood, 1978). Recent developments in the systematics of this family were summarized by Jeffrey (1995). According to this taxonomic treatment the Compositae is considered a monophyletic group with three subfamilies: Barnadesioideae, Cichorioideae and Asteroideae. One systematic group of Cichorioideae is tribus Lactuceae (Cichorieae) which is divided into eight subtribes. Tribe Lactuceae comprises about 70 genera and about 2,300 species. The genus *Lactuca* and all closely related genera belong to the subgroup Crepis within tribe Lactuceae.

Feráková (1976,1977) and Stace (1997) gave basic descriptions of the genus *Lactuca* L.. Frietema (1994) considered the common (vernacular) name of the species belonging to this genus to be "lettuce". This genus forms annual, biennial or perennial, glabrous or pubescent herbs with abundant latex, rarely shrubs, rhizomatous, sometimes with underground stolons or with fusiform tuberous roots. The genus *Lactuca* is distributed in temperate and warm regions of the northern hemisphere (Europe, Asia and Indonesia, North and Central America and Africa). The genus comprises about 100 species, which are prevalent in Asia and Africa (Lebeda, 1998b; Lebeda *et al.*, 2001a). The present distribution of European *Lactuca* spp. seems to have its centre of origin in the Mediterranean region (Feráková, 1977), where there is the greatest diversity (Lebeda *et al.*, 2001b).

Lactuca species are very diverse and occur in ecologically different habitats. *L. serriola, L. saligna* and *L. virosa* are frequent ruderal species preferring disturbed soils (Lebeda *et al.*, 2001b). *L. aurea, L. quercina* and to some extent *L. sibirica* are common in woodland habitats. Some Mediterranean species (e.g. *L. graeca, L. perennis, L. tenerrima, L. viminea*) are commonly calciphilous plants growing on rocky slopes (Lopez and Jimenez, 1974). The original habitat of *L. acanthifolia* and *L. tatarica* is on cliffs at the seashore; however, *L. tatarica* is also found as a weed in Asia and Europe (Feráková, 1977).

Opinions differ on the taxonomical position of *Lactuca*, and in this paper we will follow the modified classification of the genus *Lactuca* as summarized by Feráková (1977), Rulkens (1987), and recently in more detail by Lebeda (1998b) and Lebeda and Astley (1999), i.e. classification into nine sections and groups, respectively. The grouping is based on basic cytological, morphological, ecological and biogeographical differences. Table 1 includes *Lactuca* L. spp. and related genera (*Cicerbita* Wallr., *Ixeris* Cass., *Mycelis* Cass. (Cass.), *Steptorhamphus* Bunge, *Youngia* Cass.) which are considered in this paper from the viewpoint of interactions with *B. lactucae*, and are maintained in gene banks. Detailed information about these taxonomic groups and species was reported by Lebeda (1998b).

2.2. GERMPLASM COLLECTIONS OF *LACTUCA* SPP. AND THEIR BIODIVERSITY

There is increasing interest in *L. sativa* and wild *Lactuca* spp. for sampling, conservation, evaluation and utilization of germplasm collections (Boukema *et al.*, 1990; McGuire *et al.*, 1993; Lebeda and Křístková, 1995; Lebeda *et al.*, 2001b). Table 1 lists five internationally important wild *Lactuca* germplasm collections. In Europe: The Centre for Genetic Resources The Netherlands (CGN) in Wageningen, The Netherlands (Boukema *et al.*, 1990; Hintum van and Soest van, 1997; Soest van and Boukema, 1997); Research Institute of Crop Production Prague, Gene Bank Division in Olomouc-Holice in close cooperation with Department of Botany, Palacký University in Olomouc (Lebeda and Křístková, 1995; Lebeda and Pink, 1997; Křístková and Lebeda, 1999) in the Czech Republic and Horticulture Research International, Genetic Resources Unit (HRI, GRU) in Wellesbourne, UK (pers. comm., D. Astley, HRI, Genetic Resources Unit, Wellesbourne, UK). For the USA, detailed information on the *Lactuca* collections was reported by McGuire *et al.* (1993).

There are currently about 22 wild *Lactuca* species available in these genebanks, which belong to the primary, secondary and tertiary genepools plus five other species of related genera (*Cicerbita, Ixeris, Mycelis, Steptorhampus*) (Table 1). There are nearly 2,000 accessions available, however, this number is constantly changing. We conclude that only about 20% of the known species of the genus *Lactuca* is available in the genebanks (Lebeda *et al.*, 2001a). Mostly the species in the collections belong to the European geographic group. A very limited number of Asian, African and American species is available. For example some species (e.g. *L. azerbaijanica, L. georgica, L. scarioloides*) which are now considered as likely primary progenitors of cultivated lettuce (Zohary, 1991; Vries de, 1997) are missing in genebank collections. Also the available number of accessions of some wild taxa is not very high. The substantial part of variability in collections, represented by the number of accessions, is concentrated in only three species (*L. serriola, L. saligna* and *L. virosa*) which represent 90-95% of available accessions (Lebeda *et al.*, 2001a).

2.3. CROP-WEED COMPLEXES OF *L. SATIVA – L. SERRIOLA*, THE POTENTIAL FOR GENE FLOW BETWEEN WILD- AND CROP-PATHOSYSTEMS

The basic biosystematic concept of crop-weed complexes was developed by Pickersgill (1981). Van der Maesen (1994) defined crop-weed complexes as cultivated and related wild or weedy plants growing together and influencing each other through introgression. The phenomenon of introgression could be important when crops and wild relatives are interfertile and these crosses may result in weedy intermediates as part of "crop-weed complexes". *L. serriola* is distributed world-wide and is one of the most common weed plants in Europe (Hegi, 1987; Lebeda *et al.*, 2001b). *L. serriola* is interfertile with *L. sativa*, but the possibility of introgression is relatively low because of the autogamous character of both species. Nevertheless, cross-pollination is possible. Thompson (1933) and Thompson *et al.* (1958) demonstrated that cross-pollination to the extent of 2-3 % occurred between lettuce cultivars grown in close proximity. The occurrence of spontaneous interspecific hybridization between *L. sativa* and *L. serriola* was reported by Bohn and Whitaker (1951) and Lindquist (1960). Feráková (1977) also stated that

TABLE 1. Survey of wild species of *Lactuca* and related genera germplasm collections (pers. comm., D. Astley, HRI, Genetic Resources Unit; Boukema *et. al.* (1990); McGuire *et al.* (1993); Lebeda and Křístková (1995); Soest van and Boukema (1995, 1997); Lebeda and Pink (1997)) [*]

Lactuca spp. and related species[1]	USDA[2] Salinas	USDA Pullman	Gene Bank CGN[3] Wageningen	HRI[4] Wellesbourne	RICP[5] Olomouc
L. aculeata	+	+	+	+	+
L. altaica	+	+	+	+	+
L. biennis	+	-	+	-	+
L. canadensis	+	-	+	+	+
L. capensis	+	-	-	-	-
L. dregeana	-	+	+	-	+
L. floridana	+	-	-	-	-
L. gracoglossum (?)	+	-	-	-	-
L. graminifolia	+	-	-	-	+
L. homblei	-	-	+	+	+
L. indica	+	+	+	+	+
L. livida	+	+	-	+	-
L. perennis	+	+	+	+	+
L. quercina	+	+	+	+	-
L. saligna	+	+	+	+	+
L. serriola	+	+	+	+	+
L. taraxacifolia (?)	+	-	-	-	+
L. tatarica	-	-	+	+	+
L. tenerrima	-	-	+	-	+
L. undulata	-	+	-	-	-
L. viminea	+	+	+	+	+
L. virosa	+	+	+	+	+
Cicerbita alpina	-	-	-	+	+
C. bourgaei	-	-	-	+	-
Ixeris dentata	-	-	+	+	+
Mycelis muralis	-	-	+	+	+
Steptorhamphus tuberosus	-	-	+	-	-
Total number of:					
species	27	17	12	18	17	19
accessions: [6]	1700	250	220	680	220	300

[*] Updated survey of *Lactuca* spp. germplasm collections is available on: www.plant.wageningen-ur.nl/cgn/ildb
[1] Taxonomical classification according to Lebeda (1998b), Lebeda and Astley (1999)
[2] USDA = United States Department of Agriculture (USA); [3] CGN = Centre for Genetic Resources (The Netherlands); [4] HRI = Horticulture Research International, Genetic Resources Unit (UK);
[5] RICP = Research Institute of Crop Production Praha, Gene Bank Division Olomouc (Czech Republic)
[6] Approximate number of accessions available
+ = present, - = absent
? = taxonomical position of the species is unclear (see following remarks):
L. gracoglossum is not described under this name in the available literature, however this species could probably be a misleading description of *L. indica* L. var. *dracoglossa* (Makino) Kitam. (syn. *L. dracoglossa* Makino), mentioned also as a cultivated plant by Ohwi (1965);
L. taraxacifolia is described under this name in recent literature only by Chalkuziev (1974), for additional comments see Lebeda (1998b).

spontaneous natural hybrids between *L. sativa* and *L. serriola* could be quite common. Frietema (1994) questioned whether the recent spread of *L. serriola* can be linked to the cultivation of lettuce. Despite this information, precise experimental data have not been reported on this phenomenon. However, because there is broad genetic variation in natural populations of *L. serriola* in the occurrence of race-specific R-factors to *B. lactucae* (Crute, 1990; Lebeda, 1998b) we can predict an intensive gene flow between populations of *L. serriola* in the wild pathosystem. The same phenomenon is expected also for populations of lettuce downy mildew (*B. lactucae*) (Lebeda and Blok, 1991).

3. Characterization of parasite (*Bremia lactucae*) biodiversity

3.1. TAXONOMY, BIOGEOGRAPHY AND BIOLOGY OF *BREMIA* SPP. AND *B. LACTUCAE*

The genus *Bremia* Regel belongs to the order Peronosporales, which are considered as evolutionarily advanced oomycetes (Dick, 2000, this volume). Detailed characterization of downy mildew taxonomy and biology are summarized elsewhere (Lebeda and Schwinn, 1994; Lucas *et al.*, 1995; Hall, 1996; Dick, this volume).

The genus *Bremia* Regel is distinguished morphologically from other genera of the Peronosporales by its asexual conidiophores (Savulescu, 1962; Lebeda *et al.*, 1989). In the taxonomic designation within the genus *Bremia,* there has been much confusion as to the species divisions (Savulescu, 1962; Skidmore and Ingram, 1985). Currently only two species of *Bremia* are described; these parasitize different host families. *B. lactucae* Regel parasitizes plants of the family Compositae and *B. graminicola* Naumov is a pathogen of grass species of *Arthraxon* Beauv. (Poaceae). This species of *Bremia* is characterized by small conidia and can be clearly distinguished from *B. lactucae* isolates (Crute and Dixon, 1981). However, according to some other authors (Dick, 2000) the existence of more *Bremia* species is expected.

The distribution of *B. lactucae* is worldwide, occurring on all continents except Antarctica (Marlatt, 1974). *B. lactucae* occurs on cultivated lettuce wherever the crop is grown, especially in regions with a temperate climate (Crute and Dixon, 1981). However, there is incomplete information about the distribution of the fungus on wild *Lactuca* species. Our knowledge of the geographic distribution of the fungus is based mainly on its occurrence on lettuce (*L. sativa*), chicory (*Cichorium intybus*), endive (*C. endivia*) and *Sonchus* spp. The economic impact of lettuce downy mildew can be very high (Crute, 1992a). The epidemiological impact of *B. lactucae* on wild *Lactuca* spp. is not known. The rather common occurrence of the parasite on *L. serriola* in the territory of former Czechoslovakia was recorded by Lebeda (1984b). The first quantitative results were obtained in the Czech Republic where the occurrence of the fungus on weedy plants of *L. serriola* was studied. A high frequency (up to 79%) of naturally infected *L. serriola* populations were found. However, the degree of infection and expression of symptoms was variable (Lebeda, this volume).

3.2. HOST RANGE AND SPECIALIZATION (*FORMAE SPECIALES*) OF *B. LACTUCAE*

B. lactucae is known to infect more than two hundred species of Compositae from about 40 genera of the tribes Lactuceae, Cynareae and Arctotideae (Crute and Dixon, 1981). Some new host plants have been recorded recently (Tao Chia-Feng, 1965; Wittman, 1972; Dange *et al.*, 1976; Lucas and Dias, 1976). From cross-inoculation experiments it is evident that *B. lactucae* is highly specific and from the viewpoint of parasitism mostly limited to the same genus of plants (Lebeda and Syrovátko, 1988). On the basis of these experiments and analysis of available data eleven *formae speciales* (f.sp.) of *B. lactucae* (Table 2) have been proposed (Skidmore and Ingram, 1985; but see Dick, this volume).

TABLE 2. *Bremia lactucae* specialization at the level of *formae speciales* (modified according to Skidmore and Ingram (1985))

Forma specialis of *B. lactucae*	Host plant genera
f.sp. *centaureae*	*Centaurea* spp.
f.sp. *cirsii*	*Cirsium* spp.
f.sp. *crepidis*	*Crepis* spp.
f.sp. *hieracii*	*Hieracium* spp.
f.sp. *lactucae*	*Lactuca* spp., *Mycelis* spp.
f.sp. *lapsanae*	*Lapsana* spp.
f.sp. *leontodi*	*Leontodon* spp.
f.sp. *picridis*	*Picris* spp.
f.sp. *senecionis*	*Senecio* spp.
f.sp. *sonchi*	*Sonchus* spp.
f.sp. *taraxaci*	*Taraxacum* spp.

In a comprehensive review, Marlatt (1974) described 31 wild *Lactuca* species as hosts of *B. lactucae*. Critical review of these data confirmed that some of these species are synonyms, and some species were not recognized as naturally infected host species, but were artificially inoculated by *B. lactucae* originating from *L. sativa*. A detailed analysis of all the available published data related to the host range of *B. lactucae* showed that infection on naturally growing plants has been recorded in only a limited number of *Lactuca* spp. and closely related genera. Table 3 summarizes data on wild species of *Lactuca* and related genera which may be considered as natural hosts of *B. lactucae* f.sp. *lactucae*. However, some of these species are still questionable as hosts. Surprisingly, there is no clear information on the occurrence of *B. lactucae* on *L. saligna* which is a common species in Europe (Feráková, 1976, 1977; Hegi, 1987). Nevertheless, one questionable report of *B. lactucae* occurrence on *L. saligna* originate from Viennot-Bourgin (1956). Of the 16 wild *Lactuca* species described in Europe (Feráková, 1976) only seven are definitely known as a natural hosts of *B. lactucae* (Table 3). *L. serriola* can be considered as the most common *Lactuca* spp. weed occurring in Europe (Hegi, 1987). However, except in the Czech Republic (Lebeda, 1984b, this volume), there is no detailed information on the natural occurrence of lettuce downy mildew and its epidemiological impact on this species. For example in England the natural occurrence of *B. lactucae* on *L. serriola* has not been verified (Crute, 1990).

TABLE 3. Wild species of *Lactuca* and related genera reported as natural hosts of *B. lactucae*

Lactuca and other species	References
L. biennis	Shaw (1958)
L. canadensis	Berlese and De Toni (1888), Savulescu (1962), Marlatt (1974)
L. formosana	Sawada (1919)
L. hirsuta	Schweizer (1920)
L. indica	Ito and Tokunaga (1935), Ling and Tai (1945)
L. ludoviciana	Shaw (1958)
L. perennis	Schweizer (1920)
L. quercina	Fischer (1892), Berlese and De Toni (1888), Schweizer (1920), Savulescu (1962), Skidmore and Ingram (1985)
L. raddeana	Ito and Tokunaga (1935)
L. serriola	Fischer (1892), Erwin (1920, 1921), Savulescu (1962), Lebeda (1984b, this volume), Lebeda *et al.* (2001b)
L. sibirica	Jorstad (1964)
L. tatarica	Shaw (1958)
L. viminea	Schweizer (1920), Savulescu (1962)
L. virosa	Losa Espana (1942)
Cicerbita alpina	Jorstad (1964)
C. plumieri	Skidmore and Ingram (1985)
Ixeris dentata	Ito and Tokunaga (1935)
I. chinensis	Ling and Tai (1945)
I. japonica	Ito and Tokunaga (1935)
Mycelis muralis	Berlese (1898), Schweizer (1920), Skidmore and Ingram (1985)

Detailed information, based on cross-inoculation experiments, about the specificity and virulence variability of isolates originating from *L. serriola* and related species is not available. Only *B. lactucae* isolates originating from *L. sativa* and *L. serriola* have been studied (Table 4). More details on the variability for virulence of these isolates is given below. Ling and Tai (1945) investigated the specificity of *B. lactucae* isolates from *Lactuca indica* and *Ixeris chinensis*. All these data support the earlier opinion of Ogilvie (1943), later summarized by Lebeda and Syrovátko (1988), that wild Compositae plants (other than *Lactuca* spp.) infected by *B. lactucae* cannot serve as a source of inoculum, and therefore do not play an important role in the epidemiology of the fungus and its interaction with cultivated lettuce.

3.3. VARIATION FOR VIRULENCE (PHYSIOLOGICAL RACES) IN *B. LACTUCAE*

In host plant–downy mildew interactions, there is generally a very clear expression of compatibility or incompatibility. The classification of races is based on the pattern of compatible and incompatible reactions on differential host genotypes (Lebeda and Schwinn, 1994).Variability in virulence was first described in *B. lactucae* in the USA. Similar situations were found subsequently in most countries where lettuce is widely grown (Crute, 1987). The variation was first categorized in terms of physiological races (Crute and Dixon, 1981), later as specific virulence determinants (virulence phenotypes, virulence factors) based on the interpretation of the host-pathogen interaction in a gene-for-gene relationship (Crute, 1987). This approach provided a basis for the application of population studies in distinct pathogen populations (Lebeda and Jendrulek, 1987).

There is very limited knowledge of virulence variation of downy mildews in wild pathosystems (Renfro and Shankara Bhat, 1981). Recently such variation was decribed for *Peronospora parasitica* and *Albugo candida* on naturally growing populations of *Arabidopsis thaliana* (Holub and Beynon, 1997). For *B. lactucae* only isolates originating from natural populations of *L. serriola* have been investigated for specific virulence variation (Lebeda, 1984b, 1986, 1989, 1990a, this volume; Lebeda and Boukema, 1991). These studies showed for the first time that variation for specific virulence on wild host species existed within peronosporaceous fungi. In comparison with isolates obtained from the crop pathosystem, the isolates derived from the wild plant pathosystem were characterized by rather simple virulence phenotypes. However, in a relatively small number of isolates (eight), four different races were recognized (Lebeda, 1984b). Generally, *B. lactucae* isolates from the wild pathosystem were characterized by possessing virulence factors (v-factors) mostly matching *Dm* genes or resistance factors (R-factors) located or derived from *L. serriola*. This phenomenon was confirmed in recent studies with a larger number of isolates originating from more locations and a greater range of host resistance genes (Lebeda, this volume). From these investigations it is evident that in the wild plant pathosystem *L. serriola – B. lactucae*, differential interaction (race-specificity) occurs as in some other wild host-pathogen associations (Burdon *et al.*, 1996).

The comparison of pathogenicity (expressed by means of the frequency of sporulation occurrence and by the degree of sporulation) of isolates of *B. lactucae* originating from *L. sativa* and *L. serriola* on the same set of *L. serriola* accessions shows isolates from *L. serriola* as being significantly more pathogenic than those from *L. sativa* (Lebeda, 1986). This is in full agreement with the data on virulence variation interpreted in genetical terms (Lebeda, this volume).

3.4. SURVIVAL STRATEGIES OF *B. LACTUCAE* AND THE POTENTIAL GENE FLOW FROM THE WILD- TO THE CROP-PATHOSYSTEM

In the life cycle of *B. lactucae*, sexual reproduction has a crucial role. The product of sexual reproduction, the oospores, are important propagules for pathogen survival and the initiation of primary infection (Crute, 1992d). Michelmore and Ingram (1980) demonstrated that *B. lactucae* is predominantly heterothallic and that two sexual compatibility types exist, B1 and B2. Oospores of lettuce downy mildew were observed and studied only in *L. sativa* (Crute and Dixon, 1981). However, from experimental results it is obvious that sexual reproduction also occurs in weedy growing *L. serriola* plants. It was confirmed that isolates of *B. lactucae* from *L. serriola* are heterothallic and both compatibility types were found (Lebeda and Blok, 1990). These results suggest that sexual recombination of *B. lactucae* isolates from the wild- and crop-pathosystems may occur. More detailed field and experimental studies are required to test this assumption.

4. Specificity of interactions between wild *Lactuca* spp. and *Bremia lactucae*

Research on the specificity of interactions between plants and parasitic microorganisms is not possible without germplasm collections of the host and parasite, and a basic knowledge of their biology and evolution. General aspects related to the specificity of interactions between plants and downy mildews have been discussed and summarized by many authors during the last two decades (Crute, 1981; Lucas and Sherriff, 1988; Michelmore *et al.*, 1988; Lebeda and Schwinn, 1994; Lucas *et al.*, 1995; Holub and Beynon, 1997). This section will analyse and discuss the specificity of interaction between wild *Lactuca* spp. and *B. lactucae* on an individual, tissue and cellular level, and from the mechanistic, genetic and breeding viewpoint.

4.1. WILD GERMPLASM OF *LACTUCA* SPP. AS SOURCES OF RESISTANCE AND THEIR UTILIZATION IN BREEDING

There has been a general tendency in plant breeding programmes towards the rapid elimination of genetic variability (biodiversity), coupled with the notion that strictly uniform crop populations are the universal ideal. However, as genetically uniform cultivars have been grown over wider areas, their vulnerability to diseases has increased (Lenné and Wood, 1991). The need to broaden the genetic base of crops by use of wild germplasm resources has been widely recognized and there is a continuing requirement that the genetic diversity of crop species and their wild relatives is conserved as a resource for future programmes of genetic improvement (Crute, 1992b; Dinoor and Eshed, 1997). Wild relatives of crop plants have proved to be fruitful sources of resistance genes against different pathogens (Burdon and Jarosz, 1989). Some important examples were summarized by Lenné and Wood (1991).

The first observations and experimental studies related to the specificity of interactions between wild *Lactuca* species and *B. lactucae* occurred in the beginning of the 1920s (Erwin, 1920; Schweizer, 1920) and in the 1930s (Schultz, 1937; Schultz and Röder, 1938). These studies mostly included only *L. serriola*, however, some observations were also made with *L. canadensis, L. ludoviciana, L. quercina* (Erwin, 1920) and *L. virosa* (Schultz and Röder, 1938) after infestation with *B. lactucae*. More intensive research in this area started in the 1940s and is ongoing. All the available literature data (from 27 papers), derived from experiments using well characterized plant material and pathogen isolates and artificial inoculation in controlled conditions, related to this topic are summarized in Table 4. The description of plant genera and species is according to recent taxonomic convention (Lebeda, 1998b). A total of 15 wild *Lactuca* species and another 8 species of 5 related genera have been screened for resistance to *B. lactucae*. In most of these studies *B. lactucae* isolates originating from *L. sativa* were used. A limited number of studies were made with isolates originating from *L. serriola* or other species (Table 4). A substantial part of knowledge on the specificity of interaction between wild *Lactuca* spp. and *B. lactucae* is therefore only based on pathogen isolates from *L. sativa*. However, there is also another gap in the information in these interactions, because most of the studies concerned interactions between *B. lactucae* and accessions of *L. serriola, L. saligna* and *L. virosa* (more than 10 papers), with only a few publications describing other interactions (Table 4). Early on this research had a botanical and pure phytopathological character, but later (post 1970s)

TABLE 4. Wild species of *Lactuca* and related genera screened for resistance to *Bremia lactucae* (compiled from all available literature sources published in the period 1945-1998)

Lactuca and related species [1]	Isolates of *B. lactucae* originating from			
	L. sativa	*L. serriola*	*L. indica*	*I. chinensis*
L. aculeata	+ (3)	+ (1)	-	-
L. altaica	+ (2)	+ (1)	-	-
L. aurea	+ (1)	-	-	-
L. biennis	+ (1)	+ (1)	-	-
L. dregeana	+ (2)	+ (1)	-	-
L. indica	+ (3)	+ (1)	+ (1)	+ (1)
L. perennis	+ (6)	+ (1)	-	-
L. quercina	+ (1)	-	-	-
L. saligna	+ (16)	+ (3)	-	-
L. serriola	+ (23)	+ (2)	-	-
L. tatarica	+ (2)	+ (1)	-	-
L. tenerrima	+ (1)	+ (1)	-	-
L. undulata	+ (1)	-	-	-
L. viminea	+ (4)	+ (2)	-	-
L. virosa	+ (11)	+ (3)	-	-
Cicerbita alpina	+ (1)	+ (1)	-	-
C. bourgaei	+ (1)	-	-	-
C. plumieri	+ (3)	-	-	-
Ixeris chinensis	+ (1)	-	+ (1)	+ (1)
I. dentata	+ (1)	+ (1)	-	-
Mycelis muralis	+ (6)	+ (1)	-	-
Steptorhamphus tuberosus	+ (1)	-	-	-
Youngia denticulata	+ (1)	-	-	-

[1]Taxonomical classification according to Lebeda (1998b), Lebeda and Astley (1999)
+ = interaction studied, - = interaction not studied,
( ) = number of references recorded in the literature
References used for compilation of Table 4:
Bannerot (1980), Bonnier *et al.* (1992), Crute (1990), Crute and Davis (1977), Crute and Dickinson (1976), Crute and Norwood (1981), Eenink (1974, 1980), Farrara and Michelmore (1987), Globerson *et al.* (1980), Gustafsson (1986, 1989), Lebeda (1983, 1984c, 1986, 1990b), Lebeda and Boukema (1991), Ling and Tai (1945), Netzer *et al.* (1976), Norwood *et al.* (1980), Norwood *et al.* (1981), Ogilvie (1945), Powlesland (1954), Verhoeff (1960), Wild (1948), Zink and Duffus (1969, 1973)

there was more focus on searching for new sources of resistance and genes suitable for practical lettuce breeding.

In commercial lettuce breeding programmes only a limited number of resistance genes have been utilized and these have differed between countries (Crute, 1987, 1992a; Reinink, 1999). However, the utilization of wild *Lactuca* species, as sources of resistance in practical breeding, started relatively early. Accessions of *L. serriola* (PI 91532, PI 167150) originating from Russia and Turkey were used in the 1930s in the USA as sources of resistance against *B. lactucae*. These sources created the background for a new generation of lettuce cultivars (Imperial 410, Calmar, Valmaine) for outdoor cropping which were introduced in the 1940s and 1950s (Whitaker *et al.*, 1958). All of these cultivars have race-specific resistance (Table 5).

In Europe, the utilization of wild *Lactuca* germplasm in lettuce breeding was based on two diverse strategies (The Netherlands and Great Britain). In the 1950s, genes originating from old German and French cultivars of *L. sativa* were used mostly (Crute, 1992a). At the end of the 1960s in The Netherlands an interspecific hybrid between *L. sativa* (cv. Hilde) and an accession of *L. serriola*, which is described in the literature as H x B, Hilde x *L. serriola* was released. Resistance derived from this material is assigned to the race-specific gene *Dm*11 (Crute, 1992c) (Table 5). In the 1970s and 1980s other sources of resistance to *B. lactucae*, derived from *L. serriola* with resistance genes (factors) described as *Dm*16 and R18 (Table 5), were used in the Netherlands. All of these genes have been used frequently in breeding programmes in Europe during the last twenty years. However, resistance based on these genes is no longer effective against some *B. lactucae* isolates (Lebeda, 1998a; Lebeda and Zinkernagel, 1998, 1999). From the end of the 1980s there was increasing interest (esp. in the Netherlands and U.K.) for the utilization of resistance located in the hybrid line *L. serriola* (Swedish) x *L. sativa* (Brun Hilde) (Lebeda, 1983, 1984c, 1986) and line CS-RL (Lebeda and Blok, 1991) derived from this material. Material comparable with CS-RL was also derived and used in practical breeding programmes of crisp lettuce at Horticulture Research International, Wellesbourne (Pink and McClement, 1996; Lebeda and Pink, 1998). On the basis of this material cv. Libusa by Rijk Zwaan (The Netherlands) (pers. comm., K. Reinink, Rijk Zwaan) and cv. Miura by Clause (France) (Anonymous, 1995) were released with resistance based on a single dominant gene (Lebeda and Pink, 1998; pers. comm., S. McClement, HRI, UK).

During the last two decades, researchers and breeders have been increasingly interested in the exploitation of new sources of resistance among wild *Lactuca* species. The most important potential sources of resistance are considered to be *L. saligna* and *L. virosa* (Lebeda, 1983, 1984c, 1986; Lebeda and Boukema, 1991; Lebeda and Reinink, 1994; Lebeda and Pink, 1998; Reinink, 1999). Several studies showed that wild species may possess a new and very interesting resistance to *B. lactucae* (Lebeda and Boukema, 1991; Bonnier *et al.*, 1992; Lebeda and Reinink, 1994; Lebeda and Pink, 1998). As a result of studies in the 1990s a new lettuce cultivar Titan (Sluis & Groot) with race-specific gene *Dm*6 and resistance derived from *L. saligna* (pers. comm., K. Reinink, Rijk Zwaan, The Netherlands) was released in the Netherlands. The original *L. saligna* accession used for this cross was collected in Israel (Eenink and Roelofsen, 1977). However, this resistance is no longer effective (Lebeda, 1998a) being overcome by some *B. lactucae* isolates found in Germany (Lebeda, 1997, 1998a; Lebeda and Zinkernagel, 1998, 1999; Zinkernagel *et al.*, 1998).

4.2. RESISTANCE MECHANISMS OF WILD *LACTUCA* SPP.

The specificity of interactions between plants and microorganisms is based on many defence mechanisms which can be considered and classified from different viewpoints (phenotypic expression, mechanistic, genetic, epidemiological, durability etc.). These and other attributes serve as the most important features for the classification of resistance mechanisms in plants (Lebeda, 1984a; Crute, 1985; Heath, 1997). From both the theoretical and practical (breeding) standpoint, plant resistance mechanisms can be classified into four categories: race-specific, race-nonspecific, field and nonhost resistance. A more detailed description of these types of resistance is given elsewhere (Lebeda, 1984a; Crute, 1985; Niks, 1987, 1988; Parlevliet, 1992; Heath, 1997). These categories will be used for the characterization of current knowledge of resistance mechanisms to *B. lactucae* in wild *Lactuca* spp. and/or the expression of interaction specificity.

4.2.1. *Race-specific resistance*

It was recognized early on that the interaction between lettuce (*L. sativa*) cultivars and *B. lactucae* was clearly differential, i.e. race-specific (Jagger, 1924). Later it was confirmed that all resistances used in breeding of lettuce cultivars were race-specific (Crute and Johnson, 1976; Lebeda, 1984d). The first genetical studies (Jagger and Whitaker, 1940) showed that resistance to downy mildew was inherited simply and this was confirmed by further research (see Crute (1992c) for review). On the basis of detailed analyses of the available data, Crute and Johnson (1976) constructed a gene-for-gene model, based on 10 matching gene pairs, which explained the cultivar x isolate interaction. Further classical and molecular genetic studies have shown race-specificity based on a complementary gene-for-gene relationship governs the interaction between both *L. sativa* and *L. serriola* with *B. lactucae*. The specificity is determined by dominant resistance *Dm* genes in the host which are matched by dominant factors for avirulence in the fungus (Crute, 1992c, 1998). Resistance genes in lettuce occur in distinct clusters, and at least five linkage groups have been recognized (Norwood and Crute, 1980; Hulbert and Michelmore, 1985; Bonnier *et al.*, 1994). Current research is focused on molecular genetics and mapping of these resistance genes by using molecular markers (Kesseli *et al.*, 1994; Maisonneuve *et al.*, 1994; Witsenboer *et al.*, 1995, 1997; Meyers *et al.*, 1998; Sicard *et al.*, 1999; Jeuken and Lindhout, 2000; Chin *et al.*, 2001).

During these investigations it was also recognized that the resistance of *L. serriola* to *B. lactucae* is characterized by the expression of clear race-specificity. At least five genes (*Dm5/8, Dm6, Dm11, Dm16* and R18) for race-specificity have been introgressed from *L. serriola* to *L. sativa* cultivars. The origin and characterization of these genes is summarized in Table 5. In seven accessions of *L. serriola* eight new resistance factors (designated R23-R30) dissimilar from known *Dm* genes, were identified in the Netherlands (Bonnier *et al.*, 1994). Evidence about a new, fifth, linkage group, which contains resistance factors R23 and R25, was also obtained. Some other R-factors were not assigned to a linkage group (Table 5). From other studies (Lebeda, 1983, 1986, unpubl. results; Norwood *et al.*, 1981; Bonnier *et al.*, 1992) it is evident that numerous additional resistance genes might be present in *L. serriola* accessions. The occurrence of race-specificity will also probably be a very common phenomenon in the *L. serriola* – *B. lactucae* pathosystem (Lebeda, 1984b, 1986, this volume; Farrara and Michelmore, 1987; Lebeda and Jendrulek, 1989; Lebeda and Petrželová, 2001).

The effectiveness of the expression of some *Dm* genes located in *L. serriola* can be dependent on environmental factors. Judelson and Michelmore (1992) showed that resistance (assessed as the absence of sporulation) based on *Dm6, Dm7, Dm11, Dm15*, and *Dm16* became less effective or ineffective at lower temperatures (below 10 °C). Also histological examination indicated that the extent of fungus development was dependent on temperature. The ecological and epidemiological consequences of this effect are not known.

The occurrence of race-specificity in other wild *Lactuca* species and related genera has not been analysed and confirmed. Most investigations were focused only on related *Lactuca* species. Our recent detailed analyses of all available published data (Table 4) has shown that the occurrence of race-specific resistance in wild *Lactuca* species is a common phenomenon (Table 6). In the section Lactuca, all of the species studied express race-specificity after inoculation with isolates of *B. lactucae* from *L. sativa* and *L. serriola*. The presence of race-specific resistance in *L.saligna* is questionable because most of the screened accessions expressed complete or incomplete resistance at both the seedling and adult stage (Lebeda, 1983, 1986; Lebeda and Boukema, 1991; Lebeda, 1997). A race-specific response was also confirmed in some species from other sections of the genus *Lactuca* (*L.viminea, L. tatarica, L. quercina, L. indica*). In other species (*L. aurea, L. taraxacifolia, L. undulata, L. capensis, L. homblei, L. biennis*) there is insufficient information to confirm the existence of this phenomenon (Table 6). The clear evidence of the occurrence of a race-specific response in some species of related genera is surprising (Table 6). However, there is very little (Table 7) or no information available on the genetics of this interaction.

4.2.2. *Race-nonspecific resistance*

Race-nonspecific or horizontal (nondifferential) resistance is characterized by some effectiveness against many different races of the pathogen, however, this type of resistance does not protect plants from becoming infected (Agrios, 1997). Genotypes with this type of resistance have a certain level of nonspecific resistance which can be measured in laboratory experiments by phenotypic expression. Generally, this resistance is controlled by many genes (minor gene resistance, polygenic resistance) which contribute to the expression of resistance (Crute, 1985). Sometimes this type of resistance is considered synonymous with field (or partial) resistance (Zadoks and Schein, 1979). The same is true for interactions of *L. sativa* and *B. lactucae* (Gustafsson, 1992). However, this is not an entirely correct view. The phenomenon of field resistance is identified from the epidemiological viewpoint (see part 4.2.3.) and is generally not measurable in laboratory experiments (Lebeda and Reinink, 1991).

There is only limited information available about race-nonspecific resistance in wild *Lactuca* species. The presence of race-nonspecific resistance has only been reported in *L. serriola* (Tables 5 and 6). Currently only two *L. serriola* accessions can be considered as potential sources of this type of resistance. Norwood *et al.* (1981), Lebeda (1983, 1984c, 1986) and Lebeda and Jendrulek (1989) recognised that accessions PI 281876 and PI 281877 in the seedling stage were infected by some *B. lactucae* isolates, however the intensity of sporulation was mostly very low and in some interactions followed by an expression of a necrotic response. According to Norwood *et al.* (1981) it is improbable that the resistance of these accessions is governed by the same factors since PI 281877 showed extensive necrosis, whereas the response of PI 281876 was more discrete. This characteristic response of this accession was recently confirmed by more detailed histological study (Lebeda and Pink, 1998).

There is limited information on the genetical background of this type of resistance. Norwood *et al.* (1981) found that it was not possible to clearly interpret the segregation data from F_2 progeny because of the sparse sporulation, frequent occurrence of necrosis and difficulty in classifying the infected seedlings into discrete categories. Current thoughts are that this resistance is based on some major gene(s) and modifiers (Table 5).

TABLE 5. Race-specific resistance genes (*Dm*) or factors (R) located or derived from *Lactuca serriola*

Dm gene (R-factor)	*L. serriola* accession (line)	Origin	Occurrence in *L. sativa* cultivars[1]	Linkage group	References
Dm5	PI 167150[2]	Turkey	Valmaine	2	2,6,9,12,14,15,30
Dm5/8+10	PI 91532[3]	USSR	Sucrine	2	2,6,9,12,14,15,18,30
	PI 167150	Turkey			
Dm8	PI 91532	USSR	Avoncrisp Calmar Salinas	2	as for *Dm5* or *Dm5/8*
Dm6	PI 91532	USSR	Sabine	1	2,6,7,9,12,14,15,18
Dm7	LSE/57/15	UK	Great Lakes Mesa 659	3	6,9,12,14,15,16,18
*Dm*11	IVT Wageningen	?	Capitan	3	5,6,12,14,15,16,20, 21
*Dm*15	PIVT 1309	Netherlands	*	1	1,6,11,12,15,18,19
*Dm*16	LSE/18	Czechoslo-vakia	Saffier Titania	1	8,12,15,18,19,27
Dm7+10+13	PI 114512	Sweden	Vanguard		5,12,25,29,31
	PI 114535	UK			
	PI 125819 (+*L.virosa*	Afghanistan			
	PI 125130)	Sweden			
R17	LS 102	France ?	*	2	25
R18	LS 17	France ?	Mariska	1	4,6,10,22,23,24
R19 (R18+?)	CS-RL	Sweden	Libusa	1	4,17,18,19,22,23,24, 28
	LJ88356		Miura		
Dm7+R23	CGN 5153	USSR (Krym)	*	3,5?	3,4,22
R24+R25	CGN14255	Hungary	*	3,5	3,4,22
R24+R26	CGN14256	Hungary	*	3,4	3,4,22
R24+R27	CGN14270	Hungary	*	3,4?	3,4,22
R24+R28	CGN14280	Hungary	*	3,?	3,4,22
R24+R29	PI 491178	Turkey	*	3,?	3,4,22
R30	PI 491229	Greece	*	1	3,4,22
R? (+modi-fiers,probably RNS)	PI 281876	Iraq	*	?	17,18,19,27,28

[1] = only selected examples; [2] = PI 167150 was originally described as *L. sativa* "Marul" originating from Turkey - Karadevar, near Mersin (Plant Inventory No. 156, USDA, Washington, 1955), however Leeper *et al.* (1963) mentioned it as *L. serriola;* [3] = PI 91532 was originally described as *L. virosa* originating from USSR (Abhasia) (Plant Inventory No. 106, USDA, Washington, 1932)
? = not known or unclear, * = *Dm* gene or R-factor not yet located in *L. sativa* cultivar(s)
References (for Table 5):
1) Bannerot (1980), 2) Bohn and Whitaker (1951), 3) Bonnier *et al.* (1992), 4) Bonnier *et al.* (1994), 5) Crute (1987), 6) Crute (1990), 7) Crute and Dunn (1980), 8) Crute and Gordon (1984), 9) Crute and Johnson (1976), 10) Crute *et al.* (1986), 11) Eenink (1980), 12) Farrara *et al.* (1987), 13) Gustafsson (1989), 14) Hulbert and Michelmore (1985), 15) Ilott *et al.* (1989), 16) Johnson *et al.* (1978), 17) Lebeda (1983), 18) Lebeda (1984d), 19) Lebeda (1986), 20) Lebeda (1989), 21) Lebeda (1992), 22) Lebeda (1997, 1998b), 23) Lebeda and Blok (1991), 24) Lebeda and Boukema (1991), 25) Maisonneuve *et al.* (1994), 26) McGuire *et al.* (1993), 27) Norwood *et al.* (1981), 28) Pink and McClement (pers. comm., HRI), 29) Thompson and Ryder (1961), 30) Witsenboer *et al.* (1995), 31) Yuen and Lorbeer (1983)

A. LEBEDA, D.A.C. PINK AND D. ASTLEY

TABLE 6. Possible resistance mechanisms of wild species of *Lactuca* (according to taxonomical position) and related species to *Bremia lactucae*

Section/subsection *Lactuca* or related species	Response to *B. lactucae* from		Possible resistance mechanisms
	L. sativa	*L. serriola*	
Lactuca			
Lactuca			
L. aculeata	+/-	+/-	RS
L. altaica	+/-	+/-	RS [1,2]
L. dregeana	+/-	+/-	RS
L. livida	+	+	RS ? [3]
L. saligna	+/-	(-)	RS, NR ?
L. serriola	+/-	+/-	RS, RNS, FR
L. virosa	+/-	+/-	RS, FR
Cyanicae			
L. perennis	+/-	-	RS
L. tenerrima	+/-	+/-	RS [1,2]
Phaenixopus			
L. viminea	+/-	(-)	RS
Mulgedium			
L. taraxacifolia	+	-	? [4,5]
L. tatarica	+/-	(-)/-	RS [2,5]
Lactucopsis			
L. aurea	-	*	?
L. quercina	+/-	+/-	RS
Tuberosae			
L. indica	+/-	-	RS
Micranthae			
L. undulata	-	*	?
African group			
L. capensis	-	*	? [4]
L. homblei	-	(-)	RS? [3,5]
North American group			
L. biennis	-	+/-	RS [5]
Related species			
Cicerbita alpina	+/(-)	+	RS [4]
C. bourgaei	-	*	?
C. plumieri	+/-	*	RS
Ixeris dentata	+/-	+/-	RS [4]
I. chinensis	+	*	RS ?
Mycelis muralis	+/-	+/-	RS
Steptorhamphus tuberosus	(-)	*	?
Youngia denticulata	-	*	?

+ = compatible reaction, - = incompatible reaction, +/- = differential reaction, (-) = incomplete resistance, * = reaction not studied, ? = unknown or unclear, [1] = only compatible reactions were recorded in published papers; [2] = incompatible reaction (Lebeda, 1998b, unpubl.); [3] = Lebeda (1998b, unpubl.); [4] = Lebeda (1997), [5] = Lebeda and Petrželová (2001); RS = race-specific resistance, RNS = race-nonspecific resistance, FR = field resistance, NR = nonhost resistance

TABLE 7. Genetics of resistance to *B. lactucae* in other wild *Lactuca* species

Lactuca spp.	Accession (line)	Origin	Possible genetical background of resistance	References
L. saligna	"Bet Dagan" line (LSA/92/1)	Israel	2 recessive r-factors (cross with *L. sativa*)	Globerson *et al.* (1980), Netzer *et al.* (1976)
	LSA/92/1 and others	Israel	Indication that some acces. carry *Dm* genes	Crute *et al.* (1981), Crute and Norwood (1979)
	LSA/92/1	Israel	2 dominant R-factors (crosses with *L. serriola, L. virosa, L. aculeata*)	Pink and McClement (1998, unpubl.)
			Recessive factors (in crosses with *L. sativa*)	Pink and McClement (1998, unpubl.)
	CGN5147	Italy	*Dm* gene ? unidentified	Reinink *et al.* (1993, 1995)
	CGN5271	France	recessive	Lebeda and Reinink (1994)
	CGN5327	Spain	r-factor(s)	
	CGN9313	Israel		
	VGB1629	France		
	PI 261653	Portugal	CMV (simple Mendelian inher.), *B. lactucae* ?	Lebeda (1983, 1984c, 1986), Norwood *et al.* (1981), Robinson and Provvidenti (1963)
L. virosa	PI 261651 (line B9056)	Nether-lands	1 dominant R-factor (*L. sativa* x *L. virosa* hybrids)	Maxon Smith (1984), Maxon Smith and Langton (1989)
	LVIR/26 (line K7095)	France		
	LVIR/57/1	UK	1 dominant R-factor	Pink and McClement (1998, unpubl.)
L. aculeata	LAC/92/2	Israel	1 dominant R-factor	Pink and McClement (1998, unpubl.)

? = unknown or unclear

L. serriola (PI 281876) has been used frequently in practical breeding programmes (Lebeda and Pink, 1998).

4.2.3. *Field resistance*

Research into sources of field resistance to *B. lactucae* has focused on *L. sativa* (Crute and Norwood, 1981; Eenink, 1981; Lebeda, 1987; Lebeda and Jendrulek, 1988; Lebeda 1992). Lebeda and Jendrulek (1988) defined field resistance as the interaction of a plant population (e.g. cultivar, accession) with the pathogen population during the cultivation period. Field resistance is therefore a complex epidemiological phenomenon characterized by many different features, i.e. delayed occurrence of disease, low disease progress or epidemic rate (r), reduced degree of infection, reduced number of diseased plants and infected leaves (Lebeda and Jendrulek, 1988). In *L. sativa*, crisp-head genotypes are significantly more field-resistant than other types of lettuce. Norwood *et al.* (1985) considered this type of resistance to be genetically complex.

Wild *Lactuca* species have not been studied extensively with respect to their field resistance (Lebeda, 1990b). Only some basic data on field resistance in wild *Lactuca* spp. are available. Crute and Norwood (1981) were the first to point out that some wild *Lactuca* spp. could be potential sources of field resistance. Probably the most comprehensive experiments were carried out by Lebeda (1990b). In total thirty-one accessions of four *Lactuca* species (*L. serriola, L. saligna, L. aculeata, L. indica* /syn. *L.squarrosa/*) and one *L. serriola* x *L. sativa* hybrid (line CS-RL, see Table 5) were studied in three years of field experiments. The disease incidence was significantly different across species and accessions. There were also significant differences in the number of diseased plants and infected leaves. *L. saligna, L. aculeata* accessions and the *L. serriola* x *L. sativa* hybrid were free of infection during the observation period. This reaction implies the presence of effective unknown R-factors (see 4.2.4.) in these genotypes. Limited infection was recorded on *L. indica*. However, in the *L. serriola* group of accessions, significant differences in the level of field resistance were observed. Some accessions were highly susceptible (e.g. PI 204753, PI 253468, PI 273596, PI 273617, PI 274359), in contrast accessions PI 281876 and PI 253467 were free of disease symptoms (again implying the presence of effective unknown R-factors). However, the possible race-nonspecific resistance in PI 281876 is also likely to be expressed as field resistance.

A very high level of field resistance was found in some other accessions (Lebeda, 1990b). Gustafsson (1989) also recognized in field experiments that *L. saligna* accessions were free of disease and a few *L. serriola* accessions displayed a high level of resistance. Crute (1990) identified three resistance phenotypes in a sample of British *L. serriola* populations. *L. serriola* f. *integrifolia* was characterized by the action of previously unknown race-specific dominant resistance factor(s) and a high level of field resistance. There was clear variation between lines in this respect. Preliminary genetic studies showed that field resistance in *L. serriola* f. *integrifolia* is not under simple genetic control (Crute, 1990). However, Lebeda (1990b) clearly demonstrated that *L. serriola* genotypes with race-specific resistance genes (or R-factors) displayed variation for field resistance to *B. lactuca*. In natural *L. serriola* populations it appears that there are genotypes with both race-specific resistance and relatively high levels of field resistance. This phenomenon has also been demonstrated in a few *L. sativa* cultivars (Lebeda, 1987, 1992) carrying the race-specific gene *Dm11*. However, field resistance in lettuce cultivars with this gene is generally low (Lebeda, 1992). Populations of *L. serriola* in Britain were commonly found to be homogeneous for their downy mildew resistance phenotype. There was no evidence for resistance gene pyramiding or plant population heterogeneity as a defence strategy (Crute, 1990), which has been found in some other wild plant pathosystems, e.g. *Senecio vulgaris* – *Erysiphe fischeri* (Bevan *et al.*, 1993).

4.2.4. *Nonhost resistance*

The definition of nonhost resistance is linked with the development of the basic compatibility/incompatibility concept for plant-microorganism interactions (Lebeda,

1984a; Heath, 1985; Ride, 1985; Niks, 1987, 1988; Heath, 1997). Pathogenicity factors in a potential pathogen do not match the plant's non-specific defences. Data mostly obtained from studies with rusts show that nonhost resistance is usually the result of the sequential effects of several constitutive and inducible defences (Heath, 1985). Nonhost resistance is usually very effective, durable and not influenced by changes of environmental conditions (Ride, 1985).

We can draw some tentative preliminary conclusions on the existence of nonhost resistance from the knowledge about host range. From Table 3 it is evident that 14 wild *Lactuca* species are known as natural hosts of *B. lactucae,* and from experimental studies it is known (Tables 4 and 6) that 14 wild *Lactuca* species can be artificially infected by *B. lactucae.* Currently there is only evidence that *L. saligna* may possibly posses nonhost resistance. *L. saligna* is quite a common species in Southern Europe and the Mediterranean area (Feráková, 1977; Lebeda *et al.*, 2001b) but this species has never been clearly recorded as a natural host of *B. lactucae.* After *L. serriola, L. saligna* is the most studied wild species from the viewpoint of resistance to *B. lactucae* (Table 4). The first studies with this species (Eenink, 1974; Netzer *et al.*, 1976; Crute and Davis, 1977) showed that it is characterized by a very high level of resistance; in only a few cases was limited sporulation observed. Many other studies (Norwood *et al.*, 1980, 1981; Lebeda, 1983, 1984c, 1986; Gustafsson, 1986, Farrara and Michelmore, 1987; Gustafsson, 1989; Lebeda, 1990b; Lebeda and Boukema, 1991; Bonnier *et al.*, 1992; Lebeda, 1997) confirmed the previous results using a larger numbers of accessions and *B. lactucae* isolates with a broader virulence spectrum. In some cases a compatible reaction was observed (e.g. Lebeda, 1986; accession PIVT 1168), however later it was found that this material was not well defined taxonomically and was not *L. saligna* (Lebeda and Boukema, 1991). On the basis of these results Lebeda (1986) suggested that in *L. saligna* a system of basic incompatibility may be in operation. However, some authors have discussed (Crute *et al.*, 1981; Lebeda and Reinink, 1994; Reinink *et al.*, 1995) the possibility that some *L. saligna* accessions may have race-specific resistance. A clear differential host/parasite response was recorded in the *L. saligna/Erysiphe cichoracearum* interaction (Lebeda, 1994). Recent experimental results (Lebeda, 1997, 1998a) with new highly virulent isolates of *B. lactucae* originating from *L. sativa* have not confirmed the presence of race-specific resistance in *L. saligna.* Seedlings of some accessions (e.g. PI 261653, CGN 05305, CGN 09313, CGN 10883) displayed slight sporulation after inoculation by isolate DEG2 (Lebeda, 1997, 1998 unpubl.) but all of the accessions of *L. saligna* can be considered as resistant. In contrast the resistance introgressed from *L. saligna* to lettuce cultivar Titan is no longer effective against some of these isolates (Lebeda, 1997, 1998a; Lebeda and Zinkernagel, 1998, 1999). However, the original accession of *L. saligna* has remained resistant to the isolates. It therefore probably possesses additional resistance which may be the nonhost resistance associated with *L. saligna* as a species. This result also implies that it may be difficult to exploit this type of resistance in lettuce by conventional breeding techniques. Histological and cytological studies indicated that the resistance mechanism in *L. saligna* appears to differ from that in some other wild *Lactuca* species (Lebeda and Reinink, 1994; Lebeda, 1995; Lebeda and Pink, 1998; Lebeda *et al.*, 2001c; Sedlářová *et al.*, 2001b). This type of resistance (designated also as non-*Dm* gene mediated resistance) has been patented by Sandoz (Anonymous, 1994).

4.3. HISTOLOGICAL AND CYTOLOGICAL ASPECTS OF RESISTANCE

Host-parasite specificity is determined by many different factors and expressed at various levels of biological organization. The theoretical and evolutionary background of this phenomenon have been discussed by Heath (1997). Information related to the histological and cytological aspects of plant-downy mildew interaction were summarized recently (Mansfield, 1990; Crute, 1992d; Lebeda and Schwinn, 1994; Lucas et al., 1995; Heath and Skalamera, 1997; Mansfield et al., 1997; Lebeda et al., 2001c). Despite the fact that the first histological studies of host-parasite interaction appeared at the beginning of this century, knowledge in this area is still limited. Most studies have concentrated on economically important crops. Wild relatives have been studied in only a few cases where they have been used as sources of resistance genes (Lucas et al., 1995). The crucial questions related to this topic were summarized recently by Mansfield et al. (1997) and in a modified version are be presented here:
1) Is resistance always associated with the hypersensitive response (HR)?
2) Does the plant's resistance response occur only in cells in contact with the pathogen?
3) What is the cause of cell death and the relationship between the timing of HR and the restriction of the pathogen?
4) Is HR alone sufficient to account for the restriction of obligate parasites?
5) At what stage of infection are the determining signals exchanged, and once activated does the HR involve the same biochemical changes irrespective of the R (resistance) and A (avirulence) gene combination that controls recognition?

In the L. sativa – B. lactucae interaction, several histological and cytological studies on the expression of incompatibility/compatibility under the control of different Dm genes or R-factors have been conducted (summarized by Mansfield et al., 1997 and Lebeda et al., 2001c). Generally it was concluded, that the incompatible reaction is associated with the hypersensitivity (HR) of penetrated epidermal cells. The extent and speed of their response may be dependent on the identity of interacting genes, gene dosage, the genetic background of host and pathogen (Crute, 1992d) and environmental conditions (Judelson and Michelmore, 1992). Most of this information has been obtained in studies conducted on genotypes of L. sativa carrying well defined race-specific resistance genes (Lebeda and Pink, 1998; Sedlářová et al., 2001b) or with genotypes expressing different levels of field resistance (Lebeda and Reinink, 1991). Various aspects of hypersensitivity, irreversible membrane damage (IMD) and cell death, which parallel apoptosis of animal and plant cells (Heath, 1998), have been studied and discussed. In L. sativa some genotypes carrying race-specific Dm genes derived from L. serriola (Dm5/8, Dm6 and Dm7) showed substantial differences between these genes in the timing of IMD, accumulation of autofluorescent phenolics and in the progress of the formation of infection structures (Bennett et al., 1996; Mansfield et al., 1997). However, despite these studies, we only have limited information on the histological and cytological background of different resistance types in wild Lactuca spp. (Lebeda et al., 2001a).

The first histological studies on the interactions between wild Lactuca spp. and B. lactucae (Crute and Dickinson, 1976; Crute and Davis, 1977; Norwood et al., 1981) clearly demonstrated a very broad spectrum of microscopic responses at the tissue and cell level, including phenotypic variation in macroscopic expression. Various types of reactions were characteristic, both for incompatible and compatible interactions. In comparison with the responses of *L. sativa* genotypes, macroscopically visible necrotic flecking varied in its extent. This was often linked with a rapid decomposition of cotyledon tissue, a different rate of development and extent of microscopically evaluated hypersensitivity generally occurred more commonly. This variation was characteristic on a host intra- and inter-specific level, however there were also large differences in host genotype/parasite isolate interactions (Norwood *et al.*, 1981).

Recent histological studies of the interactions between wild *Lactuca* spp. and *B. lactucea* confirmed the enormous variation in the host response and development of parasite infection structures (Lebeda, 1995; Lebeda and Reinink, 1994; Lebeda and Pink, 1998; Lebeda *et al.*, 2001c; Sedlářová and Lebeda, 2001; Sedlářová *et al.*, 2001a,b) in different host-parasite interactions. The first significant differences were recorded in the development of the germ tubes of *B. lactucae*. Generally, germ tubes on the wild *Lactuca* spp. were substantially shorter than on the *L. sativa* genotypes or on the breeding lines derived from interspecific crosses *L. sativa* x wild *Lactuca* spp. Very short germ tubes (ca 17.0 µm) with limited variability were observed on *L. serriola* (PI 281876) (Lebeda and Pink, 1996). Currently very little is known about the factors triggering germ tube and appressorium formation in downy mildews. The nature of the early stages of interaction between wild *Lactuca* spp. and *B. lactucae* may be primarily affected by tissue age, physical (mechanical structure) and chemical characteristics (e.g. quality of waxes) of the leaf surface (Lebeda and Pink, 1998; Lebeda *et al.*, 2001c). Comparative histological studies of infection structure formation showed very large variation between different types of resistance and also between *Lactuca* genotypes expressing the same phenotype of resistance.

Some selected and generalized results are summarized in Table 8. In this table, examples of the differences and the variation in expression of incompatible reactions between different types of resistance in wild *Lactuca* spp. are presented in comparison with some *L. sativa* genotypes carrying effective (cv. Valmaine, *Dm5*) and non-effective (Cobham Green, R?; Dandie, *Dm3*) race-specific genes. The cultivar Iceberg was included as a control with a known high level of field resistance (Crute and Norwood, 1981). In reviewing these data (Table 8) with reference to the Mansfield *et al.* (1997) questions, we can state that the background of an incompatible reaction (resistance) is not so simple and uniform, and that HR not always was a primary role. Except for the two control susceptible *L. sativa* cultivars (Cobham Green and Dandie), only genotypes resistant at the seedling stage are included in this table. However, some genotypes (e.g. PI 281876, LVIR/57/1) express incomplete incompatibility if tested as leaf discs from adult plants (Lebeda and Reinink, 1994; Lebeda and Pink, 1998; Lebeda *et al.*, 2001c). Tissue and cellular expression of race-specific resistance determined by *Dm5* was described in detail by Mansfield *et al.* (1997). The expression is rather similar in *L. saligna* (CGN5147), but differs substantially from the expression of race-specific resistance in *L. serriola* (PIVT 1168) and *L. virosa* (LVIR/57/1) (Table 8). In PIVT 1168 fungal development is more or less halted at the stage of secondary vesicle development and is associated with very rapid and extensive HR, including the necrosis of subepidermal cells. The same tissue response was observed in LVIR/57/1, however fungus development continued until haustoria were formed. We can conclude therefore that there are at least three different mechanisms responsible for the control of race-specific resistance in wild *Lactuca* spp. Different control systems may be due to the

timing of the recognition event, i.e. some R-genes may act later in the infection process
than others.

TABLE 8. Comparison of variability in formation of *B. lactucae* infection structures and plant tissue
response of *L. sativa* and wild *Lactuca* spp. genotypes with different types of resistance mechanisms (compiled
according to Lebeda (1995), Lebeda and Reinink (1994), Lebeda and Pink (1998), Lebeda *et al.* (2001c)); all
responses were recorded 48 h after inoculation

Type of resistance *Lactuca* species/ cultivar/accession	Response to *B. lactucae*	Relative degree of occurrence of infection structures and tissue response					
		PV	SV	HY	HA	HR	SEN
Race-specific							
Cobham Green (R?)	+	4	4	4	4	3(2)	0
Dandie (*Dm3*)	+	4	4	2	1	1	1
Valmaine (*Dm5*)	-	4	1	1	0	2	0
L. saligna (CGN 5147)	-	3	1	1	0	1	0
L. serriola (PIVT 1168)	-	2	1	0(1)	0	4	1
L. virosa (LVIR/57/1)	(-)	3	3	1	1	4	2(3)
Race-nonspecific							
L. serriola (PI 281876)	(-)	4	4	2	2	4	2
Field resistance							
Iceberg	+	3	2	1(2)	1(2)	1	0
Nonhost (?)							
L. saligna (LSA/6)	-	3	2	0	0	1	1
L. saligna (CGN 5271)	(-)	3	3	3	2	1	0
L. saligna (CGN 5327)	(-)	4	4	4	3	4	3

Response to *B. lactucae*:
- = incompatible (no sporulation)
(-) = incompletely incompatible (very limited sporulation, mostly on cutt edges of leaf discs)
+ = compatible (profuse sporulation), ? = unclear
Infection structures and tissue response:
PV = primary vesicle, SV = secondary vesicle, HY = hyphae formation, HA= haustoria formation,
HR = hypersensitive (necrotic) response, SEN = subepidermal necrosis
Relative degree of occurrence of infection structures and tissue response (in comparison with
compatible/susceptible/ control):
0 = none recorded, 1 = very low frequency, 2 = low frequency, 3 = medium frequency, 4 = high frequency

A second group of *Lactuca* spp. is represented by *L. serriola* accession PI 281876
with expected race-nonspecific resistance (Table 8) which is probably combined with a
high level of field resistance. The reaction pattern is completely different from most
typical race-specific reactions. Continuous development of infection structures to the
stage of hyphae and haustoria was observed, associated with extensive HR and
subepidermal necrosis (Lebeda and Pink, 1998). The necrotic reaction is also expressed
on the macroscopic level (Norwood *et al.*,1981; Lebeda, 1986; Lebeda and Pink, 1998).
The expression of incomplete incompatibility was followed by sparse sporulation on the
cut edges of leaf discs or by the occurrence of a few sporophores on the disc surface,
mostly in necrotic spots (Lebeda and Pink, 1998), i.e. very limited fungal reproduction

occurred and was limited to damaged parts of plant tissue. Comparable fungus development and tissue response was recorded in *L. virosa* (LVIR/57/1) with race-specific resistance. *B. lactucae* development and tissue response in cv. Iceberg differ from this reaction pattern (Table 8); there was a very low frequency of HR and no subepidermal necrosis.

In most of the *L. saligna* accessions, nonhost or another unknown type of resistance was expected. Previous studies (Lebeda and Reinink, 1994) showed that at least two resistance mechanisms are present in *L. saligna*. One of these differs in histological expression from race-specific and field resistance reaction patterns resulting in highly reduced levels of sporulation in leaf discs, and the development of *B. lactucae* was affected at a later stage of the infection cycle (Lebeda and Reinink, 1994). However, Lebeda (1995) and Lebeda and Pink (1998) recognized at least three resistant reaction patterns in *L. saligna*, all differring from typical race-specific resistance, and characterized by significant differences in fungal growth, epidermal and subepidermal tissue response (Table 8). There are not only differences in the timing of fungal growth, but also differences in the sizes of various infection structures, indicating a very intimate physiological relationship between the parasite and the plant. From these results it can be concluded that resistance in *L. saligna* is a very heterogeneous phenomenon in relation to the rate of fungus development, and tissue and cellular responses of the host (Lebeda *et al.*, 2001c). The biochemical and genetical background of these responses are not known. The preliminary results for the *L. saligna* x *L. sativa* F_1 hybrids support the view that the resistance in *L. saligna* differs from typical dominant monogenic race-specific resistance found in *L. sativa/L. serriola*. Recessive inheritance of resistance in the *L. saligna* accessions (CGN 5271 and CGN 5327) was proposed (Lebeda and Reinink, 1994), however dominant factors were found by other researchers in other *L. saligna* accessions (Table 7).

Fungus development and tissue response in the interspecific hybrids of *L. sativa* with wild *Lactuca* spp. is of interest. Lebeda and Reinink (1994) observed that fungus development, i.e. the of rate of growth and size, was much greater in the F_1 generation of *L. saligna* x *L. sativa* hybrids than in both parents. A similar phenomenon was recorded in more advanced crosses and backcrosses of *L. sativa* with *L. serriola* and *L. virosa* (Lebeda and Pink, 1998). It is difficult to explain this "heterosis" effect which is expressed at the microscopic level. Large variability was observed on a macroscopic phenotypic level. More detailed histological, cytological, physiological and genetical research is required to explain this variation.

Only limited information on the biochemical and molecular background of *L. sativa* resistance to lettuce downy mildew is available. There are no experimental data on the biochemistry and molecular genetics of wild *Lactuca* spp. resistance to *B. lactucae*. Some basic ideas related to this subject were summarized by Mansfield *et al.* (1997). They concluded that the details of the cause of cell death remains unknown, however there is probably a form of programmed cell death involved. Ireversible membrane damage (IMD) is probably not regulated at the level of transcription and it is clear that deposition of autofluorescent phenolics does not itself cause this damage. The nature of the initial signalling, signal transduction and the defence response in lettuce controlled by race-specific resistance genes remain speculative. In relation to these events, the importance of direct contact between fungal intracellular structures (primary and secondary vesicles, haustoria) and the plant cell membrane, which could allow the activities of a wide range of signalling molecules, has recently been discussed (Mansfield *et al.*, 1997; Lebeda *et al.*, 2001c). Data obtained in histological studies of resistance in wild *Lactuca* spp. suggest a wide range of possibilities.

5. Conclusions

Understanding of the relationship between plants and biotrophs (including downy mildews) will be advanced by at least two lines of research: a) study of the cellular communication and physiological co-ordination between the two interacting organisms, b) genetic analyses of co-ordinated evolution of plant and fungus (Holub and Beynon, 1997). Interactions between wild *Lactuca* spp. and *B. lactucae* represent very diverse and complicated systems, and currently they are one of the best known relationships between wild higher plants and a biotroph. The understanding of this system at different levels must be based primarily on an in-depth knowledge of the individual components of the system, i.e. genera of host plant and a fungus (Lebeda *et al.*, 2001c).

The taxonomy of wild *Lactuca* spp. and related genera is currently unclear and much basic information is missing (Koopman *et al.*, 1998; Lebeda and Astley, 1999). Also, the collection, conservation and evaluation of wild *Lactuca* germplasm are poor. From approximately one hundred known *Lactuca* species, the world gene bank collections of *Lactuca* spp. and related genera comprise relatively low numbers of accessions amounting to only about 20 species. More intensive exploration, collection and research is required in this area (Lebeda, 1998c; Lebeda and Boukema, 2001; Lebeda *et al.*, 2001a,b).

Only two species have been described in the genus *Bremia*, nevertheless some authors consider existence of more species (Dick, 2000). Lettuce downy mildew (*Bremia lactucae*) can be considered as a global pathogen of cultivated lettuce (*Lactuca sativa*), however there is very limited information available about its occurrence on wild *Lactuca* spp. in natural populations. A high level of host specificity has been recognized, with eleven *formae speciales* described (Skidmore and Ingram, 1985). The host range of this fungus is very broad, although only 14 *Lactuca* spp. and 6 species from related genera are known as natural hosts of *B. lactucae*. Variation in virulence of the fungus on *L. sativa* is known, however on wild *Lactuca* spp. this phenomenon has been studied only in isolates originating from prickly lettuce, *L. serriola* (Lebeda, this volume). There is no official international culture collection of *B. lactucae* isolates maintaining the virulence diversity of this species, except some recent activities of lettuce breeders and authorities (Van Ettekoven and Van Arend, 1999).

The knowledge on the specificity of interactions between wild *Lactuca* spp. and *B. lactucae* is limited. Only a few wild *Lactuca* spp. (18 in total) and related species (8 in total) have been studied for host specificity. The wild *Lactuca* spp. studied are mostly characterized by a race-specific response. In some *L. saligna* accessions uncharacterized, but very effective, mechanisms of resistance have been discovered. In related genera (*Cicerbita, Ixeris* and *Mycelis*) race-specificity probably occurs. Very efficient sources of field resistance (rate-reducing) and possibly race-nonspecific resistance have been identified in some *L. serriola* accessions, which are being used in practical breeding (Lebeda and Pink, 1998).

Large variations in the response and the genetical background of resistance have also been recognized. At least 16 race-specific *Dm* genes and/or R-factors have been identified or derived from *L. serriola*. Some of these genes (*Dm*5/8, *Dm*6, *Dm*7, *Dm*11, *Dm*16, R18) are used frequently in commercial lettuce breeding. However, only limited information is available on the genetic control of resistance in other wild *Lactuca* spp. (e.g. *L. aculeata, L. saligna, L. virosa*). Different mechanisms of resistance have been recognized in various wild *Lactuca* spp. based on histological and cytological studies (Lebeda and Pink, 1998; Lebeda *et al.*, 2001c; Sedlářová *et al.*, 2001a,b).

Finally, we conclude that despite over 70 years research on host-fungus specificity in wild species of *Lactuca* and related genera to *B. lactucae*, current knowledge at the population, individual, histological, cytological, genetical and molecular level of these interactions is still limited. The *Lactuca* spp./*B. lactucae* system, together with the *Arabidopsis thaliana* /*Peronospora parasitica* and *Albugo candida* (Holub and Beynon, 1997), could be considered as the most important model systems to study the interaction between wild host plants and oomycetes pathogens.

6. Acknowledgements

A. Lebeda carried out main part of this work during a fellowship under the OECD Co-operative Research Programme: Biological Resource Management for Sustainable Agricultural Systems at Horticulture Research International, Genetic Resources Unit (Wellesbourne, UK). The generous support of OECD and use of HRI facilities are gratefully acknowledged. This work was partly supported by grant MSM153100010. A. Lebeda would like to express special personal thanks to Professor I.R. Crute for more than twenty years cooperation and encouragement in the field of lettuce downy mildew research.

7. References

Agrios, G.N. (1997) *Plant Pathology, Fourth Edition*, Academic Press, San Diego and London.

Allen, D.J., Lenné, J.M. and Waller, J.M. (1999) Pathogen biodiversity: Its nature, characterization and consequences, in D. Wood and J.M. Lenné (eds), *Agrobiodiversity: Characterization, utilization and management*, CABI Publishing, Wallingford, pp. 123-153.

Anonymous (1994) European patent application, Fungus resistant plants, *European Patent Office, Bulletin* 94/51 (date of publication: 21.12.1994; publication number: 0 629 343 A2), Paris.

Anonymous (1995) Showcase for world-beating technology, *Grower* **124**, 9.

Bannerot, H. (1980) Screening wild lettuce for *Bremia* resistance, in *Proceedings Eucarpia Meeting on Leafy Vegetables*, Littlehampton (UK), pp. 104-105.

Barrett, J.A. (1985) The gene-for-gene hypothesis: parable or paradigm, in D. Rollinson (ed), *Ecology and Genetics of Host-Parasite Interactions*, Academic Press, London, pp. 215-225.

Bennett, M., Gallagher, M., Fagg, J., Bestwick, Ch., Paul, T., Beale, M. and Mansfield, J. (1996) The hypersensitive reaction, membrane damage and accumulation of autofluorescent phenolics in lettuce cells challenged by *Bremia lactucae*, *The Plant Journal* **9**, 851-865.

Berlese, A.N. (1898) *Icones Fungorum ad Usum Sylloges Saccardianae Accomodatae, Phycomycetes*: Fac. I. Peronosporaceae, Padua (published by author), pp. 1-40.

Berlese, A.N. and De Toni, J.B. (1888) Saccardo Sylloge Fungorum VII, pp. 243-244.

Bevan, J.R., Clarke, D.D. and Crute, I.R. (1993) Resistance to *Erysiphe fischeri* in two populations of *Senecio vulgaris*, *Plant Pathology* **42**, 636-646.

Bohn, G.W. and Whitaker, T.W. (1951) Recently introduced varieties of head lettuce and methods used in their development, *US Department of Agriculture Circular* 881.

Bonnier, F.J.M., Reinink, K. and Groenwold, R. (1992) New sources of major gene resistance in *Lactuca* to *Bremia lactucae*, *Euphytica* **61**, 203-211.

Bonnier, F.J.M., Reinink, K. and Groenwold, R. (1994) Genetic analysis of *Lactuca* accessions with new major gene resistance to lettuce downy mildew, *Phytopathology* **84**, 462-468.

Boukema, I.W., Hazekamp, Th. and van Hintum, Th.J.L. (1990) *The CGN Collection Reviews: The CGN Lettuce Collection*, Centre for Genetic Resources, Wageningen.

Burdon, J.J. (1993) The structure of pathogen populations in natural plant communities, *Annual Review of Phytopathology* **31**, 305-323.

Burdon, J.J. and Jarosz, A.M. (1989) Wild relatives as source of disease resistance, in A.H.D. Brown, O.H. Frankel, D.R. Marshall and J.T. Williams (eds), *The Use of Plant Genetic Resources*, Cambridge University Press, Cambridge, pp. 281-296.

Burdon, J.J., Jarosz, A.M. and Kirby, G.C. (1989) Pattern and patchiness in plant-pathogen interactions-causes and consequences, *Annual Review of Ecology and Systematics* **20**, 119-136.

Burdon, J.J., Wennström, A., Elmqvist, T. and Kirby, G.C. (1996) The role of race specific resistance in natural plant populations, *Oikos* **76**, 411-416.

Chalkuziev, P. (1974) Species novae florae Schachimardanicae, *Botaniceskije Materialy Gerbarija Instituta Botaniky Akademii Nauk Uzbekskoj SSR* **19**, 58-61.

Chin, D.B., Arroyo-Garcia, R., Ochoa, O., Kesseli, R.V., Lavelle, D.O. and Michelmore, R.W. (2001) Recombination and spontaneous mutation at the major cluster of resistance genes in lettuce (*Lactuca sativa*), *Genetics* **157**, 831-849.

Crute, I.R. (1981) The host specificity of peronosporaceous fungi and the genetics of the relationship between host and parasite, in D.M. Spencer (ed), *The Downy Mildews*, Academic Press, London, pp. 237-253.

Crute, I.R. (1985) The genetic bases of relationships between microbial parasites and their hosts, in R.S.S. Fraser (ed), *Mechanisms of Resistance to Plant Diseases*, Nijhoff/Junk, Dordrecht, pp. 80-142.

Crute, I.R. (1987) The geographical distribution and frequency of virulence determinants in *Bremia lactucae*: relationships between genetic control and host selection, in M.S. Wolfe and C.E. Caten (eds), *Populations of Plant Pathogens:Their Dynamics and Genetics*, Blackwell Scientific Publications, Oxford, pp. 193-212.

Crute, I.R. (1989) Lettuce downy mildew: a case study in integrated control, in K.J. Leonard and W.E. Fry (eds), *Plant Disease Epidemiology*, McGraw-Hill, New York, pp. 30-53.

Crute, I.R. (1990) Resistance to *Bremia lactucae* (downy mildew) in British populations of *Lactuca serriola* (prickly lettuce), in J.J. Burdon and S.R. Leather (eds), *Pests, Pathogens and Plant Communities*, Blackwell Scientific Publications, Oxford, pp. 203-217.

Crute, I.R. (1992a) The role of resistance breeding in the integrated control of downy mildew (*Bremia lactucae*) in protected lettuce, *Euphytica* **63**, 95-102.

Crute, I.R. (1992b) The contribution to successful integrated disease management of research on genetic resistance to pathogens of horticultural crops, *Journal of the Royal Agricultural Society of England* **153**, 132-144.

Crute, I.R. (1992c) From breeding to cloning (and back again?): A case study with lettuce downy mildew, *Annual Review of Phytopathology* **30**, 485-506.

Crute, I.R. (1992d) Downy mildew of lettuce, in H.S. Chaube, J. Kumar, A.N. Mukhopadhyay and U.S. Singh (eds), *Plant Diseases of International Importance,Volume II. Diseases of Vegetables and Oil Seed Crops*, Prentice Hall, New Jersey, pp. 165-185.

Crute, I.R. (1998) The elucidation and exploitation of gene-for-gene recognition, *Plant Pathology* **47**, 107-113.

Crute, I.R. and Davis, A.A. (1977) Specificity of *Bremia lactucae* from *Lactuca sativa*, *Transactions of the British Mycological Society* **69**, 405-410.

Crute, I.R. and Dickinson, C.H. (1976) The behaviour of *Bremia lactucae* on cultivars of *Lactuca sativa* and on other composites, *Annals of Applied Biology* **82**, 433-450.

Crute, I.R. and Dixon, G.R. (1981) Downy mildew diseases caused by the genus *Bremia* Regel, in D.M. Spencer (ed), *The Downy Mildews*, Academic Press, London, pp. 421-460.

Crute, I.R. and Dunn, J.A. (1980) An association between resistance to root aphid (*Pemphigus bursarius* L.) and downy mildew (*Bremia lactucae* Regel) in lettuce, *Euphytica* **29**, 483-488.

Crute, I.R. and Gordon, P.L. (1984) Downy mildew of lettuce – new combinations of R-factors in lettuce, *Annual Report of the National Vegetable Research Station (Wellesbourne) for 1983*, pp. 78.

Crute, I.R. and Johnson, A.G. (1976) The genetic relationship between races of *Bremia lactucae* and cultivars of *Lactuca sativa*, *Annals of Applied Biology* **83**, 125-137.

Crute, I.R. and Norwood, J.M. (1979) Downy mildew of lettuce, resistance studies, *Annual Report of the National Vegetable Research Station (Wellesbourne) for 1978*, pp. 77.

Crute, I.R. and Norwood, J.M. (1981) The identification and characteristics of field resistance to lettuce downy mildew (*Bremia lactucae* Regel), *Euphytica* **30**, 707-717.

Crute, I.R., Norwood, J.M. and Gordon, P.L. (1981) Downy mildew of lettuce, resistance studies, *Annual Report of the National Vegetable Research Station (Wellesbourne) for 1980*, pp. 77-79.

Crute, I.R., Norwood, J.M., Gordon, P.L., Clay, C.M. and Whenham, R.J. (1986) Diseases of lettuce – biology, resistance and control, *Annual Report of the National Vegetable Research Station (Wellesbourne) for 1985*, pp. 51-53.

Dange, S.R., Jain, K.L. and Giradhana, B.S. (1976) *Launaea asplenifolia* – a new host of *Bremia lactucae*, *Indian Journal of Mycology and Plant Pathology* **5**, 202.

Dick, M.W. (2000) The Peronosporomycetes, in D.J. McLaughlin, E.C. McLaughlin and P.A. Lemke (eds), *The Mycota, Vol. VII, Systematics and Evolution Part A*, Springer-Verlag, Berlin, Heidelberg, pp. 39-72.

Dinoor, A. and Eshed, N. (1997) Plant conservation *in situ* for disease resistance, in N. Maxted, B.V. Ford-Lloyd and J.G. Hawkes (ed), *Plant Genetic Conservation, The in situ approach*, Kluwer Academic Publishers, Dordrecht, pp. 323-336.

Eenink, A.H. (1974) Resistance in *Lactuca* against *Bremia lactucae* Regel, *Euphytica* **23**, 411-416.

Eenink, A.H. (1980) Absolute and partial resistance to *Bremia lactucae* in lettuce: two strategies to improve durability of resistance, *Proceedings Eucarpia Meeting on Leafy Vegetables*, Littlehampton (U.K.), pp. 107-111.

Eenink, A.H. (1981) Partial resistance in lettuce to downy mildew (*Bremia lactucae*), 1. Search for partially resistant genotypes and the influence of certain plant characters and environments on the resistance level, *Euphytica* **30**, 619-628.

Eenink, A.H. and Roelofsen, H. (1977) Een genenbank en het verzamelen van wilde sla in Israel, *Landbouwkundig Tijdschrift* **89**, 129-132.

Ellis, M.B. and Ellis, J.P. (1997) *Microfungi on Land Plants. An Identification Handbook*, The Richmond Publishing Co., Slough.

Erwin, A.T. (1920) Control of downy mildew of lettuce, *Proceedings of the American Society for Horticultural Science* **17**, 161-169.

Erwin, A.T. (1921) Controlling downy mildew of lettuce, *Iowa Agricultural Experimental Station Bulletin* No. **196**, 307-328.

Farrara, B.F., Ilott, T.W. and Michelmore, R.W. (1987) Genetic analysis of factors for resistance to downy mildew (*Bremia lactucae*) in species of lettuce (*Lactuca sativa* and *Lactuca serriola*), *Plant Pathology* **36**, 499-514.

Farrara, B.F. and Michelmore, R.W. (1987) Identification of new sources of resistance to downy mildew in *Lactuca* spp., *HortScience* **22**, 647-649.

Feráková, V. (1976) Lactuca L., in T.G. Tutin (ed), *Flora Europaea, Vol. 4*, Cambridge University Press, Cambridge, pp. 328-331.

Feráková, V. (1977) *The Genus Lactuca L. in Europe*, Universita Komenského, Bratislava.

Fischer, A. (1892) *Die Pilze Deutschlands, Oesterreichs und der Schweiz IV – Phycomycetes*, in L. Rabenhorst (ed), *Kryptogamen-Flora von Deutschland, Oesterreich und der Schweiz*, Eduard Kummer, Leipzig, pp. 439-442.

Frietema, F.T. (1994) The systematic relationship of *Lactuca sativa* and *Lactuca serriola*, in relation to the distribution of prickly lettuce, *Acta Botanica Neerlandica* **43**, 79.

Globerson, D., Netzer, D. and Sacks, J. (1980) Wild lettuce as source for improving cultivated lettuce, *Proceedings Eucarpia Meeting on Leafy Vegetables*, Littlehampton (UK), pp. 86-96.

Glowka, L., Burhenne-Guilmin, F. and Synge, H. (1994) *A Guide to the Convention on Biological Diversity*, IUCN, Gland and Cambridge.

Guarino, L., Ramanatha Rao, V. and Reid, R. (1995) *Collecting Plant Genetic Diversity, Technical Guidelines*, CAB International, Wallingford.

Gustafsson, I. (1986) Virulence of *Bremia lactucae* in Sweden and race-specific resistance of lettuce cultivars to Swedish isolates of the fungus, *Annals of Applied Biology* **109**, 107-115.

Gustafsson, I. (1989) Potential sources of resistance to lettuce downy mildew (*Bremia lactucae*) in different *Lactuca* species, *Euphytica* **40**, 227-232.

Gustafsson, I. (1992) Race non-specific resistance to *Bremia lactucae*, *Annals of Applied Biology* **120**, 127-136.

Hall, G.S. (1996) Modern approaches to species concepts in downy mildews, *Plant Pathology* **45**, 1009-1026.

Hawksworth, D.L. (1991) The fungal dimension of biodiversity: magnitude, significance, and conservation, *Mycological Research* **95**, 641-655.

Heath, M.C. (1985) Implications of non-host resistance for understanding host-parasite interactions, in J.V. Groth and W.R. Bushnel (eds), *Genetic Basis of Biochemical Mechanisms of Disease*, APS Press, St. Paul, pp. 25-42.

Heath, M.C. (1991) Evolution of resistance to fungal parasitism in natural ecosystems, *New Phytologist* **119**, 331-343.

Heath, M.C. (1997) Evolution of plant resistance and susceptibility to fungal parasites, in G.C. Carroll and P. Tudzynski (eds), *The Mycota, Vol. V. Plant Relationships, Part B*, Springer-Verlag, Berlin, pp. 257-281.

Heath, M.C. (1998) Apoptosis, programmed cell death and the hypersensitivity response, *European Journal of Plant Pathology* **104**, 117-124.

Heath, M.C. and Skalamera, D. (1997) Cellular interactions between plants and biotrophic fungal parasites, *Advances in Botanical Research* **24**, 196-225.

Hegi, G. (ed) (1987) *Illustrierte Flora von Mitteleuropa, Band VI, Teil 4*, Verlag Paul Parey, Berlin-Hamburg.

Heywood, V.H. (ed) (1978) *Flowering Plants of the World*, Oxford University Press, Oxford.

Hintum, Th.J.L. van and Soest, L.J.M. van (1997) Conservation of plant genetic resources in the Netherlands, *Plant Varieties and Seeds* **10**, 145-152.

Holub, E.B. and Beynon, J.L. (1997) Symbiology of mouse-ear cress (*Arabidopsis thaliana*) and Oomycetes, *Advances in Botanical Research* **24**, 228-273.

Hulbert, S.H. and Michelmore, R.W. (1985) Linkage analysis of genes for resistance to downy mildew (*Bremia lactucae*) in lettuce (*Lactuca sativa*), *Theoretical and Applied Genetics* **70**, 520-528.

Ilott, T.W., Hulbert, S.H. and Michelmore, R.W. (1989) Genetic analysis of the gene-for-gene interaction between lettuce (*Lactuca sativa*) and *Bremia lactucae*, *Phytopathology* **79**, 888-897.

IPGRI (1993) *Diversity for Development, The Strategy of the International Plant Genetic Resources Institute*, International Plant Genetic Resources Institute, Rome.

Ito, S. and Tokunaga, Y. (1935) Notae mycologicae, Asie orientalis I., *Transactions of the Sapporo Natural History Society* **14**, 11-33.

Jagger, I.C. (1924) Immunity to mildew (*Bremia lactucae* Reg.) and its inheritance in lettuce, *Phytopathology* **14**, 122 (Abstr.).

Jagger, I.C. and Whitaker, T.W. (1940) The inheritance of immunity from mildew (*Bremia lactucae*) in lettuce, *Phytopathology* **30**, 427-433.

Jeffrey, C. (1995) Compositae systematics 1975-1993, Developments and desiderata, in D.J.N. Hind, C. Jeffrey and G.V. Pope (eds), *Advances in Compositae Systematics*, The Royal Botanic Gardens, Kew, pp. 3-21.

Jeuken, M. and Lindhout, P. (2000) Towards characterization of resistance in *Lactuca saligna* (wild lettuce) to *Bremia lactucae* and genome mapping in lettuce, in *Durable disease resistance; key to sustainable agriculture*, International Symposium, Ede-Wageningen, The Netherlands, November 28 – December 1, 2000, Book of Abstracts, p. 52.

Johnson, A.G., Laxton, S.A., Crute, I.R., Gordon, P.L. and Norwood, J.M. (1978) Further work on the genetics of race specific resistance in lettuce (*Lactuca sativa*) to downy mildew (*Bremia lactucae*), *Annals of Applied Biology* **89**, 257-264.

Jorstad, I. (1964) The Phycomycetous genera *Albugo, Bremia, Plasmopara*, and *Pseudoperonospora* in Norway, with an appendix containing unpublished finds of *Peronospora, Nytt Magasin for Botanik* **11**, 47-82.

Judelson, H.S. and Michelmore, R.W. (1992) Temperature and genotype interactions in the expression of host resistance in lettuce downy mildew, *Physiological and Molecular Plant Pathology* **40**, 233-245.

Kesseli, R.V., Paran, I. and Michelmore, R.W. (1994) Analysis of a detailed linkage map of *Lactuca sativa* (Lettuce) constructed from RFLP and RAPD markers, *Genetics* **136**, 1435-1446.

Koopman, W.J.M., Guetta, E., Van de Wiel, C.C.M., Vosman, B. and Van den Berg, R.G. (1998) Phylogenetic relationships among *Lactuca* (Asteraceae) species and related genera based on ITS-1 DNA sequences, *American Journal of Botany* **85**, 1517-1530.

Křístková, E. and Lebeda, A. (1999) Collection of *Lactuca* spp. genetic resources in the Czech Republic, in A. Lebeda and E. Křístková (ed), *Eucarpia Leafy Vegetables '99*, Palacký University, Olomouc, pp. 109-116.

Lebeda, A. (1983) Resistenzquellen gegenüber *Bremia lactucae* bei Wildarten von *Lactuca* spp., *Tagungs-Berichte Akademie Landwirtschafts-Wissenschaften* DDR, Berlin **216**, 607-616.

Lebeda, A. (1984a) A contribution to the general theory of host-parasite specificity, *Phytopathologische Zeitschrift* **110**, 226-234.

Lebeda, A. (1984b) Response of differential cultivars of *Lactuca sativa* to *Bremia lactucae* isolates from *Lactuca serriola*, *Transactions of the British Mycological Society* **83**, 491-494.

Lebeda, A. (1984c) Wild *Lactuca* spp. as sources of stable resistance to *Bremia lactucae* – reality or supposition, *Proceedings Eucarpia Meeting on Leafy Vegetables*, Versailles (France), pp. 35-42.

Lebeda, A. (1984d) Race-specific factors of resistance to *Bremia lactucae* in the world assortment of lettuce, *Scientia Horticulturae* **22**, 23-32.

Lebeda, A. (1986) Specificity of interactions between wild *Lactuca* spp. and *Bremia lactucae* isolates from *Lactuca serriola, Journal of Phytopathology* **117**, 54-64.

Lebeda, A. (1987) Screening of lettuce cultivars for field resistance to downy mildew, *Tests of Agrochemicals and Cultivars* No. **8** (*Annals of Applied Biology* **110**, Suppl.), 146-147.

Lebeda, A. (1989) Response of lettuce cultivars carrying the resistance gene *Dm*11 to isolates of *Bremia lactucae* from *Lactuca serriola, Plant Breeding* **102**, 311-316.

Lebeda, A. (1990a) Identification of a new race-specific resistance factor in lettuce (*Lactuca sativa*) cultivars resistant to *Bremia lactucae, Archiv für Züchtungsforschung* Berlin **20**, 103-108.

Lebeda, A. (1990b) The location of sources of field resistance to *Bremia lactucae* in wild *Lactuca* species, *Plant Breeding* **105**, 75-77.

Lebeda, A. (1992) The level of field resistance to *Bremia lactucae* in lettuce (*Lactuca sativa*) cultivars carrying the resistance gene *Dm*11, *Plant Breeding* **108**, 126-131.

Lebeda, A. (1994) Evaluation of wild *Lactuca* species for resistance of natural infection of powdery mildew (*Erysiphe cichoracearum*), *Genetic Resources and Crop Evolution* **41**, 55-57.

Lebeda, A. (1995) *Report on Scientific Programme* (9June-18August1995), Horticulture Research International, Wellesbourne (UK).

Lebeda, A. (1997) Virulence distribution, dynamics and diversity in German population of lettuce downy mildew (*Bremia lactucae*), *Report on Research Programme*, TU Munich, Department of Plant Pathology, Freising-Weihenstephan (Germany).

Lebeda, A. (1998a) Virulence variation in lettuce downy mildew (*Bremia lactucae*) and effectivity of race-specific resistance genes in lettuce, in M. Recnik and J. Verbic (eds), *Proceedings of the Conference "Agriculture and Environment"*, Bled (Slovenia), pp. 213-217.

Lebeda, A. (1998b) Biodiversity of the Interactions Between Germplasms of Wild *Lactuca* spp. and Related Genera and Lettuce Downy Mildew (*Bremia lactucae*), *Report on Research Programme*, Horticulture Research International, Wellesbourne (U.K.).

Lebeda, A. and Astley, D. (1999) World genetic resources of *Lactuca* spp., their taxonomy and biodiversity, in A. Lebeda and E. Křístková (ed), *Eucarpia Leafy Vegetables '99*, Palacký University, Olomouc, pp. 81-94.

Lebeda, A. and Blok, I. (1990) Sexual compatibility types of *Bremia lactucae* isolates originating from *Lactuca serriola, Netherlands Journal of Plant Pathology* **96**, 51-54.

Lebeda, A. and Blok, I. (1991) Race-specific resistance genes to *Bremia lactucae* Regel in new Czechoslovak lettuce cultivars and location of resistance in a *Lactuca serriola* x *Lactuca sativa* hybrid, *Archiv für Phytopathologie und Pflanzenschutz* **27**, 65-72.

Lebeda, A. and Boukema, I.W. (1991) Further investigation of the specificity of interactions between wild *Lactuca* spp. and *Bremia lactucae* isolates from *Lactuca serriola, Journal of Phytopathology* **133**, 57-64.

Lebeda, A. and Boukema, I.W. (2001) Leafy vegetables genetic resources, in L. Maggioni and O. Spellman (eds), *Report of a Network Coordinating Group on Vegetables*; Ad hoc meeting, 26-27 May 2000, Vila Real, Portugal, International Plant Genetic Resources Institute, Rome (Italy), pp. 48-57.

Lebeda, A., Doležalová, I. and Astley, D. (2001a) Geographic distribution of wild *Lactuca* spp. (Asteraceae) and their representation in world genebank collections, *Genetic Resources and Crop Evolution* (prepared for press)

Lebeda, A., Doležalová, I., Křístková, E. and Mieslerová, B. (2001b) Biodiversity and ecogeography of wild *Lactuca* spp. in some European countries, *Genetic Resources and Crop Evolution* **48**, 153-164.

Lebeda, A. and Jendrulek, T. (1987) Application of cluster analysis for establishment of genetic similarity in gene-for gene host-parasite relationships, *Journal of Phytopathology* **119**, 131-141.

Lebeda, A. and Jendrulek, T. (1988) Application of methods of multivariate analysis in comparative epidemiology and research into field resistance, *Zeitschrift für Pflanzenkrankheiten und Pflanzenschutz* **95**, 495-505.

Lebeda, A. and Jendrulek, T. (1989) Application of multivariate analysis for characterizing the relationships between wild *Lactuca* spp. and *Bremia lactucae, Acta Phytopathologica et Entomologica Hungarica* **24**, 317-331.

Lebeda, A., Jendrulek, T. and Syrovátko, P. (1989) Application of multivariate analysis in research of conidial morphology of *Bremia lactucae* originating from the family Compositae, *Directory 10th European Mycological Congress*, Tallin (USSR), pp. 51-52.

Lebeda, A. and Křístková, E. (1995) Genetic resources of vegetable crops from the genus *Lactuca, Horticultural Science* (Prague) **22**, 117-121.

Lebeda, A. and Petrželová, I. (2001) Occurrence and characterization of race-specific resistance to *Bremia lactucae* in wild *Lactuca* spp., in *EUCARPIA Section Genetic Resources "Broad Variation and Precise Characterization – Limitation for the Future"*, May 16-20, 2001, Poznan (Poland), Abstracts, p. 93.

Lebeda, A. and Pink, D.A.C. (1995) Expression of resistance to *Bremia lactucae* in wild species of *Lactuca, The Downy Mildew Fungi, An International Symposium Organised by The British Mycological Society,* Gwatt (Switzerland), Abstracts, p. 46.

Lebeda, A. and Pink, D.A.C. (1996) Variation in germ tube growth and disease initiation by *Bremia lactucae* on different wild *Lactuca* species and interspecific hybrids. *Sixth International Fungal Spore Conference,* University Konstanz (Germany), Book of Abstracts, p. 73.

Lebeda, A. and Pink, D.A.C. (1997) Wild *Lactuca* spp. as a gene pool for resistance breeding of lettuce, in F. Kobza, M. Pidra and R. Pokluda (eds), *Proceedings of the International Horticultural Scientific Conference "Biological and Technical Development in Horticulture"*, Lednice na Moravě (Czech Republic) 1997, Published by Mendel University of Agriculture and Forestry in Brno (Czech Republic), pp. 53-59.

Lebeda, A. and Pink, D.A.C. (1998) Histological aspects of the response of wild *Lactuca* spp. and *L. sativa* x *Lactuca* spp. hybrids to lettuce downy mildew (*Bremia lactucae*), *Plant Pathology* **47,** 723-736.

Lebeda, A., Pink, D.A.C. and Mieslerová, B. (2001c) Host-parasite specificity and defense variability in the *Lactuca* spp.-*Bremia lactucae* pathosystem, *Journal of Plant Pathology* **83,** (in press)

Lebeda, A. and Reinink, K. (1991) Variation in the early development of *Bremia lactucae* on lettuce cultivars with different levels of field resistance, *Plant Pathology* **40,** 232-237.

Lebeda, A. and Reinink, K. (1994) Histological characterization of resistance in *Lactuca saligna* to lettuce downy mildew (*Bremia lactucae*), *Physiological and Molecular Plant Pathology* **44,** 125-139.

Lebeda, A. and Schwinn, F.J. (1994) The downy mildews – an overview of recent research progress, *Journal of Plant Diseases and Protection* **101,** 225-254.

Lebeda, A. and Syrovátko, P. (1988) Specificity of *Bremia lactucae* isolates from *Lactuca sativa* and some Asteraceae plants, *Acta Phytopathologica et Entomologica Hungarica* **23,** 39-48.

Lebeda, A. and Zinkernagel, V. (1998) Evolution of virulence in German population of lettuce downy mildew, *Bremia lactucae,* 7th *International Congress of Plant Pathology, Edinburgh, (Abstracts).*

Lebeda, A. and Zinkernagel, V. (1999) Durability of race-specific resistance in lettuce against lettuce downy mildew (*Bremia lactucae*), in A. Lebeda and E. Křístková (ed), *Eucarpia Leafy Vegetables '99,* Palacký University, Olomouc, pp. 183-189.

Lenné, J.M. and Wood, D. (1991) Plant disease and the use of wild germplasm, *Annual Review of Phytopathology* **29,** 35-63.

Leeper, P.W., Thompson, R.C. and Whitaker, T.W. (1963) Valmaine, a new downy mildew immune romaine lettuce variety, *Texas A and M University Agricultural Experimental Station* L-610.

Lindqvist, K. (1960) On the origin of cultivated lettuce, *Hereditas* **46,** 319-350.

Ling, L. and Tai, M.C. (1945) On the specialisation of *Bremia lactucae* on Compositae, *Transactions of the British Mycological Society* **28,** 16-25.

Lopez, E.G. and Jimenez, A.C. (1974) *ELENCO de la Flora Vascular Española (Península y Baleares),* ICONA, Madrid.

Losa Espana, D.M. (1942) Aportacion al estudio de la flora micologica espanola, *Annales de Jardin Botanique Madrid* **2,** 87-142.

Lucas, M.T. and Dias, M.R.S. (1976) Peronosporaceae lusitaniae, *Agronomica Lusitanica* **37,** 281-299.

Lucas, J.A. and Sherriff, C. (1988) Pathogenesis and host specificity in downy mildew fungi, in R.S. Singh, U.S. Singh, W.M. Hess and D.J. Weber (eds), *Experimental and Conceptual Plant Pathology, Vol. 2. Pathogenesis and Host-Parasite Specificity,* Gordon and Breach Science Publishers, New York and London, pp. 321-349.

Lucas, J.A., Hayter, J.B.R. and Crute, I.R. (1995) The downy mildews: Host specificity and pathogenesis, in K. Kohmoto, U.S. Singh and R.P. Singh (eds), *Pathogenesis and Host Specificity in Plant Diseases, Vol. II: Eukaryotes,* Pergamon Press and Elsevier Science, Oxford, pp. 217-238.

Maesen, L.J.G. van der (1994) Systematics of crop-weed complexes, *Acta Botanica Neerlandica* **43,** 78-79.

Maisonneuve, B., Bellec, Y., Anderson, P. and Michelmore, R.W. (1994) Rapid mapping of two genes for resistance to downy mildew from *Lactuca serriola* to existing clusters of resistance genes, *Theoretical and Applied Genetics* **89,** 96-104.

Mansfield, J.W. (1990) Recognition and response in plant/fungus interactions, in R.S.S. Fraser (ed), *Recognition and Response in Plant-Virus Interactions,* ASI Series, Vol. H 41, Springer-Verlag, Berlin, pp. 31-52.

Mansfield, J.W., Bennett, M., Bestwick, Ch. and Woods-Tör, A. (1997) Phenotypic expression of gene-for-gene interaction involving fungal and bacterial pathogens: variation from recognition to response, in I.R.

Crute, E.B. Holub and J.J. Burdon (eds), *The Gene-for-Gene Relationship in Plant-Parasite Interactions*, CAB International, Wallingford, pp. 265-291.

Marlatt, R.B. (1974) Biology, morphology, taxonomy and disease relations of the fungus *Bremia, Florida Agricultural Experimental Station Technical Bulletin* **764**, 1-25.

Maxon Smith, J.W. (1984) Interspecific hybridisation in *Lactuca* with particular reference to *L. sativa* L. x *L. virosa* L., *Proceedings Eucarpia Meeting on Leafy Vegetables*, Versailles (France), pp. 21-34.

Maxon Smith, J.W. and Langton, A. (1989) New sources of resistance to downy mildew, *Grower* **112**, 54-55.

McGuire, P.E., Ryder, E.J., Michelmore, R.W., Clark, R.L., Antle, R., Emery, G., Hannan, R.M., Kesseli, R.V., Kurtz, E.A., Ochoa, O., Rubatzky, V.E. and Waycott, W. (1993) Genetic Resources of Lettuce and *Lactuca* species in California; An Assessment of the USDA and UC Collections and Recommendations for Long-term Security, Report No. **12**, 40 pp., *University of California Genetic Resources Conservation Program, Davis, CA*.

Meyers, B.C., Chin, D.B., Shen, K.A., Sivaramakrisman, S., Lavelle, D.O., Zhang, Z. and Michelmore, R.W. (1998) The major resistance gene cluster in lettuce is highly duplicated and spans several megabases, *Plant Cell* **10**, 1817-1832.

Michelmore, R.W., Ilott, T., Hulbert, S.H. and Farrara, B. (1988) The downy mildews, in D.S. Ingram and P.H. Williams (eds), *Advances in Plant Pathology, Vol. 6, Genetics of Plant Pathogenic Fungi*, Academic Press, London, pp. 54-79.

Michelmore, R.W. and Ingram, D.S. (1980) Heterothallism in *Bremia lactucae, Transactions of the British Mycological Society* **75**, 47-56.

Netzer, D., Globerson, D. and Sacks, J. (1976) *Lactuca saligna* L., a new source of resistance to downy mildew (*Bremia lactucae* Reg.), *HortScience* **11**, 612-613.

Niks, R.E. (1987) Nonhost plant species as donors for resistance to pathogens with narrow host range, I. Determination of nonhost status, *Euphytica* **36**, 841-852.

Niks, R.E. (1988) Nonhost plant species as donors for resistance to pathogens with narrow host range, II. Concepts and evidence on the genetic basis of nonhost resistance, *Euphytica* **37**, 89-99.

Norwood, J.M. and Crute, I.R. (1980) Linkage between genes for resistance to downy mildew (*Bremia lactucae*) in lettuce, *Annals of Applied Biology* **94**, 127-135.

Norwood, J.M., Crute, I.R. and Lebeda, A. (1981) The location and characteristics of novel sources of resistance to *Bremia lactucae* Regel (downy mildew) in wild *Lactuca* L. species, *Euphytica* **30**, 659-668.

Norwood, J.M., Johnson, A.G. and Crute, I.R. (1980) The utilization of novel sources of resistance to *Bremia lactucae* from wild *Lactuca* species, *Proceedings Eucarpia Meeting on Leafy Vegetables*, Littlehampton (U.K.), pp. 97-103.

Norwood, J.M., Johnson, A.G., O'Brien, M. and Crute, I.R. (1985) The inheritance of field resistance to lettuce downy mildew (*Bremia lactucae*) in the cross 'Avoncrisp' x 'Iceberg', *Zeitschrift für Pflanzenzüchtung* **94**, 259-262.

Ogilvie, L. (1943) Downy mildew of lettuce, a preliminary note on some greenhouse experiments, *Annual Report of Long Ashton Research Station*, pp. 90.

Ogilvie, L. (1945) Downy mildew of lettuce: further investigations on strains of *Bremia lactucae* occurring in England, *Annual Report of Agricultural and Horticultural Research Station 1945*, pp. 147-150.

Ohwi, J. (1965) *Flora of Japan*, Smithsonian Institution, Washington, D.C.

Parlevliet, J.E. (1992) Selecting components of partial resistance, in H.T. Stalker and J.P. Murphy (eds), *Plant Breeding in the 1990s, Proceedings of the Symposium on Plant Breeding in the 1990s*, C.A.B. International, Oxon, pp. 281-302.

Pickersgill, B. (1981) Biosystematics of crop-weed complexes, *Die Kulturpflanze* **29**, 377-388.

Pink, D.A.C. and Keane, E.M. (1993) Lettuce, *Lactuca sativa*, in G. Kalloo and B.O. Bergh (eds), *Genetic Improvement of Vegetable Crops*, Pergamon Press, Oxford, pp. 543-571.

Pink, D.A.C. and McClement, S.G. (1996) *Horticulture Research International, Annual Report 1994-1995*, p. 29.

Powlesland, R. (1954) On the biology of *Bremia lactucae, Transactions of the British Mycological Society* **37**, 362-371.

Reinink, K. (1999) Lettuce resistance breeding, in A. Lebeda and E. Křístková (ed), *Eucarpia Leafy Vegetables '99*, Palacký University, Olomouc, pp. 139-147.

Reinink, K., Groenwold, R. and Lebeda, A. (1993) Characterization of nonhost resistance to lettuce downy mildew (*Bremia lactucae*) in *Lactuca saligna*, in Th. Jacobs and J.E. Parlevliet (eds), *Durability of Disease Resistance*, Kluwer Academic Publishers, Dordrecht, p. 340.

Reinink, K., Lebeda, A. and Groenwold, R. (1995) Durable resistance to the lettuce downy mildew, *European Journal of Plant Pathology*, XIII. International Plant Protection Congress, The Hague (Abstracts), No. 469.

Renfro, B.L. and Shankara Bhat, S. (1981) Role of wild hosts in downy mildew diseases, in D.M. Spencer (ed), *The Downy Mildews,* Academic Press, London, pp. 107-118.

Ride, J.P. (1985) Non-host resistance to fungi, in R.S.S. Fraser (ed), *Mechanisms of Resistance to Plant Diseases,* Martinus Nijhoff/Dr. W.Junk Publishers, Dordrecht, pp. 29-61.

Robinson, R.W. and Provvidenti, R. (1993) Breeding lettuce for viral resistance, in M.M. Kyle (ed), *Resistance to Viral Diseases of Vegetables, Genetics & Breeding,* Timber Press, Portland, pp. 61-79.

Rulkens, A.J.H. (1987) De CGN sla-collectie: inventarisatie, paspoortgegevens en enkele richtlijnen voor de toekomst, *CGN Report: CGN-T 1987-1,* Centre for Genetic Resources (CGN), Wageningen.

Ryder, E.J. (1998) *Lettuce, endive and cichory,* CABI Publishing, Wallingford.

Savulescu, O. (1962) A systematic study of the genera *Bremia* Regel and *Bremiella* Wilson, *Revue de Biologie* **7,** 43-62.

Sawada, K. (1919) Descriptive catalogue of the Formosan fungi, *Agriculture Experimental Station of the Government Formosa, Special Bulletin* No. 19.

Schultz, H. (1937) Zur Biologie der *Bremia lactucae* Regel, des Erregers des Falschen Mehltaues des Salats, *Phytopathologische Zeitschrift* **19,** 490-503.

Schultz, H. and Röder, K. (1938) Die Anfälligkeit verschiedener Varietäten und Sorten von Salat (*Lactuca sativa* L. und *Lactuca scariola* L.) gegen den Falschen Mehltau (*Bremia lactucae* Regel), *Der Züchter* **10,** 185-194.

Schweizer, J. (1920) Untersuchungen über den Pilz des Salates *Bremia lactucae* Regel, *Mitteilungen der Thüringischer Naturforschenden Gesellschaft* **23,** 14-160.

Sedlářová, M., Binarová, P. and Lebeda, A. (2001a) Changes in microtubular alignement in *Lactuca* spp. (Asteraceae) epidermal cells during early stages of infection by *Bremia lactucae* (*Peronosporaceae*), *Phyton* **41,** 21-34.

Sedlářová, M. and Lebeda, A. (2001) Histochemical detection and role of phenolic compounds in the defense response of *Lactuca* spp. to lettuce downy mildew (*Bremia lactucae*), *Journal of Phytopathology* (in press)

Sedlářová, M., Lebeda, A. and Pink, D.A.C. (2001b) The early stages of interaction between effective and non-effective race-specific genes in *Lactuca sativa,* wild *Lactuca* spp. and *Bremia lactucae* (race NL16), *Journal of Plant Diseases and Protection* (in press)

Shaw, C.G. (1958) Host fungus index for the Pacific Northwest, I. Hosts, *Washington Agriculture Experimental Station Circular* 335.

Sicard, D., Woo, S.S., Arroyo-Garcia, R., Ochoa, O., Nguyen, D. Korol, A., Nevo, E. and Michelmore, R.W. (1999) Molecular diversity at the major clusters of disease resistance genes in cultivated and wild *Lactuca* spp., *Theoretical and Applied Genetics* **99,** 405-418.

Skidmore, D.I. and Ingram, D.S. (1985) Conidial morphology and the specialization of *Bremia lactucae* Regel (Peronosporaceae) on hosts in the family Compositae, *Botanical Journal of the Linnean Society* **91,** 503-522.

Smartt, J. and Simmonds, N.W. (1995) *Evolution of Crop Plants, Second Edition,* Longman Scientific & Technical, Harlow.

Soest, L.J.M. van and Boukema, I.W. (1995) Diversiteit in de Nederlandse genebank, *Een overzicht van de CGN collecties,* CGN, CPRO-DLO, Wageningen, pp. 9-15.

Soest, L.J.M. van and Boukema, I.W. (1997) Genetic resources conservation of wild relatives with a users' perspective, *Bocconea* **7,** 305-316.

Spencer, D.M. (ed) (1981) *The Downy Mildews,* Academic Press, London.

Stace, C. (1997) *New Flora of the British Isles, Second Edition,* Cambridge University Press, Cambridge.

Tao Chia-Feng (1965) A preliminary study on the primary sources of downy mildew of *Lactuca* caused by *Bremia lactucae* Regel in Chengtu, *Acta Phytotaxonomica Sinica* **4,** 15-20.

Thompson, R.C. (1933) Natural cross-pollination in lettuce, *Proceedings of the American Society of Horticultural Science* **30,** 545-547.

Thompson, R.C. and Ryder, E.J. (1961) Description and pedigrees of nine varieties of lettuce, *US Department of Agriculture Technical Bulletin* 1244.

Thompson, R.C., Whitaker, T.W., Bohn, G.W. and van Horn, C.V. (1958) Natural cross-pollination in lettuce, *Proceedings of the American Society for Horticultural Science* **72,** 403-409.

Van Ettekoven, K. and Van der Arend, A.J.M. (1999) Identification and denomination of "new" races of *Bremia lactucae,* in A. Lebeda and E. Křístková (ed), *Eucarpia Leafy Vegetables '99,* Palacký University, Olomouc, pp. 171-175.

Verhoeff, K. (1960) On the parasitism of *Bremia lactucae* Regel on lettuce, *Tijdschrift over Plantenziekten* **66,** 133-203.

Viennot-Bourgin, G. (1956) *Mildious, oidiums, caries, charbons et rouilles des plantes de France,* Encyclopedia Mycologique Vol. 26, Paris, p. 161.

Vries, I.M. de (1997) Origin and domestication of *Lactuca sativa* L., *Genetic Resources and Crop Evolution* **44,** 165-174.

Whitaker, T.W., Bohn, G.W., Welch, J.E. and Grogan, R.G. (1958) History and development of head lettuce resistant to downy mildew, *Proceedings of the American Society for Horticultural Science* **72,** 410-416.

Wild, H. (1948) Downy mildew disease of the cultivated lettuce, *Transactions of the British Mycological Society* **31,** 112-125.

Witsenboer, H., Kesseli, R.V., Fortin, M.G., Stanghellini, M. and Michelmore, R.W. (1995) Sources and genetic structure of a cluster of genes for resistance to three pathogens in lettuce, *Theoretical and Applied Genetics* **91,** 178-188.

Witsenboer, H., Vogel, J. and Michelmore, R.W. (1997) Identification, genetic localization, and allelic diversity of selectively amplified microsatellite polymorphic loci in lettuce and wild relatives (*Lactuca* spp.), *Genome* **40,** 923-936.

Wittman, W. (1972) Falscher Mehltau an Strohblumen (*Bremia lactucae*), *Pflanzenarzt* **25,** 70.

Yuen, J.E. and Lorbeer, J.W. (1983) A new gene for resistance to *Bremia lactucae, Phytopathology* **73,** 159-162.

Zadoks, J.C. and Schein, R.D. (1979) *Epidemiology and Plant Disease Management*, Oxford University Press, Oxford.

Zink, F.W. and Duffus, J.E. (1969) Relationship of turnip mosaic virus susceptibility and downy mildew (*Bremia lactucae*) resistance in lettuce, *Journal of American Society for Horticultural Science* **94,** 403-407.

Zink, F.W. and Duffus, J.E. (1973) Inheritance and linkage of turnip mosaic virus and downy mildew (*Bremia lactucae*) reaction in *Lactuca serriola, Journal of American Society for Horticultural Science* **98,** 49-51.

Zinkernagel, V., Wegener, D., Hecht, D. and Dittebrandt, R. (1998) Die Population von *Bremia lactucae,* Erreger des falschen Mehltau an Salat in Deutschland, *Nachrichtenblatt des Deutschen Pflanzenschutzdienstes* **50,** 13-17.

Zohary, D. (1991) The wild genetic resources of cultivated lettuce (*Lactuca sativa* L.), *Euphytica* **53,** 31-35.

Zohary, D. and Hopf, M. (1994) *Domestication of Plants in the Old World, Second Edition,* Oxford Science Publications, Oxford.

CHEMICAL CONTROL OF DOWNY MILDEWS

U. GISI

*Syngenta Crop Protection, Research, Product Biology, WRO-1060
CH-4002 Basel, Switzerland*

1. Introduction

The downy mildews are among the most devastating plant diseases caused by
pathogens. In spite of cultural practices and breeding for resistant cultivars, chemical
control is the most effective and economic measure currently used to protect crops from
these diseases. In 1996, the sales value of the global fungicide market was about 7.2
billion SFr, of which about 1.2 billion (16.7 %) was represented by chemicals for
downy mildew control. The downy mildew market can be analysed on the basis of
crops (Figure 1a) or products (Figure 1b) (Syngenta internal data). By far the biggest
crop is represented by grape vine (54 %) with *Plasmopara viticola* as pathogen. This is
followed by: cucurbits (10 %) with mainly *Pseudoperonospora cubensis*; other
vegetables (8 %), e.g. lettuce with *Bremia lactucae*, leek and onion (5 %) with
Peronospora spp.; tobacco and ornamentals (each 4 %) with different *Peronospora*
spp.; peas, brassicas and sugar beet (each 3 %) with *Peronospora* spp.; soybean and
maize (each 1.5 %) with *Peronospora* spp., and in maize the systemic pathogens
Peronosclerospora and *Sclerophthora* spp.; and finally hops and sunflower (each 1 %)
with *Pseudoperonospora humuli* and *Plasmopara halstedii*, respectively. Not included
in this list are the crops infected by *Phytophthora* spp., of which potato with
Phytophthora infestans is the most prominent segment (about 70 % of the grape/
P. viticola market). *Phytophthora* and *Pythium* species of the Pythiales are mentioned in
this review in many places, because most of the chemicals controlling downy mildews
(Peronosporales and Sclerosporales) are also active against *Phytophthora* spp., and the
modes of action and resistance of these compounds have often been studied with
Phytophthora and *Pythium* spp., that are sometimes easier to handle under laboratory
and glasshouse conditions than the biotrophic downy mildews. When the downy
mildew market is analysed according to chemical products sold in 1996, six groups
(Table 1) account for about 90 % of total sales (Figure 1b). The most important group is
represented by products containing phenylamide fungicides. These are not used alone
but in mixture with other compounds, mostly contact fungicides. About the same
market share is taken by dithiocarbamates, mainly mancozeb, followed by thiram,

P.T.N. Spencer-Phillips et al. (eds.), Advances in Downy Mildew Research, 119–159.
© 2002 *Kluwer Academic Publishers. Printed in the Netherlands.*

propineb and maneb. An important part is represented by products containing cymoxanil in mixture with (mainly) dithiocarbamates. Also copper formulations are still widely
used, historically famous because of the Bordeaux mixture described by Millardet in
1885. The contact fungicide chlorothalonil and the systemic fosetyl-Al represent about
the same part of the downy mildew market, followed by smaller products containing
hymexazol, fentins, dimethomorph, propamocarp, fluazinam, or phthalimides (mainly
folpet). Several new chemical groups (Table 1) are not included in this list, because they
were announced for commercial use only few years ago, such as strobilurins (1992-98),
amino acid amide carbamates (1998) and benzothiadiazoles (1996). They will take a
certain market share from existing products within the next years.

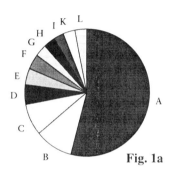 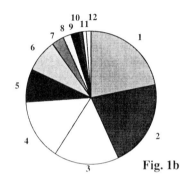

Figure 1. The downy mildew market in 1996 (total of 1.2 billion SFr worldwide) analysed according to crops
(Fig. 1a) and fungicides (Fig. 1b).   **A,** grape vine; **B,** cucurbits; **C,** other vegetables; **D,** leek and onion;
E, tobacco; **F,** ornamentals; **G,** peas; **H,** brassicas; **I,** sugar beet; **K,** soybeans and corn; **L,** hop, sunflower and
others. **1,** phenylamides (mainly metalaxyl); **2,** dithiocarbamates (mainly mancozeb); **3,** cymoxanil; **4,** copper;
5, chlorothalonil; **6,** fosetyl-Al; **7,** hymexazol; **8,** fentins; **9,** dimethomorph; **10,** propamocarb; **11,** fluazinam;
12, phthalimides (folpet) (Syngenta internal data).

For many decades, the copper formulations, the dithiocarbamates, fentins,
chloronitrines and phthalimides were the only fungicides available for the control of
downy mildews. Since the last review on chemical control of downy mildews has been
published (Schwinn, 1981), new and mostly systemic compounds have been discovered
and widely used for disease control (Table 1) such as cymoxanil (1976), fosetyl-Al
(1977), phenylamides (1977-83), propamocarb (1978), dimethomorph (1988), and
fluazinam (1992). In contrast to the contact fungicides, that control pathogens mainly
during early developmental stages at the surface of the host plant, systemic compounds
affect pathogens also at later stages and within the host tissue. Some of these fungicides
(e.g. phenylamides) are single site inhibitors and therefore bear a high intrinsic risk for
the selection of resistant pathogen sub-populations.

2. General properties of antifungal compounds

2.1 DISEASE TYPE AND APPLICATION

Fungicides may be applied to crops as soil and seed treatments, as foliar sprays, trunk injections or as tuber, bulb, and fruit dips, depending on the infection site in the host plant and the properties of the chemicals. Soil treatments are made as drench applications or granule incorporation against soil-borne pathogens such as *Pythium* and *Phytophthora* spp. in nurseries and glasshouses, but also against early attack of young plants by air-borne pathogens, e.g. *Pseudoperonospora humuli* in hops and *Peronospora tabacina* in tobacco. Many crops in subtropical regions suffer from systemic diseases and can therefore be protected by seed treatment, such as maize and sorghum against *Peronosclerospora* ssp.. Also in temperate areas, seed treatment is an important tool in crop protection against seed-borne and systemic diseases such as *Peronospora* spp. in brassicas, soybean or peas and *Plasmopara halstedii* in sunflower. Foliar treatment is by far the largest fungicide market in downy mildew control, and because many pathogen generations develop in a progressing epidemic, several spray applications per season are normally necessary to protect the crop adequately. Under heavy disease pressure 4 to 8 applications may be necessary to control *Plasmopara viticola* in grape vine. Trunk injection or paint treatments can effectively protect trees such as avocado, citrus and cacao against *Phytophthora* spp., whereas dip treatments are used in horticulture, ornamentals and vegetables against *Phytophthora* and *Peronospora* spp..

2.2. BIOLOGICAL MODE OF ACTION

Fungicides are often grouped according to their biochemical mode of action (Table 1). Equally important, especially for designing strategies for use, is the distinction based on their biological mode of action in the disease cycle. A fungicide can either control simultaneously all diseases occurring in one crop (broad spectrum) or be primarily active against one single disease (narrow spectrum). The duration of antifungal activity should be at least as long as the length of one disease cycle. Fungicides can be used as a prevent(at)ive (protectant) application, if they control the pathogen development only prior to its penetration into the host tissue. Contact fungicides are considered to act preventively, because they do not penetrate into the plant and are non-systemic; examples are the dithiocarbamates. If a fungicide also controls later stages in the disease cycle, it is either a curative compound (i.e. active between penetration of pathogen and appearance of first symptoms in host), or it is an eradicative compound (i.e. active after first symptoms have appeared, e.g. against sporulation and further mycelial growth of pathogen.) Curative and eradicative activities can be attribute only to fungicides that are taken up by the host tissue and express a certain systemicity in plants. Phenylamide fungicides, cymoxanil and fosetyl-Al are examples with strong curative and eradicative (antisporulant) activities.

Systemic fungicides exhibit either apoplastic (acropetal) mobility (i.e. translocation within the free space, cell walls and xylem elements of plant tissue governed by diffusion and the rate of transpiration), or they exhibit symplastic (basipetal, sometimes also acropetal) mobility (i.e. translocation through plasmodesmata from cell to cell, involving uptake and distribution from source to sink via the phloem) (Neumann and Jacob, 1995). The end point of transpiration is the surface of leaves, therefore xylem-mobile compounds are often accumulated at tips and edges of the leaf. Because of the source/sink behaviour of phloem transport, there is no export of fungicides from young leaves, whereas maturing leaves (about two thirds of final size) become "source leaves", that express phloem-loading and can export fungicides to other plant parts. Depending on the intensity of systemic behaviour in plants, fungicides can be fully systemic (e.g. phenylamides, propamocarb, cymoxanil, fosetyl-Al) or partially systemic, being either locally systemic (locosystemic, e.g. dimethomorph) or mesostemic (e.g. trifloxystrobin; Margot et al., 1998). A mesostemic fungicide has a high affinity with the plant surface, it is absorbed by the waxy layers of the plant and is redistributed at the plant surface by superficial vapour movement and rediposition; it penetrates plant tissue, exhibits translaminar activity but no transport within the vasular system of the plant.

Several physico-chemical properties of compounds are responsible for their systemic behaviour in plants such as their solubility in water and the lipophilic/hydrophilic balance characterised by the log P_{ow} value (log of octanol/water partition coefficient). The absorption by roots is largely controlled by diffusion, and therefore a high solubility of fungicides in water as well as low log P_{ow} values (optimum at 0.5; Neumann and Jacob, 1995) favour uptake and apoplastic translocation. After spray application of leaves uptake and translocation of several pesticides were strongest at log P_{ow} values between -0.5 and 3.0 with an optimum at about 1.7 (Stevens et al., 1988). More lipophilic compounds (log P_{ow} > 3) are retained in the cuticle and cannot be translocated. Redistribution is then only possible if compounds express a tendency for volatilization. Chloroneb has an extremely high vapour pressure of 400 mPa and thus an important vapour-phase activity for disease control. Also metalaxyl and some strobilurins with much lower vapour pressures are claimed to control diseases over short distances in the canopy partially through vapour-phase activity. Compounds that are phloem-mobile have low log P_{ow} values and are in addition slightly acidic (low pK_a value) like the phosphonate fosetyl-Al and some members of the benzothiadiazoles (Table 1). Besides differences in physico-chemical properties of compounds, the uptake into leaves is largely dependent on the spray coverage of the plant surface, but also on how long the spray deposit remains wet. The addition of surfactants and other ingredients in the formulation can enhance the uptake rate significantly.

3. Chemical groups and biochemical mode of action

Within the oomycetes, there are several orders comprising important plant pathogens, e.g. the Peronosporales and Sclerosporales (downy mildews), as well as the Pythiales and Saprolegniales. Amongst many metabolic differences. the oomycetes, unlike most fungi, do not produce chitin but cellulose as a fibrillar cell wall component, and do not synthesise their own functional sterols (e.g. ergosterol) but take them from the growth medium (e.g. host tissue), before converting and incorporating them into their membranes. The specificity of antifungal compounds might be based on such metabolic differences between oomycetes and true fungi. Not all chemical groups described below are active exclusively against downy mildews, but also control pathogens of closely related orders such as Pythiales (e.g. *Pythium* and *Phytophthora* spp.), or in addition pathogens belonging to classes outside the oomycetes. The following fungicides were arranged according to their biochemical site of action, from single to multisite inhibition and from RNA to cell wall synthesis inhibition (Table 1). As a general rule, there is no cross resistance between members of different chemical groups, but there is a positive correlation within and amongst chemical classes.

3.1. PHENYLAMIDES

The phenylamide fungicides (PAs) inhibit ribosomal RNA synthesis, specifically RNA polymerisation (polymerases). During RNA synthesis, three polymerases are involved: Polymerase I (or A) synthesizes 45 S pre r(ibosomal) RNA that represents by far the major part of the cellular RNA. Polymerase II (or B) produces m(essenger) RNA and polymerase III (or C) is responsible for t(ransfer) RNA and 5 S rRNA synthesis. Polymerase II is known to be highly sensitive to the fungal toxin alpha-amanitin, whereas polymerase I is insensitive and polymerase III slightly sensitive to the toxin. In mycelium of *Phytophthora megasperma*, metalaxyl affected primarily rRNA synthesis, whereas mRNA was much less sensitive; therefore, inhibition of rRNA synthesis is considered as the site of action of phenylamides (Davidse, 1995). PAs control almost exclusively members of the Peronosporales and Sclerosporales (downy mildews), as well as Pythiales and Saproleginales with some few exceptions: *Pythiopsis cymosa* but not *Aphanomyces euteiches* of the Saprolegniales is sensitive to PAs. All members of the Pythiales including *Pythium, Phytophthora* and *Lagenidium* species are sensitive to PAs (Fuller and Gisi, 1985; Schwinn and Staub, 1995). Especially in the genus *Pythium*, however, certain species express different intrinsic sensitivities to some members of the PAs. Species of other "zoosporic fungi" such as *Plasmodiophora* spp. of the Plasmodiophorales, *Olpidium* spp. of the Chytridiomycetes, *Rhizidiomyces* ssp. and *Hyphochytridium* spp. of the Hyphochytridiomycetes, as well as *Zoophagus* ssp. and *Rhizopus* spp. of the Zygomycetes and all higher fungi are insensitive to PAs (Bruin and Edginton, 1983; Fuller and Gisi, 1985). Major target pathogens for PAs are downy mildews of the genera *Albugo, Bremia, Peronospora, Peronosclerospora, Plasmopara,*

Pseudoperonospora, Sclerophthora and *Sclerospora*. The genus *Aphanomyces* of the Saprolegniales is insensitive to PAs, but sensitive to hymexazol and propamocarb (see below).

Within the phenylamides (Table 1), the specific compounds express differential levels of intrinsic activities: Metalaxyl-M (mefenoxam), the R-enantiomer (D-alaninate analogue) of metalaxyl provides at half rate the same level of activity as (racemic) metalaxyl (DL-alaninate analogue) and is degraded in soil more easily than metalaxyl (Nuninger *et al.*, 1996). Metalaxyl-M is the most active, versatile and broadly used molecule within PAs against a wide range of foliar diseases in crops like potato (late blight), vegetables (*Phytophthora* spp., downy mildews), tobacco (blue mold), grapes (*Plasmopara viticola*), citrus (*Phytophthora* spp.), against soil-borne diseases in turf and ornamentals (*Pythium* spp., *Phytophthora* spp.), tobacco (black shank), avocado (*Phytophthora cinnamomi*) and against seed-borne diseases, e.g. in maize against *Peronosclerospora* spp.. Furalaxyl is specifically recommended against soil-borne *Phytophthora* and *Pythium* species in ornamentals, oxadixyl is mainly used against potato late blight, grape downy mildew and some seed-borne pathogens (e.g. *Peronospora* on peas and *Phytophthora* on cotton), whereas benalaxyl and ofurace are of particular interest against several downy mildews (e.g. on grapes and vegetables) and potato late blight, respectively.

Phenylamides exhibit strong preventive and curative activity (Figure 3). They affect especially hyphal growth (inside and outside the plant tissue) as well as haustorium and spore formation. Although not fully utilized for resistance management reasons, PAs also exhibit strong eradicative and antisporulant activity in the disease cycle of target pathogens. On the other hand, PAs do not inhibit the early stages in the disease cycle like zoospore release, spore germination and penetration of the host tissue (Schwinn and Staub, 1995). Since spores contain many ribosomes to support early growth stages, RNA synthesis is fully operating only after spore germination; later development stages are therefore most sensitive to PAs. As a consequence of RNA inhibition, the precursors of RNA synthesis (i.e. nucleoside triphosphates) are accumulated; they activate β-1,3-glucansynthetases which are involved in cell wall formation. Metalaxyl-treated hyphae often produce thicker cell walls than untreated ones. In addition to RNA polymerase inhibition, benalaxyl, but not metalaxyl and oxadixyl, affected also uridine uptake into fungal cells. This observation (Davidse, 1995) implied a second mode of action for some PAs, but it was expressed only at about 10 fold higher fungicide concentrations than those needed to inhibit RNA synthesis and is probably not of practical relevance.

Phenylamides are rapidly taken up by roots and foliage and are easily translocated acropetally (apoplastically) within leaves and from leaf to leaf. Translaminar movement is substantial. Also, some basipetal (symplastic) translocation was described for most PAs, especially within the leaf and to a certain degree also from treated leaves to lower plant parts, but the molecules did not reach the roots (Table 2). Metalaxyl-M is the most systemic and ofurace the least systemic PA, with other PAs in between the two. The solubility in water is 26000, 8400, and 146 mg a.i./l and log P_{OW} values 1.71, 1.75, and

1.39 for metalaxyl-M, metalaxyl, and ofurace, respectively. Oxadixyl shows a similarly strong systemicity behaviour as metalaxyl (solubility 3400 mg/l, log P_{OW} 0.7), whereas benalaxyl behaves quite differently (solubility 29 mg/l, log P_{OW} 3.54), although strong systemicity is also claimed for the latter compound. Metalaxyl and metalaxyl-M can be distributed within plants and to a certain degree over short distances also in the canopy by vapour phase. Based on their strong systemicity in plants, PAs also protect untreated as well as newly grown plant parts through translocation from the treated parts. They provide long lasting disease control, although the buildup of resistant subpopulations limits extended spray intervals to a maximum of 10 to 14 days and the number of applications to 2 to 4 per season. It is generally assumed that apoplastic transport is less pronounced in senescing plant tissue, and for that reason the translocation of PAs is less strong in maturing crops. Therefore, best control levels are observed when PAs are applied early in the season and during periods of active vegetative growth of the crop.

3.2. HETEROAROMATICS

The isoxazole fungicide hymexazol (Table 1) has been used since 1974 mainly as a seed dressing or soil drench and incorporation against *Pythium* spp.; in addition, *Fusarium* spp. and *Aphanomyces* ssp. in sugar beet, vegetables and ornamentals are controlled (Kato *et al.*, 1990). Hymexazol is taken up by roots (solubility in water 65 g/l, log P_{OW} 0.48) and can be considered as a partially (locally) systemic fungicide (Table 2). The primary site of action was claimed to be inhibition of DNA synthesis, whereas RNA and protein synthesis are less sensitive to hymexazol (Nakanishi and Sisler, 1983) The most sensitive developmental stages of target pathogens are mycelial growth and sporulation. The (1,2,4)-thiadiazole fungicide etridiazole (Table 1) controls *Pythium* and *Phytophthora* spp. in ornamentals, cotton, vegetables and turf, but also the primary infection of sunflower by *Plasmopara halstedii*. Spore germination is not affected, but growth of germ tubes and subsequent hyphal growth in the tissue is inhibited (Viranyi and Oros, 1991). Etridiazole is considered as a contact fungicide (solubility in water 117 mg/l and log P_{OW} 3.37) and is used for soil and seed treatment applications. Although chemically closely related to hymexazol, it is claimed that etridiazole does not affect DNA synthesis but disrupts lipid peroxidation as it is known for aromatic hydrocarbons (Lyr, 1995; see below).

3.3. COMPLEX III RESPIRATION INHIBITORS

Fungicides inhibiting mitochondrial respiration at the enzyme complex III (ubiquinol-cytochrome c oxidoreductase complex) in the respiratory pathway are grouped together as complex III respiration inhibitors (Hewitt, 1998) (Table 1). The antifungal potential of complex III respiration inhibitors for effective disease control in agriculture is very

TABLE 1. Classification of compounds controlling diseases caused by oomycetes

Chemical/Cross Resistance Group (Site of Action)	Chemical Class	Common Name of Compound
ANTIFUNGAL COMPOUNDS		
Phenylamides (RNA synthesis)	Acylalanines	Benalaxyl, Furalaxyl, Metalaxyl, Metalaxyl-M (Mefenoxam)
	Oxazolidinones	Oxadixyl
	Butyrolactones	Ofurace
Heteroaromatics (DNA synthesis)	Isoxazoles	Hymexazol
	(1,2,4)-Thiadiazoles	Etridiazole
Compl. III respiration inhibitors, QoI (compex III respiration)	Methoxyacrylates	Azoxystrobin
	Methoxycarbamates	Pyraclostrobin (BAS 500F)
	Oximinoacetates	Kresoxim-methyl, Trifloxystrobin
	Oximinoacetamides	Metominostrobin (SSF 126)
	Oxazolidinediones	Famoxadone
	Imidazolinones	Fenamidone (RPA 407213)
Compl. III respiration inhibitors, QiI	Cyanoimidazoles	Cyazofamid (IKF 916)
Organo-tins (Fentins) (ATP production)		Fentin acetate, Fentin hydroxide
Dinitroanilines (ATP production)		Fluazinam
Morpholines (sterol biosynthesis)		Tridemorph
Aromatic hydrocarbons (lipid peroxidation)		Chloroneb
Carbamates (membrane permeability)		Propamocarb, Prothiocarb
Cinnamic acids (cell wall synthesis)		Dimethomorph
Cyano-acetamide oximes (unknown)		Cymoxanil
Amino acid amide carbamates (unknown, amino acid metabolism ?)		Iprovalicarb
Phosphonates (unknown)		Fosetyl-Al, Phosphorous acid
Miscellaneous (unknown, micotubules?)		Zoxamide (RH-7281)
Multisites	Inorganics	Copper (Cu-oxychloride,-sulfate, -hydroxide)
	Dithiocarbamates	Mancozeb, Maneb, Zineb, Metiram, Propineb, Thiram, Ziram, Ferbam
	Phthalimides	Captan, Captafol, Folpet
	Sulphamides	Dichlofluanid
	Chloroisophthalonitriles	Chlorothalonil
HOST PLANT DEFENSE INDUCERS		
Salicylic acid pathway	Isonicotinic Acids	Isonicotinic acid derivatives
	Benzothiadiazoles	Acibenzolar-S-methyl
Jasmonic acid pathway		(Methyl-)Jasmonate, *Pseudomonas*
Unknown pathway		Fosetyl-Al, Fatty acids, Amino butyric acids, *Bacillus*

big because the spectrum of activity can cover all important classes of plant pathogens (oomycetes, ascomycetes, deuteromycetes, basidiomycetes). This was described especially for pyraclostrobin (Ammermann et al., 2000), trifloxystrobin (Margot et al., 1998; Hermann et al., 1998), kresoxim-methyl (Ammermann et al., 1992; Ypema and Gold, 1999) and azoxystrobin (Godwin et al., 1992; Baldwin et al., 1996), as well as to a certain degree also for famoxadone (Joshi and Sternberg, 1996). On the other hand, metominostrobin (SSF 126) seems to be especially active against rice blast (Mizutani et al., 1996), whereas IKF 916 (Mitani et al., 1998) and fenamidone (RPA 407213; Mercer et al., 1998) primarily control members of the oomycetes; fenamidone is also active against *Alternaria* spp. and *Mycosphaerella* ssp..

In mitochondrial respiration of fungi and plants, six key enzyme complexes (I to VI, Figure 2) are involved in electron transport from NADH and $FADH_2$ to molecular oxygen and thereby producing ATP: In complex I , electrons are transferred from NADH (reduced nicotinamide-adenine dinucleotide) to ubiquinone Q, catalysed by NADH-ubiquinone oxidoreductase, whereas in complex II, electrons flow from $FADH_2$ (reduced flavin adenine dinucleotide) to ubiquinone Q, controlled by the enzyme succinate-ubiquinone oxidoreductase. The ubiquinone/ubiquinol (Q/QH_2) pool then feeds electrons into the cytochrome bc1 enzyme complex (complex III) that contains two active centers, the ubiquinol oxidizing (outer, Qo) site and the ubiquinone reducing (inner, Qi) site controlled by ubiquinol-cytochrome c oxidoreductase. The Qo site is formed by cytochrome b (heme b_L) and an iron-sulfur protein, whereas the Qi site contains cytochrome b (heme b_H) (Geier et al., 1992). Electrons flow from ubiquinol to cytochrome c either through the linear Qo chain or via the cyclic Qi route, the latter allowing feedback reactions (Gasztonyi and Lyr, 1995). Cytochrome c then delivers electrons through cytochrome a-a3 (terminal) oxidase (complex IV) to the final acceptor O_2. Under special circumstances, not well understood in fungi, electrons can bypass the regular respiration pathway and are transferred from the ubiquinone pool to O_2; this pathway represents the cyanide-insensitive, alternative respiration catalysed by the alternative oxidase (complex V). During the respiration pathway, protons liberated at several locations are essential for the convertion of ADP+Pi to ATP by the oxidative phosphorylation involving ATP synthases (complex VI, Figure 2).

Several antifungal inhibitors are known for all six enzyme complexes. The insecticide rotenone and the fungicide fenaminosulf (Dexon, active against soil-borne pathogens including *Pythium* and *Phytophthora* species) are complex I inhibitors, whereas the carboxamide fungicides (mainly active against basidiomycetes) are complex II inhibitors (Schewe and Lyr, 1995). As shown in Figure 2, complex III is composed of two active centers. Inhibitors at the Qo site (QoIs) are the "antibiotic"myxothiazol and several agronomically important fungicides (Table 1), whereas antimycin and some newer compounds such as IKF 916 (Mitani et al., 1998) affect the Qi site (Jordan et al., 1999). Cross-resistance between all compounds of the Qo-classes (methoxyacrylates,

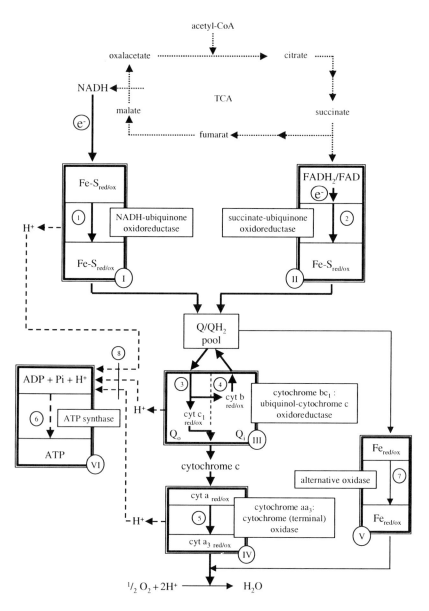

Figure 2. Antifungal inhibitors of enzyme complexes (I to VI) involved in mitochondrial respiration, electron transport and ATP production. Inhibitors are **1**: fenaminosulf; **2**: carboxamides; **3**: myxothiazol and strobilurins (QoIs); **4**: antimycin (and IKF 916) (QiIs) ; **5**: cyanide; **6**: oligomycin; **7**: SHAM; **8**: fluazinam and fentins. Solid arrows indicate flow of electrons (modified based on information of Schewe and Lyr, 1995; Jordan *et al.*, 1999).

methoxycarbamates, oximinoacetates, oximinoacetamides, oxazolidinediones, imidazo-linones, Table 1) has been observed in many pathogens such as *Erysiphe graminis* (Sierotzki *et al.*, 2000a), *Mycosphaerella fijiensis* (Sierotzki *et al.*, 2000b) and *Plasmopara viticola* (Heaney et al., 2000), but not between Qo- and Qi inhibitors (including the sulfamoyl-cyanoimidazole compound IKF 916, FRAC internal data). Inhibitors of complex IV are cyanide and azide, whereas SHAM (salicyl-hydroxamic acid) affects the alternative oxidase (Watanabe *et al.*, 1996). Inhibitors of the oxidative phosphorylation without reducing electron transport (i.e. ATP uncouplers) are nitrophenol fungicides (e.g. dinocap, active against powdery mildews), drazoxolon (soil and seed treatment fungicide against *Fusarium* and *Pythium* species), as well as the organo-tin and dinitroaniline fungicides (e.g. fluazinam, see below). They all affect the proton gradient through mitochondrial membranes but not ATP synthase activity; the latter is inhibited by oligomycin (Figure 2).

Current information and the recommended use rates for field application of complex III respiration inhibitos indicate that the strongest intrinsic activity against oomycetes is expressed by famoxadone, fenamidone, pyraclostrobin, and IKF 916 (n rate 50-125 g a.i./ha), followed by azoxystrobin and trifloxystrobin (100-250 g/ha), and kresoxim-methyl (250-400 g/ha). Most downy mildews (e.g. *P. viticola, Pseudoperonospora* spp., *Peronospora* spp.) can be controlled with lower application rates of QoI fungicides than *P. infestans*. The interval between applications is 7 to 12 days depending on disease pressure. Best disease control by QoI fungicides is achieved through preventive applications, because the most sensitive developmental stages of target pathogens are spore germination, zoospore release and motility. Later stages cannot be controlled and curative activity against oomycetes is very weak. Under glasshouse conditions, the incubation period (time between inoculation and appearance of first symptoms) for *P. viticola* is about 5 to 6 days: the more systemic members of the QoI fungicides (azoxystrobin and fenamidone) are curatively active during the first 1 to 2 days only (up to 30 % of latent period; Godwin *et al.*, 1997) and do not control later stages unlike cymoxanil or phenylamides (Figure 3). On the other hand, QoI fungicides may exhibit curative activity against some other pathogens such as *Venturia inaequalis* and powdery mildews which penetrate only the surface of the host tissue.

Some QoI fungicides have an extremely low solubility in water (famoxadone 0.05, IKF 916 0.12, trifloxystrobin 0.61, others 2-10 mg/l,), therefore the formulation is of major importance for reliable disease control. Azoxystrobin and fenamidone have log P_{OW} values enabling the molecules to be translocated in the plant (2.64 and 2.8, respectively). They are taken up by roots and slowly translocated acropetally into the shoot and within leaves, they also express strong translaminar movement, but no basipetal translocation. The strong translaminar activity of azoxystrobin might explain the significant antisporulant action against *P. viticola* (Godwin *et al.*, 1997). Azoxy-strobin and fenamidone can be considered as partially systemic molecules (Table 2). The log P_{OW} values of IKF 916, kresoxim-methyl, trifloxystrobin and famoxadone are 3.2, 3.4, 4.5 and 4.7, respectively; the molecules are hardly taken up by any plant part. It has been observed that trifloxystrobin is continuously absorbed in the waxy layers of leaves and is redistributed through the vapour phase. This unique migration behaviour on and in plant tissue (Table 2) has been defined as "mesostemic" (Margot et al., 1998). The term "quasi-systemic" or "surface systemic" has been used for kresoxim-methyl (Ypema and Gold, 1999). Thus, these members of QoIs can be clearly differentiated from non-systemic fungicides (e.g. dithiocarbamates) and fully systemic fungicides (e.g. phenylamides).

3.4. ORGANO-TINS AND DINITROANILINES

The organo-tins, or fentins, or triphenyl tins are represented by the two fungicides fentin acetate and fentin hydroxide (Table 1), which inhibit oxidative phosphorylation in mitochondria, in a very similar way as the dinitroaniline fungicide fluazinam (Table 1). They uncouple phosphorylation from electron transport by disrupting the proton gradient, and as a consequence ATP production is blocked (Figure 2). Fluazinam and fentins inhibit zoospore release and germination of spores. The fungicides of both chemical groups are non-systemic, contact fungicides (Table 2) without any curative activity (Figure 3). Best disease control is achieved in a strictly preventive spray program with intervals between applications of about 7 days. The use of fentins is limited in some crops because of their phytotoxic potential. The main target pathogens for fluazinam and fentins are *P. infestans* and *Alternaria solani* in potato. They control *P. infestans* not only on foliage but equally well on tubers and are used preferentially at the end of the season. Fentins also control *Pseudoperonospora* spp. and *Peronospora* spp. in hops, garlic and onions, as well as *Cercospora* and *Ramularia* in sugar beet, *Septoria* spp. in celery and many other pathogens in a variety of crops. Fluazinam is also used against *P. cubensis*, as well as against *Botrytis cinerea* (and *P. viticola* as a secondary effect) in grapes, apple scab and *Colletotrichum* and *Sclerotinia* in a range of crops (Anema et al., 1992).

3.5. MORPHOLINES

Although primarily active against powdery mildew, it has been claimed that the morpholine fungicide tridemorph (Table 1) can control also *P. infestans* and *P. halstedii* (Oros et al., 1988). Tridemorph inhibits the Δ 8/7- isomerase (and probably also the Δ 14-reductase) in sterol biosynthesis and because of the deficiency of a complete sterol pathway in oomycetes, the activity of tridemorph against *P. infestans* and *P. halstedii* is somewhat surprising. Most sensitive development stages are the release, motility and encystment of zoospores and also the germination of cystospores, but not any other stages. Probably because of these shortcomings and in addition due to its phytotoxic potential, tridemorph has not been used in practice.

3.6. AROMATIC HYDROCARBONS

The chemical group of the aromatic hydrocarbon fungicides contains several members such as chloroneb (Table 1), which show activity against several soil- and seed-borne pathogens in cotton, vegetables and sugar beet such as *Pythium* and *Phytophthora* species. Also, some pathogens of the *Rhizoctonia* soil-borne disease complexes are controlled. The compounds disrupt lipid peroxidation and cause aberrations in membranes, mitochondria, the endoplasmatic reticulum and nuclear envelope (Lyr, 1995). Also, cell wall synthesis is disorganized and some mutagenic effects have been observed in target fungi. Chloroneb displays some systemic activity and can be redistributed through its strong vapour phase (vapour pressure is 400 mPa, i.e. about 20 x higher than that of fenpropidin).

3.7. CARBAMATES

The spectrum of activity of the carbamate fungicides propamocarb and prothiocarb (Table 1) includes many *Phytophthora* and *Pythium* species (e.g. *P. infestans* in potato, *P. cactorum* in strawberries, *Pythium* spp. in carrots), as well as downy mildews like *Peronospora* spp. (e.g. in tobacco), *Bremia* spp. (e.g. in lettuce), *Pseudoperonospora* spp. (e.g. in hops), but also *Aphanomyces* spp. (in the family Saprolegniaceae). All other pathogens outside the oomycetes are insensitive to carbamate fungicides. Propamocarb is very soluble in water (higher than 900 g a.i./l) and the log P_{ow} is extremely low (-2.6); it is taken up easily by roots. Within the shoot and leaves, propamocarb is translocated acropetally (Table 2) and can be considered as a fully systemic fungicide. It does not exhibit sufficient curative activity for disease control. It can be applied to the soil or used as a dip treatment (for bulbs and tubers) or seed treatment, and as a foliar spray in late-season applications. Propamocarb affects the permeability of plasma membranes and as a consequence, leakage of cell constituents such as phosphate and carbohydrates have been observed. Lipid synthesis seems to be especially sensitive: the synthesis of neutral lipids was increased, whereas several phospholipid components were diminished in the presence of propamocarb (Reiter *et al.*, 1996). It affects primarily mycelial growth in the tissue and germ tube elongation of spores (Viranyi and Oros, 1991), as well as sporulation (Reich *et al.*, 1992; Reiter *et al.*, 1996).

3.8. CINNAMIC ACIDS

The cinnamic acid fungicide dimethomorph (Table 1) controls exclusively pathogens of the families Peronosporaceae (e.g. *Bremia* in lettuce, *Peronospora* spp., especially in tobacco, pea and onion, *Plasmopara* spp. in grape and sunflower, *Pseudoperonospora* spp.), and Pythiaceae (e.g. many *Phytophthora* spp. in potato, tomato, pineapple and onion) except the entire genus *Pythium*, which is insensitive as are all other pathogens outside the oomycetes. Dimethomorph inhibits most stages during development of *P. infestans* and *P. viticola* except zoosporogenesis and zoospore motility (Albert *et al.*, 1991). Zoospore encystment, cystospore germination, sporangium and oospore formation, as well as apical growth of hyphae are particularly sensitive to dimetho-morph (Albert *et al.*, 1991; Cohen *et al.*, 1995). Because all stages involving cell wall formation are affected, it is concluded that dimethomorph inhibits biosynthesis and assembly of fibrillar wall components in target pathogens (Albert and Heinen, 1996).

Foliar application of dimethomorph results in translaminar and weak acropetal trans-location in leaves, but it does not move from leaf to leaf (Table 2). After a soil drench under growth chamber conditions, dimethomorph was taken up by roots and controlled *P. infestans* in tomato leaves, and *P. viticola* in grape leaves, but did not protect cucumbers against *P. cubensis* (Cohen *et al.*, 1995). A solubility in water lower than 50 mg a.i./l and a log P_{OW} of 2.7 support the classification for dimethomorph as being a partially (locally) systemic fungicide. In addition to a long lasting preventive activity, dimethomorph also exhibits some curative (Figure 3) and strong antisporulant activity (Albert *et al.*, 1988; Cohen *et al.*, 1995).

3.9. CYANO-ACETAMIDE OXIMES

The cyano-acetamide oxime cymoxanil (Table 1) plays an important role in the control of foliar pathogens in potato and tomato (e.g. *P. infestans*), grape vine (*P. viticola*) and against downy mildews (e.g. *Peronospora* spp., *Pseudoperonospora* spp.) in a range of crops. Cymoxanil does not control any other pathogens outside the oomycetes, although an artificial mutant of *Botrytis cinerea* was used to study the biochemical mode of action (Leroux *et al.*, 1987). Soil-borne *Phytophthora* and *Pythium* species are either insensitive or not controlled because of rapid decomposition of cymoxanil in soil (Klopping and Delp, 1980). In plant tissue (and fungal mycelium) cymoxanil is degraded to glycine within 2 to 4 days, depending on temperature; therefore, cymoxanil is mostly used in combination with multisite inhibitors (e.g. mancozeb, maneb, folpet, chlorothalonil, copper and others) to improve its preventive activity through synergistic interactions (Gisi, 1996), and to extend the interval between applications from 4 (when used alone) to about 8-10 days (as fungicide mixture). Also some 3-way mixtures including systemic compounds are available (e.g. cymoxanil + mancozeb + oxadixyl) to extend the spray interval even longer (12 days) and as an enforced anti-resistance strategy to protect both cymoxanil and phenylamides (Gisi *et al.*, 1985). The curative activity of cymoxanil is quite pronounced, especially for downy mildew control in grape vine where it can last from 2 days at high temperature, up to 4 days at lower temperature (Genet *et al.*, 1997; Figure 3). Against *P. infestans*, the curative activity of cymoxanil is limited to 1-2 days, and an eradicative treatment is generally not recommended.

Cymoxanil is taken up by roots and is rapidly translocated acropetally in the shoot and leaves. It is translocated acropetally after foliar or stem application, and within leaves also shows basipetal and translaminar movement (Samoucha and Gisi, 1987a; Cohen and Gisi, 1993). Based on these observations resulting from bioassays, but also according to the physico-chemical properties (solubility in water, 890 mg a.i./l; log P_{OW}, 0.59), cymoxanil can be considered as a fully systemic fungicide (Table 2). The rapid degradation of cymoxanil in plant tissue limits its systemic potential for disease control to rather short time periods in untreated plant parts. Cymoxanil affects growth of intercellular hyphae and the formation of haustoria, as well as production of sporangia. It has been argued that cymoxanil inhibits RNA and amino acid synthesis (Leroux *et al.*, 1987), but other studies (Ziogas and Davidse, 1987) failed to identify the primary site of action.

3.10. AMINO ACID AMIDE CARBAMATES

A novel aminoacid amide (or carbamoylmethly) carbamate fungicide, iprovalicarb (SZX 722, Table 1), was presented in 1998 (Stenzel *et al.*, 1998) for the control of diseases caused by pathogens of the two families Peronosporaceae and Pythiaceae, excluding the entire genus *Pythium* as it is true for dimethomorph. Iprovalicarb is especially active against *P. viticola* on grape vine, but it also controls *P. infestans* on potato and tomato, *Pseudoperonospora* spp. on cucumber and hop, *P. tabacina* on tobacco, and *B. lactucae* on lettuce, as well as some soil-borne *Phytophthora* species. The compound affects germination of cystospores and sporangia, as well as growth of mycelium and sporulation, resulting in long lasting preventive control of *P. viticola* and *P. infestans* (similar levels as dimethomorph) and strong curative (similar to cymoxanil) as well as eradicative activities. After drench or foliar application, iprovalicarb is rapidly translocated throughout the plant and is uniformly distributed in less than one day. It can be considered as a fully systemic fungicide (Table 2), despite a rather low solubility in water (7-11 mg a.i./l) and a log P_{OW} value of 3.2. Specific adjuvants in the formulation have a strong impact on uptake and translocation behaviour leading to an enhanced efficacy of iprovalicarb (Stenzel *et al.*, 1998). Although the biochemical site of action is not yet known, iprovalicarb seems to alter the composition and reduce the pool of free amino acids in cells of treated pathogens.

3.11. PHOSPHONATES

Within the chemical group of phosphonates, fosetyl-Al and its breakdown product phosphorous acid (H_3PO_3) (Table 1) exhibit quite extraordinary translocation properties. With a solubility in water of 120 g a.i./l (extremely high) and a log P_{OW} value of -2.7 (extremly low), combined with its acidic nature, fosetyl-Al is easily taken up by roots and foliage and is translocated within the phloem (symplastically), in both acropetal and basipetal direction to all plant parts. Therefore, it is a fully systemic fungicide (Table 2). Based on the special systemicity behaviour and combined with an interesting spectrum of activity (Williams *et al.*, 1977), fosetyl-Al exhibits excellent properties for spray application to control *P. viticola* on grape vine (as early and late season treatments) with additional suppression of *Phomopsis viticola* (cane and leaf spot), *Guignardia bidwelli* (black rot), and *Pseudopeziza tracheiphila* (Rotbrenner*)* on the same crop. In addition to a long lasting preventive control, curative activity against *P. viticola* that is about as strong as that of dimethomorph has been described (Figure 3). As spray, dip or drench application, fosetyl-Al controls a range of diseases caused by pathogens of the oomycetes such as *Bremia* on lettuce, *Pseudoperonospora* spp. in hops and cucumbers, *Pythium* spp. in turf and many different *Phytophthora* species in strawberry, pepper, ornamental and nursery crops, as well as *P. cinnamomi* in avocado and *P. nicotianae* in pineapple. Against root rot and stem cancer (*Phytophthora* spp.) trunk injections of trees (avocado, palm, citrus) are recommended with fosetyl-Al or phosphorous acid. Surprisingly, *P. infestans* on potato and tomato is not adequately controlled by fosetyl-Al applications.

Fosetyl-Al inhibits the formation of sporangia, oospores and chlamydospores in *Phytophthora* spp. (Farih *et al.*, 1981), as well as release, motility, encystment of zoospores and germination of cystospores in *P. halstedii* (Viranyi and Oros, 1991); mycelial growth is much less sensitive. The antifungal activity of fosetyl-Al is affected by high levels of phosphate present in many culture media used for *in vitro* studies (Fenn and Coffey, 1984). Furthermore, fosetyl-Al is converted to H_3PO_3 more rapidly *in planta* than *in vitro*. H_3PO_3 is believed to be the more active form against fungi both *in vitro* and *in planta* tests as compared to the Al salt. These observations may have resulted in an underestimation of the direct antifungal properties of fosetyl-Al in earlier studies; the final proof was provided by a successful production of laboratory mutants of *P. palmivora* resistant to H_3PO_3 and fosetyl-Na (Dolan and Coffey, 1988). Probably as a result of changes in phosphorylated sugars and nucleotide phosphates, alterations in cell wall architecture and composition were observed in *P. palmivora* after a fosetyl-Al treatment (Grant *et al.*, 1993). Nevertheless, the primary and direct site of action of fosetyl-Al in target pathogens is not yet elucidated.

In addition to the antifungal potential, there is also evidence that fosetyl-Al and H_3PO_3 induce defense reactions in treated and infected plants (Figure 3). Fosetyl-Al is believed to stimulate the production of phytoalexins (stilbenes, flavonoids) and thus, enhance the protection against *P. viticola* in treated grape leaves (Langcake, 1981); in tobacco infected with *P. parasitica*, fosetyl-induced accumulation of the phytoalexin capsidiol was suggested to be associated with hypersensitive reactions (Guest, 1984). In addition, the accumulation of phenolic compounds and an increased production of necrotic defense reactions around infection sites of *P. capsici* was observed in treated tomato leaves (Bompeix *et al.*, 1980). Recent studies suggest a combination of direct antifungal activity of fosetyl-Al and H_3PO_3 and an enhanced response of host plant resistance to the compounds and against H_3PO_3-modified pathogens (Dercks and Creasy, 1989; Smillie *et al.*, 1989; Nemestothy and Guest, 1990). Phosphonates appear to produce changes in pathogens, e.g. changes in lipid and cell wall composition that might cause reduced virulence of the pathogens (Dustin *et al.*, 1990).

3.12. MISCELLANEOUS

Recently, the novel preventive fungicide zoxamide (RH-7281) was announced (Egan *et al.*, 1998) for foliar use in potato against *P. infestans*, in grape vine against *P. viticola* and in vegetables against *P. cubensis*. Under glasshouse and field conditions it also controls other pathogens such as *B. lactucae, Peronospora parasitica, Phytophthora capsici, P. erythroseptica, Botrytis cinerea, Cercospora arachidicola* and *Sphaerotheca fuliginea,* and *in vitro* a range of additional pathogens. Zoxamide exhibits a long lasting preventive activity similar to that of dimethomorph, but no curative control was observed (Figure 3). Spray intervals of 7 to 10 days are recommended providing similar control levels as other contact fungicides such as mancozeb, fluazinam, chlorothalonil and folpet. Zoospore motility, encystment and germination were not affected, but germ tube elongation, mycelial growth and penetration into plant tissue were suppressed. Since the solubility in water is extremely low (0.7 mg a.i./l) and log P_{OW} rather high (3.76), zoxamide can be considered as a non-systemic fungicide. It is claimed that zoxamide arrests nuclear division by binding to the β-subunits of tubulin and disrupts the microtubular cytoskeleton in target fungi.

3.13. MULTISITES

The multisite inhibitors (Table 1) are non-systemic, preventive fungicides (see F-H, K, L in Figure 3) forming a protectant barrier at the surface of plants against pathogens. They inhibit fungal development prior to the penetration into the tissue and interact mostly unspecifically with many biochemical steps in fungal metabolism. The inorganic copper fungicides such as Bordeaux mixture, $Ca(OH)_2 + CuSO_4$, and copper oxy-chloride, $Cu_2Cl(OH)_3$, and copper hydroxide, $Cu(OH)_2$, are still used singly, or in combination with systemic fungicides, to control diseases in grape vine *(P. viticola)*, potato and tomato *(P. infestans)*, hops *(P. humuli)*, and many other crops against a range of pathogens. Copper as Cu^{2+} accumulates in sensitive pathogens and forms complexes with enzymes possessing sulphydryl-, hydroxyl-, amino-, or carboxyl-groups; as a consequence, the enzymes are inactivated leading to a general disruption of metabolism and breakdown of cell integrity. In addition to the fungicidal effect, copper also exhibits a retardation of plant growth and a hardening of foliage and berries, which is considered as an additional benefit for crops like grape vine (protection against secondary pathogens and climatic stress).

Within the organic dithiocarbamate fungicides, thiram is the oldest compound; it was introduced to the market in 1937 and is still used as a broad-spectrum seed-dressing fungicide. Ferbam and ziram are iron and zinc salts, respectively, of the dimethyl-dithiocarbamates and are used as foliar fungicides against downy mildews of grape vine and certain vegetables. The most important members are the zinc and manganese salts of the ethylene-bis-dithiocarbamates such as zineb and metiram or maneb, respectively, and the combined Zn-Mn-complex mancozeb, as well as the Zn-propylene-bis-dithiocarbamate complex propineb. They are important broad-spectrum contact fungicides used as foliar, soil and seed treatments against pathogens in a range of crops including *V. inaequalis* in apple, *P. viticola* in grape vine, *Phytophthora, Botrytis, Alternaria* and *Septoria* species in vegetables, *P. infestans* and *A. solani* in potato and many others. In 1995, mancozeb, copper based fungicides and chlorothalonil represented a global sales value of 295, 280, and 256 million dollar, respectively (about 20 % of total fungicide sales; Hewitt, 1998). The toxophor is probably the R_1R_2-N-C-S_2-anion part of the molecule, which interfers with several metal-containing enzymes or sulphydryl- groups within fungal metabolism.

The phthalimide fungicides such as captan, captafol and folpet, and the closely related sulphamide fungicide dichlofluanid provide preventive control of many pathogens when applied as foliar or seed treatments. The common toxophor is the R_1-N-S-C-Cl_2-R_2-moiety of the molecules which reacts with sulphydryl- and amino-groups or other enzymes in fungal metabolism. Dichlofluanid and especially folpet control *P. viticola* and also *B. cinerea* in grape vine, when used in mixture with other fungicides, as well as downy mildews in hops and vegetables and many pathogens in other crops. The chloroisophthalonitrile fungicide chlorothalonil is a major preventive fungicide used alone and in mixture against a wide range of pathogens such as downy mildews in vegetables, *P. infestans* in potato and tomato, as well as *Septoria* spp. in cereals, *Mycosphaerella* spp. In peanuts, vegetables and banana, *Botrytis* spp., *Alternaria* spp. and *Cercospora* spp. in vegetables, field crops and ornamentals. Chlorothalonil binds to sulphydryl and mercapto groups in the fungal cell.

3.14. RELATIVE COMPARISON OF COMPOUNDS

When preventive and curative activities of selected compounds are compared, quite remarkable differences can be found (Figure 3). For *P. viticola* control under glasshouse conditions, most compounds exhibited a preventive activity of 10 days and more, except for fluazinam (8 d), famoxadone (7 d), and cymoxanil (3 d). Curative activity was observed for azoxystrobin (1 d), dimethomorph and fosetyl-Al (2 d), as well as for cymoxanil, iprovalicarb (3 d) and phenylamides (at least 4 d), but not for famoxadone, fluazinam, and contact fungicides like mancozeb, folpet and copper. The translocation behaviour in plant tissue differentiates compounds active against downy mildews very significantly from each other (Table 2). Full systemicity can be attributed to the phenylamides, iprovalicarb and fosetyl-Al, but also to propamocarb and cymoxanil (especially strong in leaves). Partially systemic are dimethomorph (not systemic in shoot), azoxystrobin and fenamidone, as well as hymexazol; these compounds can be considered as locally systemic. Special migration behaviour was attributed to some strobilurins: trifloxystrobin was defined as being mesostemic, and kresoxim-methyl as

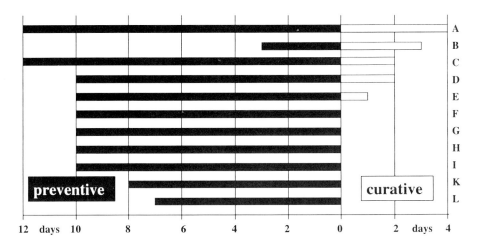

Figure 3. Duration (days) of preventive activity (black bars) and curative activitiy (open bars) of selected fungicides against *Plasmopara viticola* under glasshouse conditions. **A**: metalaxyl-M; **B**: cymoxanil; **C**: fosetyl-Al; **D**: dimethomorph; **E**: azoxystrobin; **F**: mancozeb; **G**: folpet; **H**: copper; **I**: zoxamide; **K**: fluazinam; **L**: famoxadone (Data consolidated from different sources).

TABLE 2. Translocation behaviour in plants of fungicides active against downy mildews

compound	root/shoot acro-petal	translocation in leaf			translocation leaf/leaf		redistri-bution via vapour
		acro-petal	basi-petal	trans-laminar	acro-petal	basi-petal	
phenylamides	+++	+++	++	+++	+++	+(+)	(+)
hymexazol	++	+	-	?	?	?	?
Qo inhibitors:							
azoxystrobin, fenamidone	++	+	-	++(+)	-	-	++
famoxadone, trifloxystrobin	+	-	-	+	-	-	++
fluazinam	-	-	-	-	-	-	-
chloroneb	++	?	?	?	?	?	+++
propamocarb	++	+	-	?	++	+	-
dimethomorph	++	+	-	+(+)	-	-	-
cymoxanil	+(+)	++	++	++	++	-	-
fosetyl-Al	+++	++	++	?	++	+++	-
iprovalicarb	+++	+++	-	?	?	?	?
mancozeb, folpet, chlorothalonil, zoxamide	-	-	-	-	-	-	-

+++, ++, +, - : relative intensity of translocation providing disease control in untreated plant parts
?: not known or unclear results
Data consolidated from different sources

quasi-systemic (see paragraph 3.2). All other compounds controlling downy mildews including famoxadone, fluazinam, folpet, mancozeb, chlorothalonil, zoxamide can be considered as non-systemic.

3.15. HOST PLANT DEFENSE INDUCERS

Defense mechanisms of plants against pathogen attack can be constitutive or inducible, and within the latter, local or systemic (Mauch-Mani, this volume). Chemicals and organisms inducing such reactions have been described as host plant defense inducers and several pathways are known to be involved (Figure 4). The best documented mechanism is based on the salicylic acid pathway resulting in systemic (or local) acquired disease resistance (SAR, LAR; Kessmann et al., 1994). A compound is considered as an inducer (or activator) of SAR if it does not exhibit any direct antimicrobial activity in vitro (in absence of host plant), but induces expression of the same plant genes that are also induced during biologically activated defence reactions. In addition, resistance can be activated against a broad range of pathogens including fungi, bacteria and viruses as occurs in plants after a predisposing infection with a necrotrophic pathogen. Gene induction results in a stimulated accumulation of anti-fungal proteins in and on the surface of plants such as the pathogenesis related (PR-) proteins β-1,3-glucanases and chitinases (Kessmann et al., 1994).

3.15.1 Commercial plant activator

The first non-commercial compound providing reproducible SAR was 2,6-dichloro-isonicotinic acid and its methylester (INA derivatives), followed by the functional analogue acibenzolar-S-methyl (CGA 245 704), a member of the benzothiadiazole (BTH) class of chemicals (Table 1). So far, acibenzolar-S-methyl is the only commercially available host plant defence inducer which acts as a plant activator through SAR. In tobacco, acibenzolar-S-methyl (BTH) activates enhanced resistance against *Cercospora nicotianae, Phytophthora parasitica* (black shank), *Peronospora tabacina* (blue mold), as well as the bacterial pathogen *Pseudomonas syringae* pv. *tabaci* and tobacco mosaic virus (TMV). However, it does not affect pathogens such as *Alternaria alternata* and *Botrytis cinerea in planta*. In addition, it does not control a range of 18 different fungi *in vitro* (Friedrich et al., 1996). In *Arabidopsis thaliana*, the compound activates resistance against turnip crinkle virus (TCV), *Pseudomonas syringae* pv. *tomato* and *Peronospora parasitica* (Lawton et al., 1996). A seed treatment with acibenzolar-S-methyl activated resistance against the air-borne pathogen *Peronospora parasitica* and to a lesser degree also the soil-borne pathogen *Rhizoctonia solani* in *Brassica* seedlings (Jensen et al., 1998). Under field conditions, a single spray application of 30 g a.i./ha activates longlasting resistance against powdery mildew in cereals, whereas repeated applications (12 g a.i./ha) are needed to protect tobacco against *P. tabacina* (Ruess et al., 1995, 1996).

Acibenzolar-S-methyl and its primary metabolite, benzothiadiazole-carboxylic acid, are highly mobile (parent compound and metabolite: solubility in water, 7.7 and >4000 mg/l; log P_{OW}, 3.1 and 0.5, respectively). They are weakly acidic and translocated acropetally and basipetally throughout the plant. It has been shown that BTH stimulates the SAR signal pathway downstream of the salicylic acid target (Figure 4), and induces SAR by activation of signal transduction rather than by signal release (Friedrich et al., 1996; Lawton et al., 1996). Salicylic acid is not the systemically translocated signal molecule (Vernooij et al., 1994), the nature and translocation behaviour of this molecule is still unknown. Transgenic tobacco and Arabidopsis plants that express a bacterial salicylate hydroxylase (nahG) gene were not able to accumulate significant amounts of salicylic acid and as a consequence, the plants were not able to express a SAR response after pathogen infection (Delaney et al., 1994). Treatment of these plants with BTH resulted in an activation of SAR, demonstrating that it acts downstream of salicylic acid in activation of SAR. Mutants of Arabidopsis that cannot be induced (non-inducible immunity, nim) did not activate SAR as a response to BTH treatment or pathogen infection (Lawton et al., 1996). This is strong evidence that BTH, salicylic acid and pathogens activate SAR through a common signalling pathway (Ryals et al., 1996; Figure 4).

It is assumed that SAR acts through a combination of several antifungal PR-proteins and other defense reactions such as formation of papillae (Cohen et al., 1987; Görlach et al., 1996) in a similar manner as during incompatible host-pathogen interactions. As for defense reactions after a predisposing infection with a necrotrophic pathogen, chemically induced SAR also needs a reaction time of several days between application of the chemical and the expression of sufficient protection against the pathogen. As a consequence, disease control induced by SAR is only successful if plant defense reactions are established faster than pathogenesis. Therefore, applications of acibenzolar-S-methyl are recommended in a strictly preventive schedule with one single application in cereals and repeated treatments with an interval of 8-10 days against blue mold in tobacco (Ruess et al., 1996). The speed and intensity of SAR may vary between cultivars and affect pathogen populations on different levels. In acibenzolar-S-methyl treated tobacco plants, especially the early stages such as penetration structures of Peronospora tabacina are affected (Knauf-Beiter et al., 1997).

3.15.2 Research approaches
As described above resistance can be induced by a predisposing infection with necrotrophic pathogens or compounds such as acibenzolar-S-methyl; both act through the salicylic acid pathway (SAR). Additional pathways are known for induced defense reactions in plants: through the jasmonic acid pathway, resistance can be activated by colonization of the rhizosphere with selected plant growth-promoting rhizobacteria (PGPR) such as Pseudomonas fluorescens (Table 1). This latter mechanism is generally referred to as induced systemic resistance, ISR (Kloepper et al., 1992; Pieterse et al., 1998; Figure 4). Treatment of Arabidopsis with PGPR activated an enhanced resistance

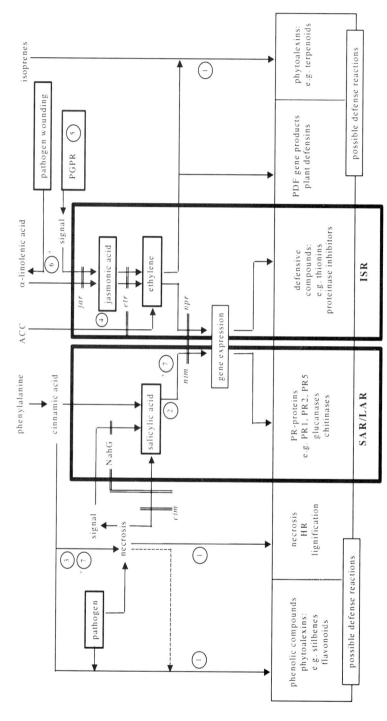

Figure 4. Possible signal transduction pathways for induced resistance in plants and sites of action for defense inducers. Defense inducers indicated are: **1,** phosphonates; **2,** isonicotinic acid, salicylic acid, benzothiadiazoles; **3,** arachidonic acid; **4,** jasmonates; **5,** *Pseudomonas* spp.; **6,** *Bacillus* spp.; **7,** β-amino butyric acid. Mutants of *Arabidopsis* indicated are: *jar,* jasmonate responsive; *etr,* ethylene responsive; *cim,* constitutive immunity, no lesions; *nim,* non inducible immunity; NahG, salicylic acid non-accumulating plants with salicylate hydroxylase gene; *npr,* non-expressor of PR genes (information modified based on Ryals *et al.,* 1996; Pieterse *et al.,* 1998).

against *Pseudomonas syringae* pv *tomato* without accumulation of salicylic acid or detectable activation of SAR gene expression (Pieterse *et al.*, 1996). In cucumber, PGPR induced protection against a range of pathogens (Wei *et al.*,1991). Jasmonic acid (JA) and jasmonic methylester (JME) (Table 1) are plant lipid derivatives that are claimed to be involved in plant wound and infection responses.Recently, JA and JME were described as inducers of local and systemic resistance in potato and tomato plants against *P. infestans* after a foliar spray under glasshouse conditions (Cohen *et al.*, 1993). Jasmonates did not inhibit fungal growth and did not induce phytoalexin production, but enhanced the level of two proteinase inhibitors. In tobacco and cucumber JA and JME induced defense reactions different from those described for SAR (Kessmann *et al.*, 1994). Therefore, a salicylic acid-independent pathway perhaps similar to that described for ISR might be involved (Figure 4).

A number of non-commercial agents (defined chemical compounds and microbial extracts) have been reported to induce plant defense reactions, and many of them may cause more or less visible local necroses triggering resistance responses (Table 1). This might be the case for unsaturated fatty acids such as arachidonic and eicosapentaenoic acid (Figure 4), well known as elicitors of phytoalexins and recently described to induce systemic resistance in potato plants against *P. infestans* (Cohen *et al.*, 1991) and *A. solani* (Coquoz *et al.*, 1995). Systemic resistance was reported to be induced in monocot and dicot plants by metabolites of selected *Bacillus subtilis* isolates (Steiner and Schönbeck, 1996). The treated plants exhibited a reduced susceptibility against biotrophic pathogens (rusts, powdery and downy mildews), but not against necrotrophic fungi and bacteria. Induced resistance was triggered by acropetal translocation of the bacterial metabolite or alternatively of a yet unknown signal induced by the metabolite (Figure 4). The chemical structures of the metabolite and the signal are unknown.

A non-commercial inducer of local and systemic resistance is also DL-3-amino-n-butyric acid (beta aminobutyric acid, BABA) (Cohen, 1994, 1996; Table 1). BABA is an amino acid. In treated tomato plants, germination and penetration of *P. infestans* was not affected, but mycelial growth was strongly inhibited. In tomato, the R-enantiomer of BABA was 2-3 times more active than the S-enantiomer (the latter was completely inactive in tobacco); other butyric acids such as 2-amino- (α-amino-), 2-amino*iso*-, 3-amino*iso*- and 4-amino- (γ-, GABA) butyric acid did not induce systemic resistance. The spectrum of BABA activity is quite remarkable ranging from *P. infestans* (mainly in tomato), to *P. cubensis* in cucumber and melon, *Peronospora tabacina* and *Phytophthora parasitica* in tobacco, *Plasmopara* spp. in grape and sunflower, *Peronospora parasitica* and *Alternaria brassicicola* in broccoli and cauliflower, as well as *Fusarium, Verticillium* and *Aphanomyces* species in vegetables and cotton (Papavizas, 1964; Cohen, 1996; Kalix *et al.*, 1996; Cohen *et al.*, this volume). When applied to the roots, as stem injection or as foliar spray, BABA was translocated mainly in acropetal and translaminar directions and was incorporated especially in the cell walls of young leaf tissue (Cohen and Gisi, 1994). The protection of the plant against pathogens was related not only to the amount of BABA in the tissue, but also to enhanced hypersensitive reaction (HR) and lignification, as well as an accumulation of PR proteins including P14a, β-1,3-glucanase, chitinase and peroxidase (in tomato plants after root and foliar treatment, in tobacco plants after foliar treatment; Cohen *et al.*, 1994). The mode of action of BABA is not fully understood but is probably different from that of acibenzolar-S-methyl (Figure 4) (see also Mauch-Mani, this volume).

The examples of chemically induced resistance described above demonstrate that it probably involves several different defense mechanisms. They also suggest that specificity rather than universality of molecules is expressed through either the chemical structure of the inducer, or the nature of the signal transduction or the effectiveness of the defense responses against the specific pathogen. As for defense responses in incompatible host-pathogen interactions, the final mode of action of induced resistance in plants remains speculative, but fungal cell wall-degrading PR proteins and elevated hydrogen peroxide levels may be involved (Hain and Schreier, 1995).

4. Pathogen (Fungicide) Resistance

4.1. APPROACHES

Pathogen resistance to fungicides (fungicide resistance) describes the presence of resistant individuals (isolates) naturally occurring in pathogen populations and their increase in frequency over time (seasonal, from year to year) to the extent that disease control may be reduced. At the start of fungicide use, this increase is due to a selection process imposed by the fungicide, which reduces the sensitive sub-population and allows less sensitive and resistant individuals to multiply. They may become dominant in the population as long as the selection process persists, but often decrease in frequency once fungicide applications are terminated. In such situations resistant individuals are less fit than the original wild-type population. In a second step not related to their sensitivity to fungicides, individuals are favoured with higher fitness attributes such as shorter incubation period and sporulation time, or larger lesion size development and sporulation capacity. By coincidence, rather than genetic linkage, resistance to a fungicide may be present along with higher fitness in the same individual. Mutations for resistance, and resistance build-up through fungicide selection may happen locally and resistance is then distributed to other fields by migration of the pathogen. More likely, the same mutation for resistance may occur at different locations recurrently, and then resistance build-up is a response to local conditions and strategies of fungicide use. Since pathogen inoculum can migrate through spore dispersal, or be transported in infested plant material during shipment of seeds, tubers, bulbs, cuttings and young plants, new races (pathotypes, genotypes) that are resistant (or sensitive) may appear at locations not exposed to the fungicide selection. Population structure and dynamics as well as the frequency of sexual recombination and fitness attributes of the pathogen play an important role for stability of resistance. Since the oomycetes are diploid, inheritance of resistance can be described in most cases by Mendelian rules. Sexual crosses have been undertaken mainly with *Phytophthora* species and it is assumed that segregation patterns in the biotrophic downy mildews should be similar as in *Phytophthora*.

Resistance originates from rare mutation events in a few individuals of a pathogen species that is intrinsically sensitive to the fungicide. After repeated fungicide treatment, sensitive individuals are largely controlled and resistant individuals can increase in frequency through the fungicide selection process. In contrast, insensitivity (or tolerance) occurs when all individuals of a species do not respond to fungicide treatment. A sensitive species is always composed of individuals expressing a range of sensitivity levels; (sensitivity profile). The sensitivity of wild-type populations may range within a factor of 10 to 100 between the most and the least sensitive individual (isolate) (Gisi *et al.*, 1997). Cross-sensitivity and cross-resistance behaviour in a pathogen is expressed when the majority of individuals have similar sensitivity reactions to compounds with the same biochemical mode of action (mostly also same chemical group or class, see Table 1). Conversely, in many cases the mode of action of a new compound is elucidated by the cross-resistance behaviour to known fungicides using a number of isolates (typically 10-50) representing the entire sensitivity range.

Because the mode of resistance is not necessarily linked to the biochemical site of action, and the outbreak of resistance in the field is not only a response of target site mutations, different approaches are needed to understand the significance of pathogen resistance to fungicides. Firstly, molecular investigations should elucidate whether one or several mutations are responsible for resistance and where in the genome mutations occur. The amino acid sequence of the target enzyme (or resistance locus) is analysed and compared in sensitive and resistant isolates. The segregation pattern of resistance after sexual crossing may give insights, whether resistance is inherited mono- or polygenically and if dominant genes are involved. Secondly, biochemical and physiological studies in cell-free systems and at the cellular level may reveal how detrimental mutations are to cell metabolism and the degree of sensitivity, but also whether the fungicide is still taken up and reaches the target site. Thirdly, population biology aspects such as pathogenicity, virulence, aggressiveness and fitness of sensitive and resistant field isolates, and epidemiological model studies help to predict whether resistance will be a relevant problem affecting fungicide performance under field conditions. Resistance risk can be estimated only if most of this information is available (Gisi and Staehle-Csech, 1988; Brent and Hollomon, 1998). Appropriate strategies to delay resistance build-up can then be designed and enforced (see below). Several organisations support this goal including the industry based network FRAC (fungicide resistance action committee), consisting of special working groups for all important fungicide groups such as phenylamides (PAs), DMIs and strobilurins (Qo inhibitors).

4.2. PHENYLAMIDES

In 1979, two years after metalaxyl has been introduced (without any companion fungicide), the first PA-resistant isolates of *P. cubensis* were detected in Israel on treated cucumbers grown under plastic (Reuveni *et al.*, 1980). In 1980, PA-resistant isolates were observed in *P. infestans* on field-grown potatoes in Ireland, the Netherlands and Switzerland (see review Gisi and Cohen, 1996), and shortly afterwards also in *P. tabacina* on tobacco in USA (Bruck *et al.*, 1982), in *P. viticola* on grape vine in France (Clerjeau and Simone, 1982; Moreau *et al.*, 1987) and Switzerland (Bosshard and Schüepp, 1983), in *Pythium* spp. on turfgrass in USA (Sanders, 1984), and in *B. lactucae* on lettuce in the UK (Crute, 1987), followed by many other pathogens of the oomycetes on a range of crops. Associated with the detection of resistant isolates was, in most cases (although not strictly correlated to treatment history), a decline in disease control. As a consequence, strict recommendations for use of PAs have been designed and enforced by PA-FRAC to prevent or further delay resistance build-up. These involve the preventive use of prepacked mixtures of PAs with well-defined rates of non-phenylamide fungicides, a limited number of applications per crop and season (max. 2-4 consecutive treatments with 10 to 14 day intervals), and no soil use of PAs for the control of air-borne pathogens. These recommendations have been successfully followed and products containing phenylamides are still the most important fungicides for the control of disease caused by oomycetes, although PA-resistant isolates can be found in all regions and crops of the world. Key elements are the quantification (rather than just detection) of resistant isolates in local populations over time (season, year to year) and the performance of PA-containing products under practical conditions.

Resistance to PAs originated from naturally occurring isolates existing at a very low proportion before the local populations had been exposed to the fungicide. Resistant isolates in *P. infestans* already existed in 1977 in northern Germany (Daggett *et al.*, 1993) several years before commercial use of PAs, and in *P. viticola* in 1978 in a Swiss vineyard never treated with PAs (Bosshard and Schüepp, 1983). These are probably not the only cases of resistant isolates detected prior to the first product usage, and spontaneous mutations may have occured many times in several places. Phenylamide resistance is a monogenic trait (Shattock, 1986): In crossing experiments, the majority of the F1 progeny from metalaxyl-resistant (r) and metalaxyl-sensitive (s) parental isolates of *P. infestans* were intermediate in sensitivity (i) to metalaxyl. Crosses between two F1 isolates with intermediate sensitivity yielded a 1s:2i:1r ratio of progeny in the F2 generation. This Mendelian segregation pattern reflects a single-gene controlled (monogenic) resistance (Shattock, 1988). Analyses of inheritance of resistance to metalaxyl in F1, F2 and backcrosses (frequencies of phenotypes, isozymes and sensitivity to PAs) suggested that a single, incompletely dominant gene was involved (Shaw and Shattock, 1991). This was again confirmed recently for *P. infestans* field isolates from Israel (Kadish and Cohen, in Gisi and Cohen, 1996).

Resistance to metalaxyl was reported to be controlled by a single incompletely dominant gene also in *P. capsici* (Lucas *et al.*, 1990), *P. sojae* (Bhat *et al.*, 1993) and *B. lactucae* (Crute and Harrison, 1988). Recent investigations with *P. infestans* suggest that either one or two semi-dominant loci, *MEX*, and several minor loci (Fabritius *et al.*, 1997; Judelson and Roberts, 1999) or one dominant gene (plus minor genes) (Lee *et al.*, 1999) may be involved in PA-resistance. Thus, it is not clear whether more than one resistance gene exists; whether a combination with minor genes controls resistance to metalaxyl in field populations; whether reported differences in sensitivity (resistance) levels are attributed to variations at the *MEX* loci alone or in combination with minor loci. Although many investigations on the mode of action and mechanism of resistance to PAs have been undertaken over the last 20 years, the PA-resistance gene(s) and the site of mutation(s) in the genome have not yet been mapped.

In biochemical studies, endogenous RNA polymerase activity of isolated nuclei of *P. megasperma* and *P. infestans* was highly sensitive to metalaxyl in sensitive isolates but insensitive in resistant isolates, suggesting that a target site mutation is responsible for resistance (Davidse, 1988). This hypothesis was further supported by the observation that (^{3}H)-metalaxyl binds to cell-free mycelial extracts of sensitive but not of resistant isolates. In spite of the observation that the four phenylamides, metalaxyl, oxadixyl, benalaxyl and ofurace, exhibit different levels of intrinsic activity and rRNA polymerase inhibition (Davidse, 1988), positive cross-resistance behaviour was observed for all four molecules (Diriwächter *et al.*, 1987) with correlation coefficients of higher than 0.8. Nevertheless, the size of resistance factors is specific for each compound and isolate.

Field populations may contain individuals with sensitive, intermediate and resistant responses to PAs and the sensitivity range can be over 1000 fold, as exemplified by *P. viticola* populations collected in 1995 in vineyards of southwestern France (Figure 5). Although PAs are considered to bear a high intrinsic resistance risk (Gisi and Staehle-Csech, 1988), they failed to eliminate the sensitive sub-population from nature, even after 20 years of intensive use. After a drastic decline in the late 1980's, the proportion of sensitive *P. viticola* isolates in French populations has recovered and equilibrated to around 40 % in the last 4 years; conversely, the proportion of resistant isolates decreased in the last 10 years from 60-80 % initially to about 25 % in 1998. Isolates with an intermediate response to PAs increased in frequency (Figure 6). Whether intermediates represent a mixture of sensitive and resistant individuals or are genetically "true" intermediates is unclear, but their appearance and steady increase in frequency after a selection period of many years indicate that they originate from sexual recombination. Resistant (r) isolates of *P. viticola* sporulated within shorter intervals compared to sensitive (s) isolates, especially when temperatures were below 20°C, but were less competitive in s+r mixtures (9+1) than sensitive isolates when incubated on untreated plants for 2 days at 30°C followed by 6 days at 22°C, and completely disappeared after 3 generations (Piganeau and Clerjeau, 1985). The proportion of resistant isolates is obviously low at the beginning and high at the end of the season.

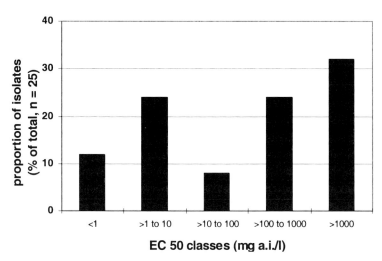

Figure 5. Sensitivity (EC$_{50}$) profile to oxadixyl of *Plasmopara viticola* field isolates collected in French commercial vineyards in 1995 (sensitive, intermediate, and resistant isolates are in EC$_{50}$ classes <1-100, >100-1000, and <1000, respectively; EC$_{50}$, effective concentration of fungicide controlling growth of specific isolate at 50 %; sensitive reference isolate is in EC$_{50}$ class >1-10; Syngenta internal data).

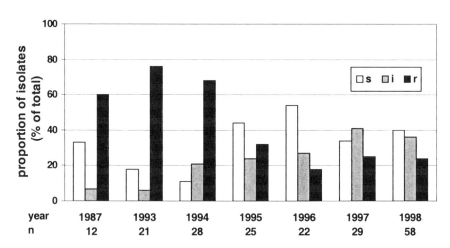

Figure 6. Proportion of *Plasmopara viticola* field isolates (% of total) collected in French vineyards between 1987 and 1998 exhibiting a sensitive (s), intermediate (i) and resistant (r) response to phenylamide fungicides (Syngenta intenal data; no data available for 1988-1992).

A similar behaviour was observed for *P. infestans* populations, where the proportion of PA-resistant isolates fluctuated from year to year and also within the season. It increased within the season, more rapidly in fields treated with PAs than in untreated fields, started to decline at the end of the season and was significantly lower at the beginning of the next season compared to the proportion at the end of the previous year (Gisi and Cohen, 1996). This behaviour demonstrates on one hand, that resistant isolates are selected by the use of PAs, but also that many must exhibit a higher aggressiveness and are more competitive than sensitive isolates (Kato *et al.*, 1997) resulting in an increase during disease epidemics. However, this increase is compensated by a decline during the overwintering period possibly because survival of resistant isolates in tubers is reduced at low temperatures between the seasons (Kadish and Cohen, 1988, 1989; Kadish *et al.*, 1990; Walker and Cooke, 1990; Cooke, 1991; Gisi *et al.*, 1997). The currently used sensitivity test methods (Sozzi *et al.*, 1992) provide a resistant response when as few as 1-5 % resistant sporangia are present in the test sample. Therefore, the importance of resistance in field populations is often overestimated, and PA containing products can be used in spite of the presence of resistant sub-populations, if strictly used according to label recommendations.

4.3. COMPLEX III RESPIRATION INHIBITORS

The fungicides inhibiting respiration at the Qo site (QoIs) (Table 1) are used commercially in plant protection since 1996 and therefore, pathogen populations have been exposed to selection for only a short period. Nevertheless, it was reported recently that efficacy of kresoxim-methyl against *Erysiphe graminis* on wheat was less than expected in certain areas, and that a significant proportion of the tested isolates were resistant to QoIs (Reschke, 1999). Also in other pathogens including downy mildews, such as *P. viticola* and *P. cubensis*, resistant field isolates have been identified (Gisi *et al.*, 2000; Heaney *et al.*, 2000). Only few quantitative results are available describing the distribution and significance of resistant isolates in treated fields (Chin *et al.*, 2001). Nevertheless, the inherent resistance risk for QoIs is estimated to be rather high and probably more pronounced than that of DMIs (demethylation inhibitors like triazoles). The existence of mutants resistant to myxothiazol and strobilurins has been known since many years, especially in the yeast *Saccharomyces cerevisiae* (Di Rago *et al.*, 1989). In this fungus and other organisms, resistance to QoIs was described as point mutations in the cytochrome b gene causing single amino acid (aa) exchanges in two regions of the cytochrome bc1 enzyme complex at positions between aa 127 and 147 and between aa 275 and 296 of the Qo center (Geier *et al.*, 1992). Within QoIs, cross-resistance was observed for pathogens such as *Erysiphe graminis*, *Mycosphaerella fijiensis*, *Plasmopara viticola* and *Venturia inaequalis* (Sierotzki *et al.*, 2000a, b; Heaney *et al.*, 2000; Steinfeld *et al.*, 2001) when tested at the enzyme and cellular level as well as *in planta*. No cross-resistance, however, was found between QoIs and the Qi inhibitor antimycin (Sierotzki *et al.*, 2000b). For the Qo site, many mutations affecting inhibitor activity have been described in a range of organisms (Brasseur *et al.*, 1996), of which a single aa exchange, G143A, seems to be specific for QoI resistance in all plant pathogens tested so far (Sierotzki *et al.*, 2000a, b; Gisi *et al.*, 2000; Heaney *et al.*, 2000). Because of mitochondrial inheritance, resistance to QoIs will be donated after sexual recombination only if it is present in the maternal crossing partner. In field populations however, the actual frequency of resistant mitochondria cannot be predicted yet. Besides target site mutations in the cytochrome b gene, the stimulation of alternative oxidase activity was proposed as a possible resistance mechanism for *Septoria tritici* (Ziogas *et al.*, 1997), but it is unlikely, that this is a primary effect. However, other resistance mechanisms (Steinfeld *et al.*, 2001) and additional mutations are likely to exist in certain plant pathogens and future research is needed to understand the evolution of QoI resistance.

4.4. CYANO-ACETAMIDE OXIMES

For about 20 years, cymoxanil has been used primarily in mixture with contact fungicides to control downy mildews (e.g. in grape vine) and *P. infestans* in potato and tomato without significant control failures. Cymoxanil is degraded in plants within few days and thus might not exhibit an important selection pressure for resistant populations. Sensitivity profiles of *P. infestans* populations to cymoxanil are more or less unimodal but rather extended and can range over a factor of 40 to 500 between the most and the least sensitive isolate (Power *et al.*, 1995; Gisi *et al.*, 1997), depending on the test method, but resistant isolates have not been observed in this pathogen. For *P. viticola*, however, sensitivity profiles can be very large (up to 1000). Between 1995 and 1997, French and Italian populations contained an important proportion of isolates at the less sensitive end of the sensitivity distribution (Genet and Vincent, 1999). They were much less sensitive than those from Portugal and non-European populations (e.g. from USA) and may be considered as cymoxanil-resistant. In some Italian vineyards, cymoxanil failed to control the severe downy mildew attacks in the years 1993 to 1995, and cymoxanil-resistant isolates were identified at some of the sites expressing resistance factors of up to 100 compared to sensitive populations (Gullino *et al.*, 1997). The possible presence of resistant isolates in *P. viticola* populations limits the curative potential of cymoxanil, and preventive treatments provide more reliable disease control.

4.5. MISCELLANEOUS INHIBITORS

For several fungicides with activity against members of the Peronosporales, it has been reported that resistant mutants can be generated under laboratory conditions. However, no reports have been published for these fungicides on disease control failures or the widespread occurrence of resistant field isolates. For technical reasons, mutants have been produced mainly in *Phytophthora* and *Pythium* species. When isolates of *Phytophthora parasitica* were exposed to UV light and increasing concentrations of dimethomorph, resistant mutants were detected. After several transfers to non-amended agar, the resistant mutants remained stable and were equally pathogenic as wild type isolates (Chabane *et al.*, 1993). The analysis of progeny derived from crosses between sensitive and resistant mutants suggested that resistance to dimethomorph is controlled by a single dominant gene; therefore, the inherent resistance risk for dimethomorph was estimated to be rather high (Chabane *et al.*, 1996). In fact, dimethomorph-resistant isolates of *P. viticola* were claimed to be present in French vineyards in 1994 (Albert, pers. comm. in Chabane *et al.*, 1996). As an anti-resistance strategy, but also to improve fungicidal activity, dimethomorph is recommended in mixture with other fungicides.

As described above, phosphonate-resistant mutants of *Phytophthora palmivora* have been produced and used to elucidate the antifungal mode of action of fosetyl-Al (Dolan and Coffey, 1988). After extensive use of fosetyl-Al over many years, the existence of fosetyl-Al-resistant field isolates has not been described, except for one case in French horticulture, where a resistant isolate of *Phytophthora cinnamomi* has been detected in *Chamaecyparis lawsoniana* (Vegh *et al.*, 1985). In *Phytophthora parasitica*, mutants resistant to chloroneb have been produced (Chang and Ko, 1990), but no reports on resistant field isolates were published.

For plant defense inducers it is generally assumed that pathogens cannot develop resistance very easily against defense responses such as production of PR proteins, phytoalexins and structural barriers. Nevertheless, detoxification and efflux through ABC (ATP binding cassette) membrane pumps are known to be effective mechanisms in pathogens for protection against plant defense products (Denny *et al.*, 1987; Urban *et al.*, 1999). Defense reactions require a certain expression time to provide sufficient protection against disease and they have to be more rapid than the infection process by the pathogen. In host-pathogen interactions, plants will always select for new pathotypes that are able to overcome plant resistance reactions. As a consequence, more aggressive pathotypes with increased fitness may be favoured in pathogen populations after a certain selection time. This may make disease control by defense inducers more difficult, especially in situations of heavy pathogen attack and climatic conditions which favour epidemics.

5. Fungicide mixtures and alternations

5.1. PURPOSE

There are three main reasons why different fungicides are combined either in mixture or in alternation as part of a treatment programme:
(I) In most crops several diseases and disease complexes occur simultaneously, which in most cases cannot be controlled by a single fungicide. In addition, the control of one disease often allows a second one to colonize the empty niche. Alternation, and more effectively mixtures of fungicides, extend the spectrum of activity of single components. Important examples are mixtures of rather specific fungicides like

cymoxanil and phenylamides with multisite inhibitors (mancozeb, folpet, copper) to control late and early blight in potato, and downy mildew as well as *B. cinerea* and secondary diseases in grape vine.

(II) In mixtures, interactions occur between the components resulting in either antagonistic, additive or synergistic activities against the target disease. Antagonistic interactions should be avoided (loss in efficacy), whereas additive effects ensure disease control levels to be expected as the sum of the single components. Mixtures providing synergistic interactions result in an improved level of disease control compared to the single components alone, or alternatively allow use of reduced amounts of active ingredients in the mixture without losing efficacy.

(III) Fungicides lacking any cross-resistance behaviour can be combined in mixtures to prevent or delay the process of selection for resistant subpopulations to one or both components in the mixture.

5.2. SYNERGISTIC INTERACTIONS

Several approaches are available to describe synergistic interactions in fungicide mixtures, the most important being the Colby and the Wadley methods (De Waard and Gisi, 1995). Because of its short residual activity, cymoxanil has been used in most cases with contact fungicides resulting in an improved disease control, and synergistic interactions have been claimed for this effect (Douchet *et al.*, 1977). Also dimethomorph is used in mixture with other fungicides for the same reason. Many investigations have been undertaken for mixtures containing phenylamides and one or two other fungicides that are not cross-resistant (Gisi, 1991). When oxadixyl was combined with mancozeb and/or cymoxanil, the fungicide concentration of the single components providing 90 % control of *P. viticola* (EC_{90}) decreased significantly in the mixture compared to the fungicides alone (Table 3). For the three-way mixture, this effect was stronger for phenylamide-resistant than for sensitive isolates and resulted in synergistic interactions of 6.4 and 5.0, respectively (Grabski and Gisi, 1987). When the components were applied one after the other to control *P. infestans* on tomato, synergistic interactions decreased with increasing interval between the applications of components, and were lost after three days (Samoucha and Gisi, 1987b). Thus, the components do not necessarily need to be present simultaneously in the mixture to express synergistc interactions, but the pathogen has to be hit consecutively by the fungicides within a short period. The first fungicide may affect the pathogen (e.g. at germination stage) to a degree, that sublethal concentrations of the second component exhibit a stronger effect (e.g. inhibition of haustorium formation) than observed without the pre-treatment with the first fungicide (Gisi, 1996). The mechanisms of synergistic interactions are not fully elucidated, but may also involve enhanced uptake of the fungicides (Cohen and Gisi, 1993).

TABLE 3. Antifungal activity (EC$_{90}$, mg a.i./l) of oxadixyl (ox), mancozeb (ma) and cymoxanil (cy) alone and in mixture against a phenylamide-sensitive and -resistant isolate of *Plasmopara viticola* under growth chamber conditions and synergistic interaction (SI) between components.

Fungicide	PA-sensitive isolate			PA-resistant isolate		
	EC$_{90th}$	EC$_{90ob}$	SI	EC$_{90th}$	EC$_{90ob}$	SI
ox	-	12	-	-	452	-
ma	-	64	-	-	75	-
cy	-	27	-	-	29	-
ox+ma=1+7	42	7(1+6)	6.0	84	33 (4+29)	2.5
ox+cy=1+0.4	14	7(6.7+0.3)	2.0	87	7(6.7+0.3)	12.4
ox+ma+cy=1+7+0.4	40	8(1+6.7+0.3)	5.0	77	12(1.4+10+0.6)	6.4

Numbers in parentheses are concentrations of the individual components in the mixture.

th and ob are theoretical (calculated according to Wadley method) and observed values, respectively.

SI (EC$_{90th}$/EC$_{90ob}$) for sensitive field populations were 2.1 (ox + ma), 1.5 (ox + cy), 3.4 (ma + cy) and 2.0 (ox + ma + cy).

Data from Grabski and Gisi (1987), Samoucha and Gisi (1987b).

5.3. RETARDATION OF RESISTANCE BUILD-UP

The frequency of resistant individuals often starts increasing shortly after the intensive use of a new fungicide controlling the sensitive but selecting for the resistant sub-populations. The aim of so called "anti-resistance strategies" (Sozzi *et al.*, 1992) is the delay of this selection process before the resistant part of the population has become dominant and thus affecting disease control. Because the appearance of resistant sub-populations is hard to predict, appropriate strategies have in the past been implemented rather late and after disease control failures have been observed. Anti-resistance strategies are designed according to current knowledge on mode of action and activity profiles of compounds as well as agronomic practices. Model trials involving defined mixtures of sensitive and resistant isolates, and controlled conditions for the progress of the disease over several cycles, help to validate fungicide mixtures or alternations for the control of both the disease progress and the dynamics of resistant sub-populations. As exemplified under growth chamber conditions for *P. infestans* and *P. viticola* populations containing 1, 0.1 or 0.01 % PA-resistant sporangia initially, a resistance level of about 10 % was reached after 2, 3 and 4 pathogen cycles, respectively, each treated with one spray of an oxadixyl + mancozeb mixture (Samoucha and Gisi, 1987c). A maximum of two to four applications of phenylamide containing products per season, as recommended by FRAC, can be supported by the model tests described. Thus, the proportion of resistant individuals in the primary inoculum, as well as the type and use strategy of the fungicide are important factors controlling the progress of selection for resistance.

6. Integration into agroecosystem and future perspectives

Several models exist to predict the onset and progress of diseases caused by oomycetes. The ultimate goal of such forecasting systems is an optimal positioning of fungicide applications, whereas the expected reduction in the number of treatments remains mostly the exception. Typically, in each country specific simulation models have been developed for forecasting *P. viticola* on grape vine: in Switzerland with microcomputer programmes and warning systems (Blaise and Gessler, 1990; Siegfried *et al.*, 1992); in Germany with "Freiburger Prognosemodell" (Kassemeyer, 1996); in Italy with PLASMO (Orlandini *et al.*, 1993); in France with MILVIT (Muckensturm, 1995). The input parameters are local weather conditions such as temperature, rainfall and leaf wetness (determining infection events by sporangia), relative humidity and temperature (for incubation period), as well as the survival and germination potential of oospores. The output of such models is the prediction of primary infection, incubation period and sporulation, as well as the progress of resulting disease epidemics. The information can be used as basis for treatment recommendations and the choice of appropriate product types. Similar forecasting systems exist for *P. infestans* in potato (e.g. SIMPHYT, Gutsche, 1993; PhytoPRE, Cao et al., 1997) and for *P. tabacina* in tobacco (Davis and Main, 1986).

Antifungal compounds and plant defense inducers will remain major tools for an effective control of plant diseases and are an integral part of modern agriculture. Their effectiveness depends on combination with other important agronomical practices such as the choice of less susceptible cultivars, a balanced fertiliser input, careful sanitation programmes and the use of forecasting systems and diagnostics. All these practices contribute to an integrated crop management, which should also take into consideration ecological and economical aspects. The production of new cultivars with increased disease resistance can be achieved by classical breeding or new molecular methods (transgenic crop plants) and is often considered as an alternative to chemical control. Pathogen populations, however, can adapt very rapidly to both new resistance genes in crop plants and novel antifungal compounds. When used excessively and on a large scale, resistance genes of host plants select for avirulence genes in pathogen populations that consequently overcome resistance responses of plants, whereas antifungal compounds select for naturally occuring mutations in the fungal genome resulting in an increase of resistant sub-populations. The most effective and durable disease control requires the combination of resistance genes (less susceptible cultivars) and antifungal compounds. The successful control of downy mildews through development of adapted cultivars and balanced treatment programmes will only be achieved, however, if recommendations are agreed and followed by industry, government, advisors and end users as a co-ordinated strategy.

Acknowledgements

Many colleagues have contributed with information and suggestions to this review. Special thanks go to Khoon Chin, Ian Dalton, Jürg Jehli, Gaby Knapova, Ruth Küng, Neil Leadbitter, Antonio Lopez, Urs Neuenschwander, David Nevill, Christian Pillonel, Helge Sierotzki, Theo Staub, Ute Steinfeld, Eileen Watson, Martin Weiss and Hugo Ziegler.

References

Albert, G. and Heinen, H. (1996) How does dimethomorph kill fungal cells? - A time lapse video study with *Phytophthora infestans*, in H. Lyr, P.E. Russell and H.D. Sisler (eds), *Modern Fungicides and Antifungal Compounds*, Intercept, Andover, pp. 141-146.

Albert, G., Curtze, J. and Drandarevski,C. A. (1988) Dimethomorph (CME 151): A novel curative fungicide, *Brighton Crop Protection Conference*, 17-24.

Albert, G., Thomas, A. and Gühne, M. (1991) Fungicidal activity of dimethomorph on different stages in the life cycle of *Phytophthora infestans* and *Plasmopara viticola*, *ANPP: 3rd International Conference on Plant Diseases*, Bordeaux, 887-894.

Ammermann, E., Lorenz, G. and Schelberger, K. (1992) BAS 490 F: A broad-spectrum fungicide with a new mode of action, *Brighton Crop Protection Conference*, 403-410.

Ammermann, E., Lorenz, G., Schelberger, K., Mueller, B., Kirstgen, R. and Sauter, H. (2000) BAS 500F: The new broad-spectrum strobilurin fungicide, *Brighton Crop Protection Conference*, 541-548.

Anema, B.P. Bouwmann, J.J., Komyoji, T. and Suzuki, K. (1992) Fluazinam: a novel fungicide for use against *Phytophthora infestans* in potatoes, *Brighton Crop Protection Conference*, 663-668.

Baldwin, B.C., Clough, J.M., Godfrey, C.R.A., Godwin, J.R. and Wiggins, T.E. (1996) The discovery and mode of action of ICIA5504, in H. Lyr, P.E. Russell and H.D. Sisler (eds), *Modern Fungicides and Antifungal Compounds*, Intercept, Andover, pp. 69-77.

Bhat, R.G., McBlain, B.A. and Schmitthenner, A.F. (1993) The inheritance of resistance to metalaxyl and to fluorophenylalanine in matings of homothallic *Phytophthora sojae*, *Mycological Research* **97**, 865-870.

Blaise, P. and Gessler, C. (1990) Development of a forecast model of grape downy mildew on a microcomputer, *Acta Horticulturae* **276**, 63-70.

Bompeix, G., Ravisé, A. Raynal, G., Fettouche, F. and Durand, M.C. (1980) Modalités de l'obtention des nécroses bloquantes sur feuilles détachées de tomates par l'action du tris-O-éthyl phosphonate d'aluminium (phoséthyl d'aluminium), hypothèses sur son mode d'action *in vivo*, *Annales de Phytopathologie* **12**, 337-351.

Bosshard, E. and Schüepp, H. (1983) Variability of selected strains of *Plasmopara viticola* with respect to their metalaxyl sensitivity under field conditions, *Zeitschrift für Pflanzenkrankheiten und Pflanzenschutz* **90**, 449-459.

Brasseur, G., Saribas, A.S. and Daldal, F. (1996) A compilation of mutations located in the cytochrome *b* subunit of the bacterial and mitochondrial *bc*1 complex, *Biochimica Biophysica Acta* **1275**, 61-69.

Brent, K.J. and Hollomon, D.W. (1998) Fungicide Resistance: The Assessment of Risk, *FRAC Monograph No. 2*, Global Crop Protection Federation, Brussels.

Bruck, R.I., Gooding, Jr.G.V. and Main, C.E. (1982) Evidence for resistance to metalaxyl by isolates of *Peronospora hyoscyami*, *Plant Disease* **66**, 44-45.

Bruin, G.C.A. and Edgington, L.V. (1983) The chemical control of diseases caused by zoosporic fungi, in S.T. Buczacki (ed), *Zoosporic Plant Pathogens: A Modern Perspective*, Academic Press, London, pp. 193-232.

Cao, K.Q., Ruckstuhl, M. and Forrer, H.R. (1997) Crucial weather conditions for *Phytophthora infestans*: a reliable tool for improved control of potato late blight?, in E. Bouma and H. Schepers (eds.),*Workshop on the European Network for Development of an Integrated Control Strategy of Potato Late Blight*, Applied Research for Arable Farming, Lelystad, pp. 85-90.

Chabane, K., Leroux, P. and Bompeix, G. (1993) Selection and characterisation of *Phytophthora parasitica* mutants with ultraviolet-induced resistance to dimethomorph or metalaxyl, *Pesticide Science* **39**, 325-329.

Chabane, K., Leroux, P., Maia, N. and Bompeix, G. (1996) Resistance to dimethomorph in laboratory mutants of *Phytophthora parasitica*, in H. Lyr, P.E. Russell and H.D. Sisler (eds), *Modern Fungicides and Antifungal Compounds*, Intercept, Andover, pp. 387-391.

Chang, T.T. and Ko, W.H. (1990) Resistance to fungicides and antibiotics in *Phytophthora parasitica*: Genetic nature and use in hybrid determination, *Phytopathology* **80**, 1414-1421.

Chin, K.M., Chavaillaz, D., Kaesbohrer, M., Staub, T. and Felsenstein, F.G. (2001) Characterizing resistance risk of *Erysiphe graminis* f.sp. *tritici* to strobilurins, *Crop Protection* **20**, 87-96.

Clerjeau, M. and Simone, H. (1982) Apparition en France de souches de mildiou *(Plasmopara viticola)* résistantes aux fongicides de la famille des anilides (métalaxyl, milfurame). *Progrès Agricole et Viticole* **99**, 59-61.

Coquoz, J.L., Buchala, A.J., Meuwly, P. and Métraux, J.P. (1995) Arachidonic acid induces local but not systemic synthesis of salicylic acid and confers systemic resistance in potato plants to *Phytophthora infestans* and *Alternaria solani, Phytopathology* **85**, 1219-1224.

Cohen, Y. (1994) Local and systemic control of *Phytophthora infestans* in tomato plants by DL-3-amino-n-butanoic acids, *Phytopathology* **84**, 55-59.

Cohen, Y. (1996) Induced resistance against fungal diseases by aminobutyric acids, in H. Lyr, P.E. Russell and H.D. Sisler (eds), *Modern Fungicides and Antifungal Compounds*, Intercept, Andover, pp. 461-466.

Cohen, Y. and Gisi, U. (1993) Uptake, translocation and degradation of [^{14}C]cymoxanil in tomato plants, *Crop Protection* **12**, 284-292.

Cohen, Y. and Gisi, U. (1994) Systemic translocation of ^{14}C-DL-3-aminobutyric acid in tomato plants in relation to induced resistance against *Phytophthora infestans, Physiological and Molecular Plant Pathology* **45**, 441-456.

Cohen, Y., Baider, A. and Cohen, B. (1995) Dimethomorph activity against oomycete fungal plant pathogens, *Phytopathology* **85**, 1500-1506.

Cohen, Y., Gisi, U. and Mösinger, E. (1991) Systemic resistance of potato plants against *Phytophthora infestans* induced by unsaturated fatty acids, *Physiological and Molecular Plant Pathology* **38**, 255-263.

Cohen, Y., Gisi U. and Niderman, T. (1993) Local and systemic protection against *Phytophthora infestans* induced in potato and tomato plants by jasmonic acid and jasmonic methyl ester, *Phytopathology* **83**, 1054-1062.

Cohen, Y., Reuveni, M. and Baider, A. (1999) Local and systemic activity of BABA (DL-3-aminobutyric acid) against *Plasmopara viticola*, in P.T.N. Spencer-Phillips (ed), *Advances in Downy Mildew Research*, Kluwer, this volume.

Cohen, Y., Niderman, T., Mösinger, E. and Fluhr, R. (1994) β-Aminobutyric acid induces the accumulation of pathogenesis-related proteins in tomato (*Lycopertison esculentum* L.) plants and resistance to late blight infection caused by *Phytophthora infestans, Plant Physiology* **104**, 59-66.

Cohen, Y., Peter, S., Balass, O. and Coffey, M. (1987) A fluorescent technique for studying growth of *Peronospora tabacina* on leaf surface, *Phytopathology* **77**, 201-204.

Cooke, L.R. (1991) Current problems in chemical control of late blight: the Northern Ireland experience, in J.A. Lucas, R.C. Shattock, D.S. Shaw and L.R. Cooke (eds), *Phytophthora*, Cambridge University Press, Cambridge, pp. 337-348.

Crute, I.R. (1987) The occurrence, characteristics, distribution, genetics and control of a metalaxyl-resistant genotype of *Bremia lactucae* in the United Kingdom, *Plant Disease* **71**, 763-767.

Crute, I.R. and Harrison, J.M. (1988) Studies on the inheritance of resistance to metalaxyl in *Bremia lactucae* and on the stability and fitness of field isolates, *Plant Pathology* **37**, 231-250.

Dagget, S.S., Götz, E. and Therrien, C.D. (1993) Phenotypic changes in populations of *Phytophthora infestans* from eastern Germany, *Phytopathology* **83**, 319-323.

Davidse, L.C. (1988) Phenylamide fungicides: mechanism of action and resistance, in C.J. Delp (ed) *Fungicide Resistance in North America*, APS Press, St. Paul, MN, pp. 63-65.

Davidse, L.C. (1995) Phenylamide fungicides: biochemical action and resistance, in H. Lyr (ed), *Modern Selective Fungicides*, 2nd ed, Gustav Fischer, Jena, pp. 347-354.

Davis, J.M. and Main, C.E. (1986) Applying atmospheric trajectory analysis to problems in epidemiology, *Plant Disease* **70**, 490-487.

Delaney, T., Uknes, S., Vernooij, B., Friedrich, L., Weymann, K., Negrotto, D., Gaffney, T., Gut-Rella, M. Kessmann, H., Ward, E. and Ryals, J. (1994) A central role of salicylic acid in plant disease resistance, *Science* **266**, 1247-1250.

Denny, T.P., Matthews, P.S. and van Etten, H.D. (1987) A possible mechanism of non-degradative tolerance of pisatine in *Nectria haematococca*. *Physiological and Molecular Plant Pathology* **30**, 93-107.

Dercks, W. and Creasy, L.L. (1989) Influence of fosetyl-Al on phytoalexin accumulation in the *Plasmopara viticola*-grapevine interaction, *Physiological and Molecular Plant Pathology* **34**, 203-213.

DeWaard, M.A. and Gisi, U. (1995) Synergism and antagonism in fungicides, in H. Lyr (ed), *Modern Selective Fungicides*, 2nd ed, Gustav Fischer, Jena, pp. 565-578.

Di Rago, J.P., Coppée, J.Y. and Colson, A.M. (1989) Molecular basis for resistance to myxothiazol, mucidin (strobilurin A), and stigmatellin, *Journal of Biological Chemistry* **264**, 14543-14548.

Diriwächter, G., Sozzi, D., Ney, C. and Staub, T. (1987) Cross resistance in *Phytophthora infestans* and *Plasmopara viticola* against different phenylamides and unrelated fungicides, *Crop Protection* **6**, 250-255.

Dolan, T.E. and Coffey, M.D. (1988) Correlative *in vitro* and *in vivo* behavior of mutant strains of *Phytophthora palmivora* expressing different resistances to phosphorous acid and fosetyl-Na, *Phytopathology* **78**, 974-978.

Douchet, J.P., Absi, M., Hay, S.J.P., Muntan, L. and Villani, A. (1977) European results with DPX 3217, a new fungicide for the control of grape downy mildew and potato late blight, *Brighton Crop Protection Conference*, 535-540.

Dustin, R.H., Smillie, R.H. and Grant, B.R. (1990) The effect of sub-toxic levels of phosphonate on the metabolism and potential virulence factors of *Phytophthora palmivora*, *Physiological and Molecular Plant Pathology* **36**, 205-220.

Egan, A.R., Michelotti, E.L., Young, D.H., Wilson, W.J. and Mattioda, H. (1998) RH-7281: a novel fungicide for control of downy mildew and late blight, *Brighton Crop Protection Conference*, 335-342.

Fabritius, A.L., Shattock, R.C. and Judelson, H.S. (1997) Genetic analysis of metalaxyl insensitivity loci in *Phytophthora infestans* using linked DNA markers, *Phytopathology* **87**, 1034-1040.

Farih, A., Tsao, P.H. and Menge, J.A. (1981) Fungitoxic activity of efosite aluminium on growth, sporulation and germination of *Phytophthora parasitica* and *P. citrophthora*, *Phytopathology* **71**, 934-936.

Fenn, M.E. and Coffey, M.D. (1984) Studies on the *in vitro* and *in vivo* antifungal activity of fosetyl-Al and phosphorous acid, *Phytopathology* **74**, 606-611.

Friedrich, L., Lawton, K., Ruess, W., Masner, P., Specker, N., Gut Rella, M., Meier, B., Dincher, S., Staub, T., Uknes, S., Métraux, JP., Kessmann, H. and Ryals, J. (1996) A benzothiadiazole derivative induces systemic acquired resistance in tobacco, *The Plant Journal* **10**, 61-70.

Fuller, M.S. and Gisi, U. (1985) Comparative studies of the *in vitro* activity of the fungicides oxadixyl and metalaxyl, *Mycologia* **77**, 424-432.

Gasztonyi, M. and Lyr, H. (1995) Miscellaneous fungicides, in H. Lyr (ed), *Modern Selective Fungicides*, 2nd ed, Gustav Fischer, Jena, pp. 389-414.

Geier, B.M., Schägger, H., Brandt, U., Colson, A.M. and VonJagow, G. (1992) Point mutation in cytochrome *b* of yeast ubihydroquinone:cytochrome-*c* oxidoreductase causing myxothiazol resistance and facilitated dissociation of the iron-sulfur subunit, *European Journal of Biochemistry* **208**, 375-380.

Genet, J-L. and Vincent, O. (1999) Sensitivity of European *Plasmopara viticola* populations to cymoxanil, *Pesticide Science* **55**, 129-136.

Genet, J-L., Bugaret, Y., Jaworska, G. and Hamlen R. (1997) Etudes de l'action curative du cymoxanil sur le mildiou de la vigne, *ANPP: 5th International Conference on Plant Diseases*, Tours, 879-886.

Gisi, U. (1991) Synergism between fungicides for control of *Phytophthora*, in J.A. Lucas, R.C. Shattock, D.S. Shaw and L.R. Cooke (eds), *Phytophthora*, Cambridge University Press, Cambridge, pp. 361-372.

Gisi, U. (1996) Synergistic interaction of fungicides in mixtures, *Phytopathology* **86**, 1273-1279.

Gisi, U. and Cohen, Y. (1996) Resistance to phenylamide fungicides: a case study with *Phytophthora infestans* involving mating type and race structure, *Annual Review of Phytopathology* **43**, 549-572.

Gisi, U. and Staehle-Csech, U. (1988) Resistance risk evaluation of phenylamide and EBI fungicides, *Brighton Crop Protection Conference*, 359-366.

Gisi, U., Binder, H and Rimbach, E. (1985) Synergistic interactions of fungicides with different modes of action, *Transactions British Mycological Society* **85**, 299-306.

Gisi, U., Hermann, D., Ohl, L. and Steden, C. (1997) Sensitivity profiles of *Mycosphaerella graminicola* and *Phytophthora infestans* populations to different classes of fungicides, *Pesticide Science* **51**, 290-298.

Gisi, U., Chin, K.M., Knapova, G., Kueng Faerber, R., Mohr, U., Parisi, S., Sierotzki, H. and Steinfeld, U. (2000) Recent developments in elucidating modes of resistance to phenylamide, DMI and strobilurin fungicides, *Crop Protection* **19**, 863-872.

Godwin, J.R., Anthony, V.M., Clough, J.M. and Godfrey, C.R.A. (1992) ICIA5504: a novel, broad spectrum, systemic β-methoxyacrylate fungicide, *Brighton Crop Protection Conference*, 435-442.

Godwin, J.R., Young, J.E., Woodward, D.J. and Hart, C.A. (1997) Azoxystrobin: effects on the development of grapevine downy mildew *(Plasmopara viticola)*, *ANPP: 5th International Conference on Plant Diseases*, Tours, 871-878.

Görlach, J., Volrath, S., Knauf-Beiter, G., Hengy, G., Beckhove, U., Kogel, KH., Oostendorp, M., Staub, T., Ward, E., Kessmann, H. and Ryals, J. (1996) Benzothiadiazole, a novel class of inducers of systemic acquired resistance, activates gene expression and disease resistance in wheat, *The Plant Cell* **8**, 629-643.

Grabski, C. and Gisi, U. (1987) Quantification of synergistic interactions of fungicides against *Plasmopara* and *Phytophthora*, *Crop Protection* **6**, 64-71.

Grant, B.R., Griffith, J.M., Davis, A.J. and Niere, J.O. (1993) Progress towards understanding the mechanism by which phosphonates act in plants challenged by plant pathogens, *10th International Symposium on Systemic Fungicides and Antifungal Compounds*, German Phytomedical Society Series, Bd. 4, Ulmer, Stuttgart, pp. 51-60.

Guest, D.I. (1984) Modification of defense responses in tobacco and *Capsicum* following treatment with fosetyl-Al [aluminium tris (o-ethyl phosphonate)], *Physiological Plant Pathology* **25**, 125-134.

Gullino, M.L., Mescalchin, E. and Mezzalama, M. (1997) Sensitivity to cymoxanil in populations of *Plasmopara viticola* in northern Italy, *Plant Pathology* **46**, 729-736.

Gutsche, V. (1993) PROGEB: a model-aided forecasting service for pest management in cereals and potatoes, *EPPO Bulletin* **23**, 577-581.

Hain, R. and Schreier, P.H. (1995) Genetic engineering in crop protection: opportunities, risks and controversies, in M. Esters (ed), *Genetic Engineering in Agriculture*, Pflanzenschutz Nachrichten Bayer, Leverkusen, pp. 25-88.

Heaney, S.P., Hall, A.A., Davies, S.A. and Olaya, G. (2000) Resistance to fungicides in the QoI-STAR cross-resistance group: current perspectives, *Brighton Crop Protection Conference*, 755 -762.

Hermann, D., Fischer, W., Knauf Beiter, G., Steinemann, A., Margot, P., Gisi, U. and Laird, D. (1998) Behavior of the new strobilurin fungicide trifloxystrobin on and in plants, *Phytopathology* **88**, S37.

Hewitt, H.G. (1998) *Fungicides in Crop Protection*, CAB International, Wallingford.

Jensen, B.D., Latunde-Dada, A.O., Hudson, D. and Lucas, J.A. (1998) Protection of *Brassica* seedlings against downy mildew and damping-off by seed treatment with CGA 245704, an activator of systemic acquired resistance, *Pesticide Science* **52**, 63-69.

Jordan, D.B., Livingston, R.S., Bisaha, J.J., Duncan, K.E., Pember, S.O., Picollelli, M.A., Schwartz, R.S., Sternberg, J.A. and Tang X.S. (1999) Mode of action of famoxadone, *Pesticide Science* **55**, 105-118.

Joshi, M.M. and Sternberg, J.A. (1996) DPX-Je874: a broad-spectrum fungicide with a new mode of action, *Brighton Crop Protecion Conference*, 21-26.

Judelson, H.S. and Roberts, S. (1999) Multiple loci determining insensitivity to phenylamide fungicides in *Phytophthora infestans*. *Phytopathology* **89**, 754-760.

Kadish, D. and Cohen, Y. (1988) Competition between metalaxyl-sensitive and –resistant isolates of *Phytophthora infestans* in the absence of metalaxyl, *Plant Pathology* **37**, 558-564.

Kadish, D. and Cohen, Y. (1989) Population dynamics of metalaxyl-sensitive and metalaxyl-resistant isolates of *Phytophthora infestans* in untreated crops of potato. *Plant Pathology* **38**, 271-276.

Kadish, D., Grinberger, M. and Cohen, Y. (1990) Fitness of metalaxyl-sensitive and metalaxyl-resistant isolates of *Phytophthora infestans* on susceptible and resistant potato cultivars, *Phytopathology* **80**, 200-205.

Kalix, S., Anfoka, G., Li, Y., Stadnik, M. and Buchenauer, H. (1996) Induced resistance in some selected crops: prospects and limitations, in H. Lyr, P.E. Russell and H.D. Sisler (eds), *Modern Fungicides and Antifungal Compounds*, Intercept, Andover, pp. 451-460.

Kassemeyer, H.H. (1996) Integrated management of grapevine diseases and pests, *Brighton Crop Protection Conference*, 119-124.

Kato, M., Mizubuti, E.S., Goodwin, S.B. and Fry, W.E. (1997) Sensitivity to protectant fungicides and pathogenic fitness of clonal lineages of *Phytophthora infestans* in the United States. *Phytopathology* **87**, 973-978.

Kato, S., Coe, R., New, L. and Dick., W. (1990) Sensitivities of various Oomycetes to hymexazol and metalaxyl, *Journal of General Microbiology* **136**, 2127-2134.

Kessmann, H., Staub, T., Hofmann, C., Maetzke, T., Herzog, J., Ward, E., Uknes, S. and Ryals, J. (1994) Induction of systemic acquired disease resistance in plants by chemicals, *Annual Review of Phytopathology* **32**, 439-459.

Kloepper, J.W., Tuzun, S. and Kuc, J.A. (1992) Proposed definitions related to induced disease resistance, *Biocontrol Science and Technology* **2**, 349-351.

Klopping, H.L. and Delp, C.J. (1980) 2-Cyano-N-[(ethylamino)carbonyl]-2-(methoxyimino) acetamide, a new fungicide, *Journal of Agricultural Food Chemistry* **28**, 467-468.

Knauf-Beiter, G., Theiler, M., Gisi, U. and Staub, T., (1997) Cytology of SAR in tobacco against tobacco blue mold, *Phytopatholohy* **87**, 53.

Langcake, P. (1981) Alternative chemical agents for controlling plant disease, *Philosophical Transactions of the Royal Society London. Biological Sciences* **295**, 83-101.

Lawton, K.A., Friedrich, L., Hunt, M., Weymann, K., Delaney, T., Kessmann, H., Staub, T. and Ryals, J. (1996) Benzothiadiazole induces disease resistance in *Arabidopsis* by activation of the systemic acquired resistance signal transduction pathway, *The Plant Journal* **10**, 71-82.

Lee, T.Y., Mizubuti, E.S. and Fry, W.E. (1999) Genetics of metalaxyl resistance in *Phytophthora infestans, Fungal Genetics and Biology* **26**, 118-130.

Leroux, P., Fritz, R. and Despreaux, D. (1987) The mode of action of cymoxanil in *Botrytis cinerea*, in R. Greenhalgh and T.R. Roberts (eds), *Pesticide Science and Biotechnology*, Blackwell, Oxford, pp. 191-196.

Lucas, J.A., Greer, G., Oudemans, P.V. and Coffey, M.D. (1990) Fungicide sensitivity in somatic hybrids of *Phytophthora capsici* by protoplast fusion, *Physiological and Molecular Plant Pathology* **36**, 175-187.

Lyr, H. (1995) Aromatic hydrocarbon fungicides and their mechanism of action, in H. Lyr (ed), *Modern Selective Fungicides*, 2nd ed, Gustav Fischer, Jena, pp. 75-98.

Mauch-Mani, B. (1999) Host resistance to downy mildew diseases, in P.T.N. Spencer-Phillips (ed), *Advances in Downy Mildew Research*, Kluwer, this volume.

Margot, P., Huggenberger, F., Amrein, J. and Weiss, B. (1998) CGA 279202: a new broad-spectrum strobilurin fungicide, *Brighton Crop Protection Conference*, 375-382.

Mercer, R.T., Lacroix, G., Gouot, J.M. and Latorse, M.P. (1998) RPA 407213: a novel fungicide for the control of downy mildews, late blight and other diseases on a range of crops, *Brighton Crop Protection Conference,* 319-326.

Mitani, S., Araki, S. and Matsuo, N. (1998) IKF-916: a novel systemic fungicide for the control of oomycete plant diseases, *Brighton Crop Protection Conference,* 351-358.

Mizutani, A., Miki, N., Yukioka, H. and Masuko, M. (1996) Mechanism of action of a novel alkoxyaminoacetamide fungicide SSF-126, in H. Lyr, P.E. Russell, and H.D. Sissler (eds), *Modern Fungicides and Antifungal Compounds,* Intercept, Andover, pp. 93-99.

Moreau, C., Clerjeau, M. and Morzières, JP. (1987) Bilan des essais détection de souches de *Plasmopara viticola* résistantes aux anilides anti-oomycetes (métalaxyl, ofurace, oxadixyl et benalaxyl) dans le vignoble français en 1987. Pont de la Maye: Rapport G.R.I.S.P. de Bordeaux.

Muckensturm, N. (1995) Modélisation du mildiou de la vigne en Champagne: Bilan de trois ans de validation du modèle MILVIT et d'un an d'utilisation au service des avertissements agricoles, *Mededelingen Faculteit Landbouwwetenschappen Rijksuniversiteit Gent* **60**, 477-481.

Nakanishi, T. and Sisler, H.D. (1983) Mode of action of hymexazol in *Pythium aphanidermatum, Journal of Pesticide Science* **8**, 173-181.

Nemestothy, G.N. and Guest, D.I. (1990) Phytoalexin accumulation, phenylalanine ammonia lyase activity and ethylene biosynthesis in fosetyl-Al treated resistant and susceptible tobacco cultivars infected with *Phytophthora nicotianae* var. *nicotianae, Physiological and Molecular Plant Pathology* **37**, 207-219.

Neumann, St. and Jacob, F. (1995) Principles of uptake and systemic transport of fungicides within the plant, in H. Lyr (ed), *Modern Selective Fungicides*, 2nd ed, Gustav Fischer, Jena, pp. 53-73.

Nuninger, C., Watson, G., Leadbitter, N. and Ellgehausen, H. (1996) CGA 329351: introduction of the enantiomeric form of the fungicide metalaxyl, *Brighton Crop Protection Conference*, 41-46.

Orlandini, S., Gozzini, B., Rosa, M., Egger, E., Storchi, P., Maracchi, G. and Miglietta, F. (1993) PLASMO: a simulation model for control of *Plasmopara viticola* on grapevine, *EPPO Bulletin* **23**, 619-626.

Oros, G., Ersek, T. and Viranyi, F. (1988) Effect of tridemorph on *Phytophthora infestans* and *Plasmopara halstedii, Acta Phytopathologica et Entomologica Hungarica* **23**, 11-19.

Papavizas, F.G. (1964) Greenhouse control of *Aphanomyces* root rot of peas with aminobutyric acid and methylaspartic acid, *Plant Disease Reporter* **48**, 537-541.

Pieterse, C.M.J., van Wees, S.C.M., Hoffland, E., van Pelt, J.A. and van Loon, L.C. (1996) Systemic resistance in *Arabidopsis* induced by biocontrol bacteria is independent of salicylic acid accumulation and pathogenesis-related gene expression, *The Plant Cell* **8**, 1225-1237.

Pieterse, C.M.J., van Wees, S.C.M., van Pelt, J.A., Knoester, M., Laan, R., Gerrits, H., Weisbeek, P.J. and van Loon, L.C. (1998) A novel signaling pathway controlling induced systemic resistance in *Arabidopsis, The Plant Cell* **10**, 1571-1580.

Piganeau, B. and Clerjeau, M. (1985) Influence differentielle de la temperature sur la germination des sporocystes et la sporulation des souches de *Plasmopara viticola* sensibles et résistantes aux phenylamides, *Fungicides for Crop Protection*, BCPC Monograph No 31, pp. 327-330.

Power, R.J., Hamlen, R.A. and Morehart, A.L. (1995) Variation in sensitivity of *Phytophthora infestans* field isolates to cymoxanil, chlorothalonil and metalaxyl, in L.J. Dowley, E. Bannon, L.R. Cooke, T. Keane and E. O'Sullivan (eds), *Phytophthora infestans 150*, EAPR Boole Press, Dublin, pp. 154-159.

Reich, B., Buchenauer, H., Buschhaus, H. and Wenz, M. (1992) Wirkungsweise von Propamocarb gegenüber *Phytophthora infestans*, *Mitteilungen Biologische Bundesanstalt Land- und Forstwirtschaft* **226**, 423.

Reiter, B., Wenz, M., Buschhaus, H. and Buchenauer, H. (1996) Action of propamocarb against *Phytophthora infestans* causing late blight of potato and tomato, in H. Lyr, P.E. Russell and H.D. Sisler (eds), *Modern Fungicides and Antifungal Compounds,* Intercept, Andover, pp. 147-156.

Reschke, M. (1999) Strobis in Gefahr, *Deutsche Landwirtschafts-Gesellschaft (DLG)-Mitteilungen* **1**, 51.

Reuveni, M., Eyal, M. and Cohen, Y. (1980) Development of resistance to metalaxyl in *Pseudoperonospora cubensis*, *Plant Disease* **64**, 1108-1109.

Ruess, W., Kunz, W., Staub, T., Müller, K., Poppinger, N., Speich, J. and Ahl-Goy, P. (1995) Plant activator CGA 245704, a new technology for disease management, *XIIIth International Plant Protection Congress,* The Hague, 424.

Ruess, W., Müller, K., Knauf-Beiter, G., Kunz, W. and Staub, T. (1996) Plant activator CGA 245704: an innovative approach for disease control in cereals and tobacco, *Brighton Crop Protection Conference,* 53-60.

Ryals, J.A., Neuenschwander, U.H., Willits, M.G., Molina, A., Steiner, H. and Hunt, M.D. (1996) Systemic acquired resistance, *The Plant Cell* **8**, 1809-1819.

Samoucha, Y. and Gisi, U. (1987a) Systemicity and persistence of cymoxanil in mixture with oxadixyl and mancozeb against *Phytophthora infestans* and *Plasmopara viticola*, *Crop Protection* **6**, 393-398.

Samoucha, Y. and Gisi, U. (1987b) Possible explanations of synergism in fungicide mixtures against *Phytophthora infestans*, *Annals of Applied Biology* **110**, 303-311.

Samoucha, Y. and Gisi, U. (1987c) Use of two- and three-way mixtures to prevent buildup of resistance to phenylamide fungicides in *Phytophthora* and *Plasmopara*, *Phytopathology* **77**, 1405-1409.

Sanders, P.L. (1984) Failure of metalaxyl to control *Pythium* blight on turfgrass in Pennsylvania, *Plant Disease* **68**, 776-777.

Schewe, T. and Lyr, H. (1995) Mechanism of action of carboxin fungicides and related compounds, in H. Lyr (ed), *Modern Selective Fungicides*, 2nd ed, Gustav Fischer, Jena, pp. 149-161.

Schwinn, F.J. (1981) Chemical control of downy mildews, in D.M. Spencer (ed), *The Downy Mildews*, Academic Press, London, pp. 305-320.

Schwinn, F.J. and Staub, T. (1995) Oomycetes fungicides: phenylamides and other fungicides against Oomycetes, in H. Lyr (ed), *Modern Selective Fungicides*, 2nd ed, Gustav Fischer, Jena, pp. 323-346.

Shattock, R.C. (1986) Inheritance of metalaxyl resistance in the potato late blight fungus, *Brighton Crop Protection Conference*, 555-560.

Shattock, R.C. (1988) Studies on the inheritance of resistance to metalaxyl in *Phytophthora infestans*, *Plant Pathology* **37**, 4-11.

Shaw, D.S. and Shattock, R.C. (1991) Genetics of *Phytophthora infestans*: the Mendelian approach, in J.A. Lucas, R.C. Shattock, D.S. Shaw and L.R. Cooke (eds), *Phytophthora*, Cambridge University Press, Cambridge, pp. 218-230.

Siegfried, W., Bosshard, E. and Schüepp, H. (1992) Erste Erfahrungen mit *Plasmopara*-Warngeräten im Rebbau, *Schweizerische Zeitschrift für Obst- und Weinbau* **128**, 143-150.

Sierotzki, H., Wullschleger, J. and Gisi, U. (2000a) Point-mutation in cytochrome b gene conferring resistance to strobilurin fungicides in *Erysiphe graminis* f. sp. *tritici* field isolates. *Pesticide Biochemistry and Physiology* **68**, 107-112.

Sierotzki, H., Parisi, S., Steinfeld, U., Tenzer, I., Poirey, S. and Gisi, U. (2000b) Mode of resistance to respiration inhibitors at the cytochrome bc_1 enzyme complex of *Mycosphaerella fijiensis* field isolates. *Pest Management Science* **56**, 833-841.

Smillie, R., Grant, B.R. and Guest, D. (1989) The mode of action of phosphite: evidence for both direct and indirect modes of action on three *Phytophthora* spp. in plants, *Phytopathology* **79**, 921-926.

Sozzi, D., Schwinn, F.J. and Gisi, U. (1992) Determination of the sensitivity of *Phytophthora infestans* to phenylamides: a leaf disc method, *EPPO Bulletin* **22**, 306-309.

Steiner, U. and Schönbeck, F. (1996) Induced resistance against biotrophic fungi, in H. Lyr, P.E. Russell and H.D. Sisler (eds), *Modern Fungicides and Antifungal Compounds*, Intercept, Andover, pp. 511-517.

Steinfeld, U., Sierotzki, H., Parisi, S., Poirey, S. and Gisi, U. (2001) Sensitivity of mitochondrial respiration to different inhibitors in *Venturia inaequalis*, *Pest Management Science* **57**, 787-796.

Stenzel, K., Pontzen, R., Seitz, T., Tiemann, R. and Witzenberger, A. (1998) SZX 722: a novel systemic oomycete fungicide, *Brighton Crop Protection Conference,* 367-374.

Stevens, P.J.G., Baker, E.A. and Anderson, N.H. (1988) Factors affecting the foliar absorption and redistribution of pesticides. 2. Physicochemical properties of the active ingredient and the role of surfactant, *Pesticide Science* **24**, 31-53.

Urban, M., Bhargava, T. and Hamer, J.E. (1999) An ATP-driven efflux pump is a novel pathogenicity factor in rice blast disease, *The EMBO Journal* **18**, 512-521.

Vegh, I., Leruox, P., LeBerre, A. and Lanen, C. (1985) Détection sur *Chamaecyparis lawsoniana* 'Ellwoodii' d'une souche de *Phytophthora cinnamomi* Rands résistante au phoséthyl-Al, *P.H.M.-Revue Horticole* **262**, 19-21.

Vernooij, B., Friedrich, L., Morse, A., Reist, R., Kolditz-Jawhar, R., Ward, E., Uknes, S., Kessmann, H. and Ryals, J. (1994) Salicylic acid is not the translocated signal responsible for inducing systemic acquired resistance but is required in signal transduction, *The Plant Cell* **6**, 959-965.

Viranyi, F. and Oros, G. (1991) Developmental stage response to fungicides of *Plasmopara halstedii* (sunflower downy mildew), *Mycological Research* **92**, 199-205.

Walker, A.S.L. and Cooke, L.R. (1990) The survival of *Phytophthora infestans* in potato tubers: the influence of phenylamide resistance, *Brighton Crop Protection Conference*, 1109-1114.

Watanabe, M., Hayashi, K., Tanaka, T. and Uesugi, Y. (1996) Cyanide-resistant alternative respiration in *Botrytis cinerea*, in H. Lyr, P.E. Russell and H.D. Sisler (eds), *Modern Fungicides and Antifungal Compounds*, Intercept, Andover, pp. 111-15.

Wei, G., Kloepper, J.W. and Tuzun, S. (1991) Induction of systemic resistance of cucumber to *Colletotrichum orbiculare* by select strains of plant growth-promoting rhizobacteria, *Phytopathology* **81**, 1508-1512.

Williams, D.J., Beach, B.G.W., Horrière D. and Marechal, G. (1977) LS 74-783: a new systemic fungicide with activity against phycomycete diseases, *Brighton Crop Protection Conference*, 565-573.

Ypema, H.L. and Gold, R.E. (1999) Kresoxim-methyl: modification of a naturally occurring compound to produce a new fungicide, *Plant Disease* **83**, 4-19.

Ziogas, B.N. and Davidse, L.D. (1987) Studies on the mechanism of action of cymoxanil in *Phytophthora infestans*, *Pesticide Biochemistry and Physiology* **29**, 89-96.

Ziogas, B.N., Baldwin, B.C. and Young, J.E. (1997) Alternative respiration: a biochemical mechanism of resistance to azoxystrobin (ICIA 5504) in *Septoria tritici*, *Pesticide Science* **50**, 28-34.

AN ITS-BASED PHYLOGENETIC ANALYSIS OF THE RELATIONSHIPS BETWEEN *PERONOSPORA* AND *PHYTOPHTHORA*

D.E.L. COOKE, N.A. WILLIAMS, B. WILLIAMSON and J.M. DUNCAN
Scottish Crop Research Institute, Invergowrie, Dundee, DD2 5DA, UK

1. Introduction

Despite the economic and ecological importance of the downy mildews, understanding of their evolution and taxonomy is poor and the current approach to species concepts is "inadequate and potentially misleading" (Hall, 1996).

The adaptation of downy mildews to survival in an aerial environment, their biotrophic mode of nutrition, tightly defined host specificity and, in some genera, loss of the water-borne zoosporic phase has led many workers to consider them the most evolutionarily "advanced" of the oomycetes (Dick, this volume). Gaümann (1952) hypothesised that the downy mildews evolved through a succession from the water-dependent saprophytic taxa (e.g. *Saprolegnia*) through to a range of increasingly specialised parasites of plants; from *Pythium* into *Phytophthora* and lastly to the downy mildews. However this hypothesis has yet to be tested objectively.

A recurrent problem in downy mildew characterisation is their biotrophic habit that has hampered detailed morphological, physiological, biochemical and molecular studies. Taxonomic and evolutionary studies have thus relied heavily on a few morphological characters and host preference. Gaümann's 'biological species' concept (see Hall, 1996) for example, led to the idea of 'one host - one species' which, without rigorous cross-infection studies, resulted in a profusion of species based only on their host plants. In *Peronospora,* which is by far the largest genus of downy mildews, more than 83 species have been listed in the British Isles alone (Francis and Waterhouse, 1988). Classifications such as this have resulted in inconsistent, and often inadequate, species definitions (see review by Hall, 1996) and it is widely accepted that a more rigorous analysis of the downy mildews and related genera is required.

A phylogenetic analysis of appropriate DNA sequences would allow such a re-examination, but molecular analysis of these fungi is often hampered by the intimate association of host and pathogen and the difficulties of obtaining pure DNA of the pathogen. Sequence analysis of the Internal Transcribed Spacer (ITS) regions of ribosomal RNA gene tandem repeat (rDNA) has proved useful for interspecific comparisons and phylogenetic analysis of *Phytophthora* species and, to a limited extent, some downy mildew taxa (Cooke and Duncan, 1997; Cooke *et al.*, 2000a; Rehmany *et al.*, 2000).

The aims of this study were to:
- develop PCR primers for the molecular characterisation of downy mildew

161

P.T.N. Spencer-Phillips et al. (eds.), Advances in Downy Mildew Research, 161–165.

species, even in the presence of plant material;
- investigate intra- and interspecific relationships within the Peronosporales;
- discuss the wider impact of such findings on current concepts in downy mildew research.

2. Materials and Methods

Sequences of the 18S rDNA gene from a range of organisms (oomycetes, true fungi, plants and bacteria) were extracted from the EMBL database, aligned and compared. Within a segment c. 200 bp from the 3′ end of the gene a region specific to *Phytophthora* but not other fungi or plants was identified. The oligonucleotide primer DC6 (5′ GAGGGACTTTTGGGTAATCA 3′) was designed from this sequence and tested against a range of oomycetes.

Seventeen isolates of eleven species of *Peronospora* and two species of *Albugo* were examined. Detailed morphological examination was not undertaken; species are therefore named according to host association (Francis and Waterhouse, 1988) and details provided by the supplier.

Fresh infected plant material was examined by epifluoresence microscopy after staining in aniline blue (Williamson *et al.*, 1995) to confirm the presence of downy mildew, and DNA was extracted from either infected leaf pieces (c. 3 mm^2) or a centrifuged conidial suspension (flushed from sporulating lesions with a micro-pipette). The first round of PCR using primers DC6 and ITS4 (Cooke *et al.*, 2000b) and the second using primers ITS6 and ITS4 was carried out according to Cooke and Duncan (1997). Primer ITS6 (5′ GAAGGTGAAGTCGTAACAAGG 3′) lies at the 3′ end of the 18S region and is a form of ITS5 modified to allow more efficient amplification of the Peronosporales.

Direct sequencing of the PCR products was carried out with a PCR-based cycle sequencing kit (Applied Biosystems) and run on a 373 automated sequencer (Applied Biosystems) according to the manufacturer's instructions. Sequences were aligned with published ITS sequences of *Phytophthora* species (Cooke and Duncan, 1997; Cooke *et al.*, 2000b). Phylogenetic (DNA-distance based and DNA maximum likelihood) analysis of all *Peronospora* species, *Albugo candida* and 13 *Phytophthora* species (encompassing much of the total ITS diversity in the genus) was performed using the PHYLIP computer programme (Felsenstein, 1993).

3. Results

In a screen of many oomycete species the primer DC6 only resulted in an amplification product from members of the Peronosporales and Pythiales. No product was amplified from *Phaeodactylium*, *Saprolegnia*, *Thraustotheca* or *Achlya*. Re-amplification of the first round 1.2 Kb product with primers ITS6 and ITS4 resulted in a *c.* 930 bp product that was sequenced and used for phylogenetic analysis. Standard DNA extraction techniques from infected plant material and DNA released from freeze-thawed sporangia provided suitable templates for PCR.

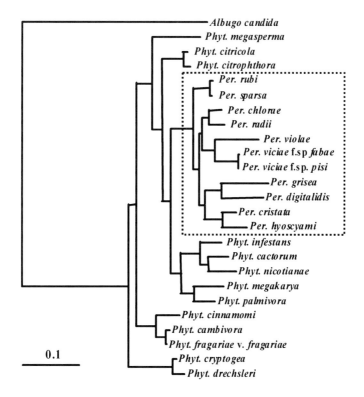

Figure 1. Phylogenetic tree produced from a DNA maximum likelihood analysis of complete ITS sequences of *Phytophthora, Peronospora* and *Albugo* species. Scale bar equals the number of nucleotide substitutions per site.

Phylogenetic analysis indicated that all 10 species of *Peronospora* examined were monophyletic, forming a single clade in that part of the *Phytophthora* tree containing papillate species of *Phytophthora* (see box in Fig.1). The genetic distances between *Peronospora* species and the nearest *Phytophthora* species were less than between non papillate and papillate *Phytophthora* species (e.g. *P. cinnamomi* and *P. infestans*). The blister rust, *Albugo candida* was not closely related to *Peronospora* but formed a distinct outgroup.

4. Conclusions

This study indicates that the genus *Peronospora* comprises a group of biotrophic *Phytophthora*-like species. Although many more taxa await examination it is clear that taxonomic revision of *Peronospora* and perhaps other genera of downy mildews is required. The fact that *Phytophthora* and *Peronospora* are so closely related has implications for the study of both genera. Co-ordination of research objectives should be encouraged. For example, comparisons of the current *Phytophthora* genomics programmes (e.g. Kamoun *et al.*, 1999) with those in *Peronospora* may help highlight the key mechanisms underlying biotrophy and host selection.

Although the intimacy of their interaction implies that the downy mildews have co-evolved with their hosts, the processes underlying host specificity and the ability of *Peronospora* species to cross-infect are poorly understood and have been little studied. Parallels exist in *Phytophthora*, for although some species have a wide host range (e.g. *P. cinnamomi* has been recovered from thousands of host species in many families), others are restricted to only a few hosts and comparable to those of *Peronospora* (e.g. *P. fragariae* var. *rubi*, which only occurs on raspberry). An understanding of host specificity and its evolution in *Peronospora* should benefit, therefore, from studies on host specificity within closely related *Phytophthora* species.

In this study there is some evidence for coevolution of host and pathogen. The hosts of the related *Per. grisea* (*Hebe*) and *Per. digitalidis* (*Digitalis*) for example, are both members of the Scrophulariaceae. Conversely, the related taxa *Per. cristata* and *Per. hyoscyami* (Fig. 1) are pathogens of the unrelated hosts, *Mecanopsis* (Papa-veraceae) and *Nicotiana* (Solanaceae), respectively. A rigorous examination of the parallels between the evolutionary relationships of more *Peronospora* species and their respective host plants will add to the understanding of coevolutionary processes. For example, the extent and rate of change in downy mildew host range will, for the first time, be open to examination.

The results support the intuitive hypothesis of an evolutionary succession amongst fungi from unspecialised saprotrophs or opportunistic invaders of dead or senescing plant material, through to those hemibiotrophic or necrotrophic pathogens with specialised haustoria and an aerial habit (*Phytophthora infestans*), to highly specialised obligate biotrophs with a narrow host range.

Such sequence analysis clearly offers an objective approach to taxonomic issues (but see caveats of sole reliance on molecular methods in Cooke *et al.*, 2000a,b). *P. sparsa* (from *Rosa* spp.) and *P. rubi* (from *Rubus* spp.) were identical in sequence, which alongside evidence of cross-infection (Breese *et al.*, 1994) is suggestive of their conspecificity. Similarly, isolates of *P. viciae* from faba bean and pea were identical in sequence. In contrast to this, isolates of *Peronospora parasitica* from *Arabidopsis* and *Brassica oleracea* differ markedly in ITS sequence (Rehmany *et al.*, 2000) suggesting reproductive isolation despite their overlapping geographic range. There are clearly many cases within the genus *Peronospora* where the establishment of a database of ITS sequences will aid the discrimination between cases where a single *Peronospora* taxa infects many plant species and perhaps conversely where multiple *Peronospora* taxa infect a single plant species.

At a more practical level, an excellent set of tools now exists for rapid PCR - based identification and detection of many downy mildew species. The PCR primer DC6 eliminates the problems of PCR contamination from plant or microbial sources. Identification on the basis of restriction digest patterns of ITS sequences is now established as a reliable method in the genus *Phytophthora* (Cooke *et al.*, 2000a,b; and see the web – based tool at www.phytID.org) and is being applied to the downy mildews. Species – specific primers have been designed for both *Phytophthora* (Bonants *et al.*, 1997) and to a limited extent *Peronospora* species (Lindqvist *et al.*, 1998) and this approach can now be extended to cover more downy mildew taxa.

5. Acknowledgements

We thank colleagues who have kindly provided samples of downy mildew and Scottish Executive Environment and Rural Affairs Department for funding this work.

6. References

Bonants, P., de Weerdt, M.H., van Gent-Pelzer, M., Lacourt, I., Cooke, D. and Duncan, J.M. (1997) Detection and identification of *Phytophthora fragariae* Hickman by the polymerase chain reaction, *European Journal of Plant Pathology* **103**, 345-355.

Breese, W.A., Shattock, R.C., Williamson, B. and Hackett, C. (1994) *In vitro* spore germination and infection of cultivars of *Rubus* and *Rosa* by downy mildews from both hosts, *Annals of Applied Biology* **125,** 73-85.

Cooke, D.E.L. and Duncan, J.M. (1997) Phylogenetic analysis of *Phytophthora* species based on the ITS1 and ITS2 sequences of ribosomal DNA, *Mycological Research* **101,** 667-677.

Cooke, D.E.L., Drenth, A., Duncan, J.M., Wagels, G. and Brasier, C.M. (2000a) A molecular phylogeny of *Phytophthora* and related Oomycetes, *Fungal Genetics and Biology* **30**, 17-32.

Cooke, D.E.L., Duncan, J.M., Williams, N.A., Hagenaar-de Weerdt, M., Bonants P.J.M. (2000b) Identification of *Phytophthora* species on the basis of restriction enzyme fragment analysis of the Internal Transcribed Spacer regions of ribosomal RNA, *OEPP/EPPO Bulletin* **30**, 519-523.

Felsenstein, J. (1993) PHYLIP: Phylogeny inference package (version 3.5c), distributed by the author, Department of Genetics, University of Washington, Seattle, USA.

Francis, S.M. and Waterhouse, G.M. (1988) List of Peronosporaceae reported from the British Isles, *Transactions of the British Mycological Society* **91**, 1-62.

Gäumann, E.A. (1952) *The Fungi. A description of their morphological features and evolutionary development,* Hafner Publishing Company, New York.

Hall, G.S. (1996) Modern approaches to species concepts in downy mildews, *Plant Pathology* **45**, 1009-1026.

Kamoun, S., Hraber, P., Sobral, S., Nuss, D. and Govers, F. (1999) Initial assessment of gene diversity for the Oomycete pathogen *Phytophthora infestans* based on expressed sequences, *Fungal Genetics and Biology* **28**, 94–106.

Lindqvist, H., Koponen, H., Valkonen, J.P.T. (1998) *Peronospora sparsa* on cultivated *Rubus arcticus* and its detection by PCR based on ITS sequences, *Plant Disease* **82**, 1304-1311.

Rehmany, A.P., Lynn, J.R, Tör, M., Holub, E.B. and Beynon, J.L. (2000) A comparison of *Peronospora parasitica* (Downy Mildew) isolates from *Arabidopsis thaliana* and *Brassica oleracea* using amplified fragment length polymorphism and Internal Transcribed Spacer 1 sequence analyses, *Fungal Genetics and Biology* **30**, 95–103.

Williamson, B., Breese, W.A. and Shattock, R.C. (1995) A histological study of downy mildew (*Peronospora rubi*) infection of leaves, flowers and developing fruits of Tummelberry and other *Rubus* spp, *Mycological Research* **99**, 1311-1316.

THE SUNFLOWER - *PLASMOPARA HALSTEDII* PATHOSYSTEM: NATURAL AND ARTIFICIALLY INDUCED COEVOLUTION

F. VIRÁNYI
Department of Plant Protection, St. Stephen University,
Gödöllő, 2103 Gödöllő, Páter K. u. 1., Hungary

1. Introduction

Downy mildew of sunflower, caused by *Plasmopara halstedii* (Farlow) Berlese et de Toni, is one of the major diseases of this crop worldwide. Since both the biotrophic fungus and its main (obligatory) host are assumed to have originated in the central portion of North America (Leppik, 1962), they are likely to have coevolved in the past (Fig.1). In addition, open-pollinated sunflower cultivars released from the 1960s were highly susceptible to this fungus so that it spread rapidly wherever the crop was grown. Though plant breeders have produced a number of resistant cultivars, soon after their introduction new virulent forms (pathotypes) of *P. halstedii* appeared (Gulya *et al.*, 1991; Virányi and Gulya, 1995). In recent years, efforts are being continued to detect new sources of resistance within both annual and perennial members of the genus *Helianthus*.

In this review, based on published data and the author's investigations, attempts are being made to illustrate how sunflower and *P. halstedii* have coevolved.

2. Materials and Methods

Field isolates of *P. halstedii*, collected from either cultivated or volunteer sunflower plants and from diseased individuals of *Xanthium strumarium*, were the subject of host range studies on a series of annual and perennial *Helianthus* species, as well as on other members of the Asteraceae. Tests for virulence and aggressiveness were carried out by using a methodology described by Gulya *et al.* (1991) and Tourvieille *et al* (2000). As for inoculum, we usually used bulk isolates but, in some cases, single spore isolates (clones) were also used.

3. Results and Discussion

Earlier investigators (Savulescu, 1941; Leppik, 1962, 1966; Orellana, 1970) reported at least 12 wild *Helianthus* species as well as several other composites as natural hosts of *P. halstedii*. Moreover, artificial inoculations of 10 species of the Asteraceae were also successful (Tuboly, 1971). In our experience, the *Xanthium* isolates could hardly infect

P.T.N. Spencer-Phillips et al. (eds.), Advances in Downy Mildew Research, 167–172.

cultivated sunflower at first but they became as pathogenic as those isolates from cultivated sunflower after being reinoculated (subcultured).

On a range of wild *Helianthus* species as well as on a few composites it was shown that besides *H. annuus*, five additional *Helianthus* species and *Artemisia vulgaris* appeared to support fungal sporulation (Table 1), as well as *Ambrosia artemisiifolia* proved to be naturally infected (Walcz *et al.*, 2000). With a few exceptions, marked in Table 1, our results are in accordance with those described in the literature (see Leppik, 1966).

TABLE 1. Sporulation of *Plasmopara halstedii* on a number of composites based on seedling tests (Virányi, 1984) and on field survey

Plant species	Sporulation	Artificial/Natural host
Annual:		
Helianthus annuus (wild)	+	A
H. argophyllus	+	A
H. debilis	+	A
H. petiolaris	+	A
Perennial:		
H. angustifolius	-	A
H. decapetalus	-	A
H. divaricatus	+	A
H. doronicoides	-	A
H. grosseserratus	+	A
H. maximiliani	-	A
H. mollis	-	A
H. multiflorus	-	A
H. nuttallii	-	A
H. occidentalis	-	A
H. resinosus	-	A
H. rigidus	-	A
H. salicifolius	-	A
H. scaberrimus	-	A
H. strumosus	-	A
H. tuberosus	-	A
Ambrosia artemisiifolia	+	N
Artemisia vulgaris	+	A
Centaurea cyanus	-	A
Xanthium strumarium	+	N

Items underlined indicate contradiction with published data.

In Table 2 the pathotypes (virulence phenotypes) of *P. halstedii,* identified to date in various laboratories, are summarized (Tourvieille *et al.*, 2000). Forty-eight records out of 66 have been obtained by either the author or T. Gulya (USDA, Fargo). In Hungary, over a period of eight years between 1989 and 1996, a total of 134 field isolates of *P. halstedii*, the majority from Hungary and some from Bulgaria, Italy, Romania and Yugoslavia, respectively, have been tested. Out of these, six pathotypes each with distinct virulence pattern have been identified (Virányi and Gulya, 1995). In addition, on a wider scale, Gulya and co-workers have identified a total of 12 pathotypes among the isolates from many countries (Gulya *et al.*, 1996).

Additionally, some of the single spore clones obtained from a bulk isolate belonging to a particular pathotype differed in their virulence character as compared to the parent isolate. This happened when the clone either had been stored deep-frozen or was regularly subcultured on susceptible sunflower seedlings prior to the virulence test.

From published data and from our experimental results it can be concluded that *P. halstedii*, a complex biological species in the past, had evolved in two steps:
1) the fungus distributed intercontinentally and had adapted to cultivated sunflower, a natural coevolution over the XVI-XX centuries (see Fig.1);
2) due to resistance breeding, fungal populations of a particular geographical region became diverse in virulence, a man-made coevolution from the 1960s to present (Table 3).

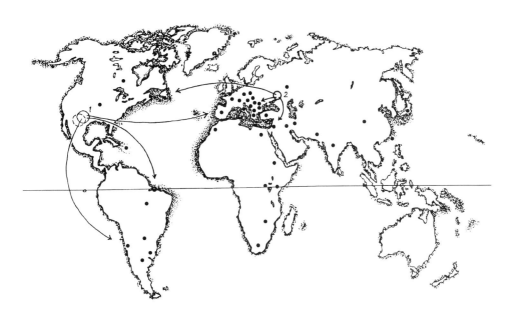

Figure 1. Dotted circle, assumed primary gene centre of cultivated sunflower; solid circles, assumed area of origin (1) and secondary centre of distribution (2) of *Plasmopara halstedii*; black dots, countries with known occurrence of downy mildew on cultivated sunflower.

TABLE 2. *Plasmopara halstedii* pathotypes/virulence phenotypes identified on a global scale

Country	Pathotype/Virulence phenotype															
	1/100	2/300	304	6/310	313	7,9/330	332	3/700	701	703	8/710	713	4/730	732	733	5/770
EUROPE																
Bulgaria	□							□								
France	□	□	□	□				□	□	□	□			□	□	□
Germany	□	□		□		□					□		□			
Hungary	□	□				□		□			□		□			
Italy	□	□				□		□			□					
Romania	□	□		□	□	□		□			□	□	□			
Spain	□	□		□		□		□		□	□		□			
U.S.S.R.	□	□						□					□			
Yugoslavia	□															
AMERICA																
Argentina						□		□								
Brazil		□														
Canada	□	□		□		□		□			□		□			□
U.S.A.		□				□		□					□			□
ASIA																
China	□	□						□								
India	□															
AFRICA																
Morocco	□					□	□	□			□		□	□		
R.S.A.	□	□						□						□		
Zimbabve													□			

□ Indicates records originating from either T. Gulya's laboratory or the author's own test

TABLE 3. Coevolution of downy mildew resistance in sunflower and the diversity of virulence in *P. halstedii*

Year	First appearance		Pathotype(s)
	Pathotype	R-gene	overcome
?	100		
1970		Pl1	100
1971		Pl3	100
1972	300	Pl2	100, 300
1974		Pl4	100, 300
1980	700		
1982		Pl5	700
1985	730		
1988	770		
1990	300, 330, 710	Pl6, Pl7, Pl8	100, 300, 700, 710, 730, 770
1990 (?)		Pl9	310
1991		Pl10	330
1993	330		

Furthermore, the fungus still retained in part its original pathogenicity to members of the Asteraceae, as was proven by local differences in host range studies (Novotel'nova 1963; Leppik, 1966; Virányi, 1984; Seiler and Gulya, 1993). Lastly, some correlation appears to exist between genetic determinants of either pathogenicity or virulence in *P. halstedii* since several pathotype-specific resistance factors (genes) have been found in wild *Helianthus* species known to be natural hosts of this fungus (Seiler, 1994).

4. References

Gulya, T.J., Sackston, W.E., Virányi, F., Masirevic, S. and Rashid, K.Y. (1991) New races of the sunflower downy mildew pathogen (*Plasmopara halstedii*) in Europe and North and South America, *Journal of Phytopathology* **132**, 303-311.

Gulya, T.J., Virányi, F., Nowell, D., Serrhini, M.N. and Arouay, K. (1996) New races of sunflower downy mildew in Europe and Africa, *Proceedings, 18th National Sunflower Association Research Workshop, Fargo*, pp. 181-184.

Leppik, E.E. (1962) Distribution of downy mildew and some other seed-borne pathogens on sunflowers, *FAO Plant Protection Bulletin* **10**, 126-129.

Leppik, E.E. (1966) Origin and specialization of *Plasmopara halstedii* complex on the Compositae, *FAO Plant Protection Bulletin* **14**, 72-76.

Novotel'nova, N.S. (1963) An overview of the species *Plasmopara* pathogenic to members of the genus *Helianthus, Proceedings of the 2nd Symposium on Problems in Investigating the Fungus and Lichen Flora of the Baltic Republics, Acad. Sci. Lithuanian SSR, Vilnius*, pp. 111-118. (in Russian)

Orellana, R.G. (1970) Resistance and susceptibility of sunflowers to downy mildew and variability in *Plasmopara halstedii, Bulletin of the Torrey Botanical Club* **97**, 91-97.

Savulescu, T. (1941) Die auf Compositen parasitierenden *Plasmopara*-Arten, *Bulletin Section Acaedemie Roumain* **24**, 45-67.

Seiler, G.J. (1994) Progress report of the working group of the evaluation of wild *Helianthus* species for the period 1991-1993, *Helia* **17**, 87-92.

Seiler, G.J. and Gulya, T.J. (1993) Wild sunflower species evaluated for downy mildew (*Plasmopara halstedii*) resistance, *Proceedings, 15th National Sunflower Association Research Workshop, Fargo*, p. 25.

Tourvieille, D. de Labrouhe, Gulya, T.J., Masirevic, S., Penaud, A., Rashid, K.Y. and Virányi, F. (2000) New nomenclature of races of *Plasmopara halstedii* (sunflower downy mildew), *Proceedings, 15th International Sunflower Conference, Toulouse*, Vol. II, I-61-66.

Tuboly, L. (1971) Investigations on the downy mildew of sunflower and on conditions of infection, *Növényvédelem Korszerűsítése* **5**, 51-62. (in Hungarian)

Virányi, F. (1984) Recent research on the downy mildew of sunflower in Hungary, *Helia* **7**, 35-38.

Virányi, F. and Gulya, T.J. (1995) Inter-isolate variation for virulence in *Plasmopara halstedii* (sunflower downy mildew) from Hungary, *Plant Pathology* **44**, 619-624.

Walcz, I., Bogár, K. and Virányi, F. (2000) Study on an *Ambrosia* isolate of *Plasmopara halstedii*, Helia **23**, 19-24.

PERONOSPORA VALERIANELLAE, THE DOWNY MILDEW OF LAMB'S LETTUCE (*VALERIANELLA LOCUSTA*)

G. PIETREK and V. ZINKERNAGEL
Institute of Phytopathologie, Technical University of Munich
D-85350 Freising-Weihenstephan, Germany

1. Introduction

Since the production of lamb's lettuce (*Valerianella locusta*) is rising in Germany and France it has become necessary to develop a strategy against the diseases which constrain its production. The downy mildew is one of the most important diseases of lamb's lettuce. The pathogen *Peronospora valerianellae* causes severe damage to the plants under favourable conditions. The leaves begin to fold downwards and under humid conditions, profuse sporulation occurs on the infected parts of the leaves. In the later stages of the infection the foliage becomes yellow and necrotic. In the yellow areas brown spots are visible which contain the oospores of the fungus.

In cooperation with breeding companies it is intended to develop rational inoculation procedures and to look for resistance sources in varieties, breeding lines and wild species against the downy mildew of lamb's lettuce.

2. Materials and Methods

2.1. PLANT MATERIAL

Seeds of different commercial varieties and breeding lines were obtained from a German breeding company. Plants were grown in trays under greenhouse conditions at temperatures of 10-15 °C. Plants were inoculated in the two leaf stage with spores of *P. valerianellae*. Twenty five plants of each variety were tested with three replications.

2.2. FUNGAL ISOLATES

Some of the isolates were obtained from diseased plants grown from oospore infected seeds of different varieties. The other ones were isolated from infected plants from different regions of Germany. The following nine isolates were tested: G, V-1, V-2, M, L, H-1, H-2, DV and T. The isolates were maintained on their respective host varieties.

P.T.N. Spencer-Phillips et al. (eds.), Advances in Downy Mildew Research, 173–177.

2.3. INOCULATION AND DISEASE ASSESSMENT

Inoculum was prepared by washing infected leaves with sporulating fungus in distilled water with gentle agitating. All inoculations were by spraying plants with the spore suspension containing 10^5 spores/ml using a chromatography sprayer. Inoculated plants were incubated in a growth chamber. Following inoculation the plants remained in darkness for 48 h at 10 °C and under a polyethene cover to provide moist conditions. After removing the cover, plants were incubated at temperatures of 10-12 °C with a 12 h photoperiod under a light intensity of approximately 130 μmol/m^{2}s. Following the incubation period the plants were sprayed with distilled water and covered again to stimulate sporulation. The sporulation was assessed after 12 h in darkness at temperatures of 10 °C using the following scale:
1 = no sporulation, 3 = a few sporophores present, 5 = < 50% of leaf covered with sporophores, 9 = profuse sporulation with > 50% of leaf covered with sporophores. To interpret the results, the sporulation intensities were calculated to get a Disease Index:

$$\text{Disease Index} = \frac{i \times 1 + j \times 3 + y \times 5 + z \times 9}{n}$$

i, j, y, z = number of plants with sporulation intensity 1, 3, 5 and 9;
n = total number of plants.

To notice the homogenity of varieties or wether plants segregate in susceptibility, the frequency of each sporulation intensity (1, 3, 5 and 9) of each variety was recorded.

2.4. INOCULATION AT DIFFERENT TEMPERATURES

In previous investigations, some incompletely resistant varieties differed in disease severity depending on the temperature. The selected varieties with resistant, intermediate and susceptible reactions were tested with different temperatures in addition to the previously used temperature of 10 °C. Using the procedure described above, six varieties were inoculated with two isolates in four growths chambers at temperatures of 10, 13, 15 and 20 °C. Since in previous investigations no sporulation occurred at 20 °C, the temperature was reduced to 10 °C three days before evaluation. After 14 days incubation, the sporulation intensities were assessed.

3. Results and Discussion

3.1. SUSCEPTIBILITY OF DIFFERENT VARIETIES

The results of the inoculations showed that there are susceptible varieties with profuse sporulation and also completely resistant reactions with no macroscopically visible symptoms. But there are also varieties with very sparse sporulation. This reaction seems to be similar to the host-pathogen relationship of lettuce and *Bremia lactucae*. Crute and Norwood (1978) termed this reaction 'incomplete resistance' or 'intermediate' response. Figure 1 shows the reaction of different varieties following inoculation with isolate G.

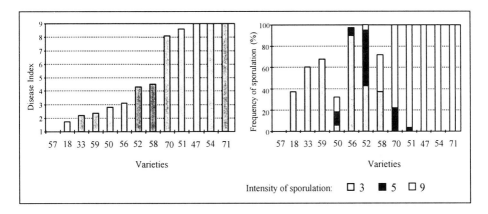

Figure 1. Susceptibility of different varieties of lamb's lettuce to *P. valerianellae*, isolate G
(3 = few sporophores, 5 = < 50%; 9 = > 50% leaf area sporulating)

The varieties 70, 51, 47, 54 and 71 are highly susceptible whereas 57 is completely resistant to isolate G. The resistant response is due to an occurrence of hypersensitive reactions in the plant tissue. A few varieties like 59 and 33 react with incomplete resistance. As shown in Figure 1, the frequency of the sporulation intensities for this varieties is very sparse. The varieties 50 and 58 include different reacting genotypes in a range from high susceptible to resistant.

3.2. INOCULATION AT DIFFERENT TEMPERATURES

The results shown in Figure 2 confirm previous investigations that there is an interaction between temperature and susceptibility. In several cases of incomplete resistant reactions the expression of resistance becomes ineffective at higher temperatures with one race (isolate G). After inoculation with isolate G the incompletely resistant varieties, e.g. 52 and 59 (Figure 2), are distinctly more susceptible at higher temperatures (15-20 °C) than at lower ones (10-13 °C) and a profuse sporulation occurs. The completely resistant varieties remain resistant, even at higher temperatures. Compatible interactions are less infected at high temperatures, with reduced sporulation intensity. At 20 °C the fungus was not able to sporulate, incubation at 10 °C for at least three days required for sporulation. The germination of sporangia is inhibited at a temperature of 25 °C.

Similar observations related to temperature dependent compatibility have been made in other host-pathogen interactions for example between lettuce and *Bremia lactucae* (Judelson and Michelmore, 1992). The reasons for the temperature sensitivity are still not completely understood, it is assumed that the temperature influence the products of avirulence or resistance genes or their interactions (Judelson and Michelmore, 1992; Balass *et al.*, 1993).

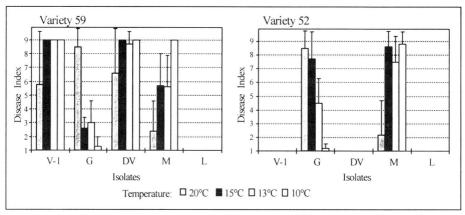

Figure 2. Interactions between temperature and susceptibility of two varieties to five isolates of *P. valerianellae.*

3.3. CHARACTERISATION OF DIFFERENT PHYSIOLOGICAL RACES

Inoculations of 45 varieties shows that six varieties could be used to characterise the physiological races.

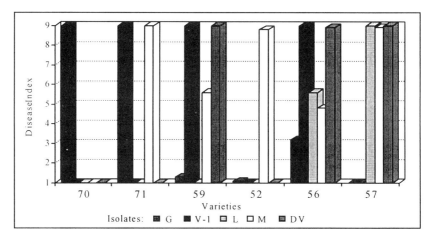

Figure 3. Susceptibility of differential varieties following inoculation with five isolates of *P. valerianellae* (incubation at 10 °C).

Figure 3 shows clearly that race specificity occurs. Every variety (except 56) is resistant to at least one race and none was completely resistant to all races. Five physiological races can be differentiated (Table 1).

TABLE 1. Characterisation of different physiological races of *P. valerianellae* with differential varieties

	Isolate				
	V-1/V-2	G/H-1/H-2/T	L	DV	M
	Race				
Variety	1	2	3	4	5
70	-	+	-	-	+
52	-	(+)	-	-	+
59	+	(+)	-	+	+
56	+	(+)	+	+	+
57	-	-	+	+	+
71	-	+	-	-	-

Reaction type: - = resistant, + = susceptible, (+) = intermediate response

It can be concluded, that there is a race specifity in Lamb's lettuce - *P. valerianellae* interaction and the resistance in *V. locusta* is assumed to be monogenic or oligogenic.

4. References

Crute, I.R. and Norwood, J.M. (1978) Incomplete specific resistance to *Bremia lactucae* in lettuce, *Annals of applied Biology* **89**, 467-474.

Judelson, H.S. and Michelmore, R.W. (1992) Temperature and genotype interactions in the expression of host resistance in lettuce downy mildew, *Physiological and Molecular Plant Pathology* **40**, 233-245.

Balass, M., Cohen, Y. and Bar-Joseph, M. (1993) Temperature-dependent resistance to downy mildew in muskmelon, *Physiological and Molecular Plant Pathology* **43**, 11-20.

OCCURRENCE AND VARIATION IN VIRULENCE OF *BREMIA LACTUCAE* IN NATURAL POPULATIONS OF *LACTUCA SERRIOLA*

A. LEBEDA
Department of Botany, Faculty of Science, Palacký University, Šlechtitelů 11, CZ-783 71 Olomouc-Holice, Czech Republic

1. Introduction

In a host-parasite interaction that follows a gene-for-gene relationship, the expression of resistance or susceptibility of the host to a particular parasite (pathogen) is conditional on the parasite genotype, and the degree of parasite virulence is conditional on the host genotype (Crute and Pink, 1996). The level of genetic variability for resistance and virulence can strongly influence the population dynamics and equilibrium of the interacting species (Simms, 1996).

Plant parasitic fungi and their populations include a very large and heterogeneous group of organisms that occupy positions of great importance in natural plant communities. Our knowledge concerning the structure and diversity of pathogen populations in wild pathosystem is poor (Burdon, 1993). The presence of race-specific resistance in natural plant-parasite associations is well established, however, its role and importance is the subject of some debate (Burdon *et al.*, 1996). Information on this subject for downy mildews and their host plants is rather limited (Lebeda and Schwinn, 1994; Drenth and Goodwin, 1999).

Only basic data are available on variation for resistance and race-specificity in wild *Lactuca* species (see Lebeda *et al.*, this volume) and variation for virulence in *Bremia lactucae* (lettuce downy mildew) isolates (Lebeda, 1984, 1986) originating from wild pathosystem *L. serriola* - *B. lactucae*. The genus *Lactuca* is distributed in temperate and warm regions of the northern hemisphere. In Europe this genus is represented by 17 species (Feráková, 1977). One of the most common species in Europe is *L. serriola* (prickly lettuce), which is a large spring annual herb growing on a wide range of soils and habitats, e.g. disturbed and ruderal (Feráková, 1977). Generally, *L. serriola* is a natural host of *B. lactucae* (Lebeda and Syrovátko, 1988). There is limited knowledge on geographical distribution and virulence variation of the pathogen in populations of *L. serriola*. The existence of physiological races of *B. lactucae* isolates originating from natural *L. serriola* f. *serriola* was first described in the former Czechoslovakia (Lebeda, 1984). Numerous *L. serriola* derived race-specific resistance genes have been used in lettuce breeding programmes (Crute, 1992). However, this resistance has been mostly "overcome" by new *B. lactucae* races (Lebeda and Schwinn, 1994; Lebeda, 1998).

The purpose of this study was to survey the occurrence and to characterize virulence variation in some recently (1995-1997) collected isolates of *B. lactucae* from natural populations of *L. serriola* f. *serriola* growing in the Czech Republic.

P.T.N. Spencer-Phillips et al. (eds.), Advances in Downy Mildew Research, 179–183.
© 2002 *Kluwer Academic Publishers. Printed in the Netherlands.*

2. Materials and Methods

The surveying and collecting trips were mostly organized in the end of August or during September (the period of peak disease development) in 1995 to 1997. Isolates were collected in the territory of the Czech Republic (mostly in the Moravia region) from populations of *L. serriola* f. *serriola* plants naturaly infected by *B. lactucae* (Table 1). One infected leaf with fresh sporophores or with yellow spots which produced sporulation after incubation was collected from each plant. The fungus was maintained on seedlings of *L. serriola* accession PI 273617 (Lebeda, 1986) susceptible to the most of the isolates. This accession also served as a susceptible control. Some isolates were also stored at –80 °C on infected seedlings. Inoculation, incubation and evaluation of sporulation intensity (0 = symptomless, 1 = isolated sporophores present, 2 = <50% and 3 = >50% cotyledon area covered with sporophores) followed procedures described previously (Lebeda and Boukema, 1991).

Virulence variation in *B. lactucae* isolates (Table 1) was determined by screening on a set of 41 *L. sativa* and *L. serriola* differential lines (accessions) with known race-specific resistance genes. Seed samples of differential lines originate from the collection at the Department of Botany, Palacký University and from the international collection of lettuce downy mildew differential series (Horticulture Research International, Genetic Resources Unit, Wellesbourne, UK). The presence of race-specific *Dm* genes or R-factors in this set has been established previously for the *Lactuca* spp. – *B. lactucae* interaction (Ilott *et al.*, 1989; Lebeda and Blok, 1991; Crute, 1992; Bonnier *et al.*, 1994; Witsenboer *et al.*, 1995). On the basis of virulence/avirulence patterns the presence of 23 virulence factors was demonstrated.

TABLE 1. Origins of *B. lactucae* isolates collected from *L. serriola* used in the virulence analyses

Year	Isolate of *B. lactucae*	Site of origin
1995	10/95	Rokytnice nad Vláři
1996	5/96	Křtiny
	6/96, 7/96	Březina
	9/96, 10/96	Ochoz u Brna
	11/96, 12/96, 13/96	Hajany
	17/96	Moravské Bránice
	20/96	Olomouc
	37/96	Olomouc-Holice
	25/96	Lutín
	27/96	Těšetice-Rataje
	28/96, 29/96	Topolany
1997	26/97	Ořechov u Brna
	42/97	Lechovice u Znojma

3. Results and Discussion

3.1. THE OCCURRENCE AND SEVERITY OF *B. LACTUCAE* INFECTION IN NATURAL POPULATIONS OF *L. SERRIOLA*

In 1995, 56 natural populations of *L. serriola* were surveyed on 50 sites. In seven sites (14%) the occurrence of infection was observed. During September 1996 77 populations of *L. serriola* on 44 sites were surveyed, and infection was observed in 57% of these. Substantial variation in severity of infection (low/1/, medium/2/, serious/3/) between different sites and/or populations were recognized: $1 = 27.2\%$, $2 = 36.4\%$, $3 = 36.4\%$. From 24 sites visited in 1997 infection occurred on 79% of the populations. The severity of infection was recorded as: $1 = 52.6\%$, $2 = 21.1\%$, $3 = 26.3\%$.

The heterogeneity in expression of macroscopic symptoms due to host response on naturally infected leaves of *L. serriola* was an interesting phenomenon. At least four different types of symptoms were recognized. The first was characterized by the occurrence of a few discrete chlorotic spots between the veins of the leaves but little or no other host response. The second type was characterized by moderate to large amounts of chlorotic spots with limited necrosis. Some plants or populations exhibited a moderate to frequent appearance of necrotic spots. However, in other populations a very different type of infection phenotype was observed. These plants had more or less completely infected leaves with profuse sporulation but without any discrete necrotic responses. Bevan *et al.* (1993) also found several forms of expression of resistance/susceptibility in *Senecio vulgaris* to *Erysiphe fischeri.*

3.2. VIRULENCE ANALYSES AND COMPARISON OF V-PHENOTYPES

The occurrence of 23 specific parasite virulence factors in the set of *B. lactucae* isolates was expressed in terms of v-factor frequencies (Lebeda, 1982). The data for the whole collecting period and all sites are summarised in Table 2. There are substantial differences in frequencies between different v-factors. Only two v-factors (v7 and v16), both matching the *Dm* genes derived from *L.serriola* (Ilott *et al.*, 1989), were present with a frequency of 1.00. Some other factors (v15, v23-30) occurred at relatively high frequency ranging between 0.61 to 0.83. These factors also match resistance genes (R-factors) located in *L. serriola* accessions (Bonnier *et al.*, 1994). In contrast v18 was not recorded at all and the remaining v-factors occurred at relatively very low frequencies.

Evaluation of variation between individual isolates showed that seven different v-phenotypes (set of v-factors in individual isolate) may be distinguished. In most of the isolates, ten to twelve v-factors were recorded. Only one isolate was characterized by a low number of v-factors. The most prevalent v-phenotype (50% of isolates) was characterized by eleven v-factors which correspond to resistance genes originating from *L. serriola*. In 15 isolates limited differences in v-phenotype structure were recorded (Table 3). Recent data are in agreement with previous results which showed that *B. lactucae* isolates originating from natural *L. serriola* populations were characterized by rather simple v-phenotypes in comparison with isolates obtained from the crop pathosystem (Lebeda, 1984, 1990).

TABLE 2. Frequency of virulence factors in 18 *B. lactucae* isolates originating from *L. serriola*

Frequency	Virulence factor (s)
0.00	*v18* **
0.05	v14 *
0.11	v3, *v6*, v10, *v11*, v13
0.17	v1, v2, v4, v12
0.33	*v5/8*
0.61	**v23 ***
0.83	**v15, v24, v25, v26, v27, v28, v29, v30**
1.00	**v7, v16**

* = v-factors matching *Dm* genes (R-factors) in *L. sativa* cultivars
** = v-factors matching *Dm* genes (R-factors) in commercial *L. sativa* cultivars,
     however derived from *L. serriola*
*** = v-factors matching *Dm* genes (R-factors) in *L. serriola*

These isolates were later recognized as also carrying v-factors matching newly described R-factors in *L. serriola* (Bonnier *et al.*, 1994). In this group of isolates frequent expression of very limited sporulation (only a few sporophores) on seedlings with the gene *Dm*11 which originated from *L. serriola* was also found. These results support previously published data (Lebeda, 1989). However, three isolates (10/95, 26/97, 42/97, see Table 3) were completely different and more like "*L. sativa* type" isolates because they have no v-factors matching the prevalent *Dm* genes originating from *L. serriola*. The isolates 26/97 and 42/97 have very similar v-phenotype to the race NL16 (Lebeda and Blok, 1991). However, a similar phenotype has been recorded previously in two *B. lactucae* isolates originating from *L. serriola* plants (Lebeda, 1984, 1990; Bonnier *et al.*, 1994). Recently some *B. lactucae* isolates with virulence to cultivated lettuce cv. Titan (resistance derived from *L. saligna*) and hybrid *L. serriola* x *L. sativa* (line CS-RL) were found in wild pathosystem (Petrželová and Lebeda, 2000).

TABLE 3. Variation in the complexity of virulence phenotypes between
     B. lactucae isolates

Isolate of *B. lactucae*	Virulence phenotype (v-factors)
10/95	v1*, v2, v4, v7, v12, **v16 ***
5/96, 11/96	**v7, v15, v16, v24, v25, v26, v27, v28, v29, v30**
9/96	*v5/8***, **v7, v15, v16, v24, v25, v26, v27, v28, v29, v30**
6/96, 10/96, 13/96, 17/96, 20/96, 27/96, 28/96, 29/96, 37/96	**v7, v15, v16, v23, v24, v25, v26, v27, v28, v29, v30**
26/97, 42/97	v1, v2, v3, v4, *v5/8*, *v6*, v7, v10, *v11*, v12, v13, **v16**
12/96	*v5/8*, **v7**, v14, **v15, v16, v24, v25, v26, v27, v28, v29, v30**
7/96, 25/96	*v5/8*, **v7, v15, v16, v23, v24, v25, v26, v27, v28, v29, v30**

*, **, ***, see Table 2.

4. Acknowledgements

The author evaluated the results during a fellowship under the OECD Co-operative Research Programme: Biological Resource Management for Sustainable Agricultural Systems at Horticulture Research International (Wellesbourne, UK). This work was partly supported by the grant MSM153100010.

5. References

Bevan, J.R., Clarke, D.D. and Crute, I.R. (1993) Resistance to *Erysiphe fischeri* in two populations of *Senecio vulgaris, Plant Pathology* **42**, 636-646.

Bonnier, F.J.M., Reinink, K. and Groenwold, R. (1994) Genetic analysis of *Lactuca* accessions with new major gene resistance to lettuce downy mildew, *Phytopathology* **84**, 462-468.

Burdon, J.J. (1993) The structure of pathogen populations in natural plant communities, *Annual Review of Phytopathology* **31**, 305-323.

Burdon, J.J., Wenström, A., Elmquist, T. and Kirby, G.C. (1996) The role of race specific resistance in natural plant populations, *Oikos* **76**, 411-416.

Crute, I.R. (1992) The role of resistance breeding in the integrated control of downy mildew (*Bremia lactucae*) in protected lettuce, *Euphytica* **63**, 95-102.

Crute, I.R. and Pink, D.A.C. (1996) Genetics and utilization of pathogen resistance in plants, *The Plant Cell* **8**, 747-1755.

Drenth, A. and Goodwin, S.B. (1999) Population structure of Oomycetes, in J.J. Worrall (ed) *Structure and Dynamics of Fungal Populations*, Kluwer Academic Publishers, Dordrecht, pp. 195-224.

Feráková, V. (1977) *The Genus Lactuca L. in Europe*, Universita Komenského, Bratislava.

Ilott, T.W., Hulbert, S.H. and Michelmore, R.W. (1989) Genetic analysis of the gene-for-gene interaction between lettuce (*Lactuca sativa*) and *Bremia lactucae, Phytopathology* **79**, 888-898.

Lebeda, A. (1982) Population genetic aspects in the study of phytopathogenic fungi, *Acta Phytopathologica Academiae Scientiarum Hungaricae* **17**, 215-219.

Lebeda, A. (1984) Response of differential cultivars of *Lactuca sativa* to *Bremia lactucae* isolates from *Lactuca serriola,Transactions of the British Mycological Society* **83**, 491-494.

Lebeda, A. (1986) Specificity of interactions between wild *Lactuca* spp. and *Bremia lactucae* isolates from *Lactuca serriola, Journal of Phytopathology* **117**, 54-64.

Lebeda, A. (1989) Response of lettuce cultivars carrying the resistance gene *Dm*11 to isolates of *Bremia lactucae* from *Lactuca serriola, Plant Breeding* **102**, 311-316.

Lebeda, A. (1990) Identification of a new race-specific resistance factor in lettuce (*Lactuca sativa*) cultivars resistant to *Bremia lactucae, Archiv für Züchtungsforschung* Berlin **20**, 103-108.

Lebeda, A. (1998) Virulence variation in lettuce downy mildew (*Bremia lactucae*) and effectivity of race-specific resistance genes in lettuce, in M. Recnik and J. Verbic (eds), *Proceedings of the Conference "Agriculture and Environment"*, Bled (Slovenia), pp. 213-217.

Lebeda, A. and Blok, I. (1991) Race-specific resistance genes to *Bremia lactucae* Regel in new Czechoslovak lettuce cultivars and location of resistance in a *Lactuca serriola* x *Lactuca sativa* hybrid, *Archiv für Phytopathologie und Pflanzenschutz* **27**, 65-72.

Lebeda, A. and Boukema, I.W. (1991) Further investigation of specificity of interactions between wild *Lactuca* species and *Bremia lactucae* isolates from *Lactuca serriola, Journal of Phytopathology* **133**, 57-64.

Lebeda, A. and Schwinn, F.J. (1994) The downy mildews – an overview of recent research progress, *Journal of Plant Diseases and Protection* **101**, 225-254.

Lebeda, A. and Syrovátko, P. (1988) Specificity of *Bremia lactucae* isolates from *Lactuca sativa* and some Asteraceae plants, *Acta Phytopathologica et Entomologica Hungarica* **23**, 39-48.

Petrželová, I. and Lebeda, A. (2000) New knowledge on the occurrence and virulence variation of *Bremia lactucae* in natural populations of *Lactuca serriola*, in *Proceedings of the XVth Czech and Slovak Plant Protection Conference in Brno*, September 12.-14., 2000, pp. 197-198.

Simms, E.L. (1996) The evolutionary genetics of plant-pathogen systems, *BioScience* **46**, 136-145.

Witsenboer, H., Kesseli, R.V., Fortin, M.G., Stanghellini, M. and Michelmore, R.W. (1995) Sources and genetic structure of a cluster of genes for resistance to three pathogens in lettuce, *Theoretical and Applied Genetics* **91**, 178-188.

OUTCROSSING OF TWO HOMOTHALLIC ISOLATES OF *PERONOSPORA PARASITICA* AND SEGREGATION OF AVIRULENCE MATCHING SIX RESISTANCE LOCI IN *ARABIDOPSIS THALIANA*

N.D. GUNN, J. BYRNE and E.B. HOLUB
Horticulture Research International, Wellesbourne, Warwick, CV35 9EF, UK

1. Introduction

Peronospora parasitica (Pers. ex Fr.) Fr. (downy mildew) in *Arabidopsis thaliana* L. Heyn. has become an important eukaryotic parasite for investigating naturally variable parasite recognition genes (so-called R-genes) and other defence-related genes identified downstream by artificial mutation. These studies are contributing to an understanding of the molecular basis of disease resistance in plants (see Mauch-Mani, this volume). More than twenty *RPP* specificities (**R**ecognition of *P. parasitica*) have been identified in *Arabidopsis*, and more than twelve downstream mutations have been characterised which at least partially influence the expression of downy mildew resistance (McDowell *et al.*, 2000; Shapiro, 2000; Holub, 2001). Molecular genetic analyses of *P. parasitica* will be essential for advancing this research by revealing parasite gene products responsible for triggering the host defence response, as well as compatibility genes required for a susceptible interaction.

Like its wild host, *P. parasitica* from *Arabidopsis* is diploid and is self-fertile (homothallic) in its vegetative stage. Studying avirulence in this oomycete presents a further challenge because of its obligate biotrophic nature. Two isolates, *Emoy2* and *Maks9*, were selected for outcrossing which between them carry **A**rabidopsis **t**haliana **r**ecognised (*ATR*) avirulence genes that correspond with six known *RPP* genes, including five that have been molecularly characterised as encoding pathogen receptor-like genes: *RPP1* (Chromosome III), *RPP4* (Chr. IV), *RPP5* (Chr. IV), *RPP8* (Chr. V) and RPP13 (Chr. III) (see review by Holub, 2001). A **c**o-dominant **a**mplified **p**olymorphic **s**equence (CAPS) marker (Konieczny and Ausubel, 1993) and pathotype characterisation using an *RPP*-gene differential set of inbred *Arabidopsis* lines were used to identify an F_1 isolate and for subsequent segregation analysis in an F_2 population.

2. Materials and Methods

P. parasitica was cultured and inoculum prepared as described by Dangl *et al.* (1992). Homozygosity of specific *ATR* loci was determined by inoculating host accessions containing combinations of *RPP* genes with progeny of the parental isolates *Maks9* and

185

P.T.N. Spencer-Phillips et al. (eds.), Advances in Downy Mildew Research, 185–188.
© 2002 *Kluwer Academic Publishers. Printed in the Netherlands.*

Emoy2. The F_1 was produced by inoculating 10-day old seedlings of an accession compatible to both *Maks9* and *Emoy2*. Two weeks after inoculation, the leaf tissue was harvested and dried. Oospores were left to mature for one month before asexual progeny were recovered.

Dried leaf tissue containing the mature oospores was ground to a fine powder using a pestle and mortar. Oospore inoculum was sprinkled onto the surface of compost in 4 cm pots and seed of a susceptible genotype sown on top. Pots were watered and stored for one week at 4 °C to break any remaining seed dormancy. Seed was incubated in sealed propagators at 18-20 °C with a 10 h photoperiod and a photon flux density of 150-250 μE m^{-2} s^{-1}. Seedlings were inspected daily for asexual conidiosporangia from 5 days post-incubation. Individual infected seedlings bearing conidiosporangiophores were harvested and the asexual inoculum was bulked on susceptible seedlings prior to testing.

Putative F_1 progeny were single-spored from conidiosporangia and confirmed as hybrids using host accessions differential for specific *RPP* genes and a PCR-based CAPS marker. The F_2 population was derived from selfing the F_1 and recovering progeny as described. The F_2 progeny were single-spored prior to testing on host genotypes differential for the six *RPP* genes complemented by putative *ATR* genes segregating among the F_2 progeny.

Amplified fragment length polymorphisms (AFLP) (Vos *et al.*, 1995) were used as an efficient PCR-based method for generating large numbers of molecular markers for *P. parasitica*. Linkages between molecular markers are being assembled into a genetic map using Joinmap® Version 2 (Stam, 1993) computer programme, and phenotypic data for avirulence segregation has been included to identify linkage groups that contain each predicted ATR locus.

3. Results and Discussion

Fifty-two asexual progeny were isolated from the oospore population; 15 progeny exhibited a hybrid phenotype, 14 exhibited an *Emoy2*-like phenotype and 23 exhibited a *Maks9*-like phenotype following inoculation of the host differentials. A PCR-based CAPS marker was produced which detected a dimorphism between the parent isolates, and all parental bands were present in the putative F_1 progeny. One of these was selected as the parental for F_2 production. Initially, eleven F_2 progeny were analysed for dimorphism using the CAPS marker, which segregated in the F_2 with a ratio of 2:6:3 (X^2 = 0.27, p = 0.60). The results described here provide the first demonstration that genetic analyses with two homothallic *P. parasitica* isolates is possible in *Arabidopsis*.

A larger mapping population consisting of ninety-four F_2 progeny was generated; and each F_2 was used to inoculate a set of *Arabidopsis* lines containing different homozygous combinations of known *RPP* genes. The resulting phenotypes were used to predict the combination of *ATR* alleles inherited by each F_2 isolate. Segregation data for *ATR1*, *ATR4*, *ATR5* and *ATR13* provided evidence of avirulence controlled in each case by a semi-dominant allele at a single locus. Avirulence at each of these loci segregated independently, demonstrating that they are unlinked, including *ATR4* and *ATR5* which interestingly correspond with different resistance alleles at the *RPP4/5* locus in *Arabidopsis*. *ATR8* appeared to be skewed, giving a ratio of 9:1 (avirulent:virulent), which could be explained either by *ATR8* being linked to a trait that affects oospore maturation and/or germination, or that *ATR8* is controlled by alleles at two closely-linked loci. Segregation of *ATR21* was not possible because the progeny exhibited a continuous distribution of sporulation phenotypes (i.e., discrete avirulent or virulent classes could not be distinguished).

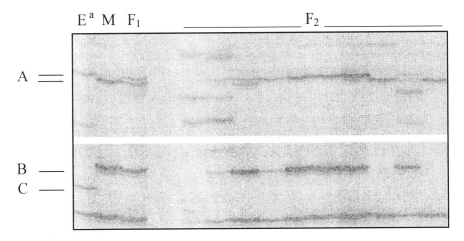

[a] A single primer pair (MseI-CAT/EcoRI-AA) incorporated radiolabelled ^{33}P in amplified bands in *Maks9* (M), *Emoy2* (E), F$_1$ hybrid (F$_1$) and 10 F$_2$ progeny (F$_2$). Both panels from the same AFLP gel to show examples of various banding patterns (see text for A,B and C).

Figure 1. AFLP autoradiograph of parental *Peronospora parasitica* isolates *Emoy2* (E) and *Maks9* (M), the F$_1$ hybrid (F$_1$) and 10 F$_2$ progeny

Our aim for generating a genetic map in *P. parasitica* has been to produce an AFLP map of marker density similar to one published for *Phytophthora infestans* consisting of 183 markers that segregated among 73 individuals (Van der Lee *et al.*, 1997). The genome size estimated for these two closely related pathogens is ca. 250 and 80 Mb for *Ph. infestans* and *P. parasitica* (Rehmany *et al.*, 2000), respectively. Figure 1 shows the AFLP profiles obtained from the two parent isolates *Emoy2*, *Maks9*, the F$_1$ hybrid and ten F$_2$ progeny. Numerous bands were dimorphic (present or absent) between the parental isolates and present in the F$_1$, providing useful markers for detecting segregation among F$_2$ progeny and for generating a map (markers A and B in Figure 1). However, a large number of parental bands were absent from the F$_1$, suggesting that these marker loci are heterozygous in the donor parent but were not inherited in F$_1$ used to generate the F$_2$ mapping population (example C in Figure 1).

A preliminary genetic map has already been assembled from the first 32 F$_2$ individuals and 130 AFLP markers. Incorporation of avirulence data identified several markers that define map intervals for *ATR1* and *ATR8*, each in a different linkage group.

Only a single marker was identified that was linked to *ATR13*. Avirulence data for *ATR4* and *ATR5* was not available for this preliminary analysis.

The clear segregation of several single ATR loci, with linkage to AFLP markers, indicates that these are five suitable targets for map-based cloning of avirulence genes from *P. parasitica*. We are currently working in collaboration with Anne Rehmany and Jim Beynon (HRI) to define a fine map interval for *ATR1*, *ATR8* and *ATR13* using bulked-segregant AFLP analysis as a prelude to gene isolation.

4. References

Dangl, J.L., Holub, E.B., Debener, T., Lehnackers, H., Ritter, C. and Crute, I.R. (1992) Genetic definition of loci involved in *Arabidopsis*-pathogen interactions, in C. Koncz, N-H. Chua and J. Schell (eds), *Methods in Arabidopsis Research*, World Scientific Publishing, pp. 393-418.

Holub, E.B. (2001) The arms race is ancient history in *Arabidopsis*, the wildflower, *Nature Reviews Genetics* **2**, 516-527.

Konieczny, A. and Ausubel, M. (1993) A procedure for mapping *Arabidopsis* mutations using co-dominant ecotype-specific PCR-based markers, *The Plant Journal* **4**, 403-410.

McDowell, J.M., Cuzick, A., Can, C., Beynon, J., Dangl, J.L. and Holub, E.B. (2000) Downy mildew (*Peronospora parasitica*) resistance genes in *Arabidopsis* vary in functional requirements for NDR1, EDS1, NPR1 and salicylic acid accumulation, *Plant Journal* **22**, 523-529.

Rehmany, A., Lynn, J.R., Tör, M., Holub, E.B. and Beynon, J.L. (2000) A comparision of *Peronospora parasitica* (downy mildew) isolates from *Arabidopsis thaliana* and *Brassica oleracea* using amplified fragment length polymorphism and internal transcribed spacer 1 sequence analyses, *Fungal Genetics and Biology* **30**, 95-103.

Shapiro, A.D. (2000) Using *Arabidopsis* mutants to determine disease signalling pathways, *Canadian Journal of Plant Pathology* **22**, 199-216.

Stam, P. (1993) Construction of integrated genetic linkage maps by means of a new computer package: JoinMap, *Plant Journal* **3**, 739-744.

Van der Lee, T., De Witte, I., Drenth, A., Alfonso, C. and Govers, F. (1997) AFLP linkage map of the oomycete *Phytophthora infestans*, *Fungal Genetics and Biology* **21**, 278-291.

Vos, P., Hogers, R., Bleeker, M., Reijans, M., Van der Lee, T., Hornes, M., Frijters, A., Pot, J., Peleman, J., Kuiper, M. and Zabeau, M. (1995) AFLP: a new technique for DNA fingerprinting, *Nucleic Acids Research* **23**, 4407-4414.

EPIDEMIOLOGY AND CONTROL OF PEARL MILLET DOWNY MILDEW, *SCLEROSPORA GRAMINICOLA*, IN SOUTHWEST NIGER

E. GILIJAMSE[1] and M.J. JEGER[2]
[1] *Rijk Zwaan Zaadteelt en Zaadhandel b.v., P.O. Box 40, 2678 ZG De Lier, the Netherlands*
[2] *Imperial College at Wye, University of London, Wye, Ashford, Kent, TN25 5AH, United Kingdom*

1. Introduction

Pearl millet, *Pennisetum glaucum*, is one of the most important staple crops, both for human consumption and animal fodder, in the drier regions of the Sahel area. Due to droughts, low soil fertility, pests and diseases, yields are low. One of the main pearl millet diseases is downy mildew, caused by the oomycete *Sclerospora graminicola*. The pathogen causes chlorosis, starting at the base of a leaf ('half-leaf' or 'partial leaf' symptom), but gradually covering a greater proportion of leaves until the entire leaf is chlorotic (Williams, 1984). Severely infected plants are barren and produce malformed heads, so-called 'green ears' or 'crazy tops'. Crop losses of up to 50% have been reported (Werder and Manzo, 1992), although in general more conservative estimates are between 0 and 20% (Mbaye, 1992; Gilijamse *et al.*, 1997; Jeger *et al.* 1998).

Initial infection of plants is caused by oospores which are thick-walled resting spores, produced at the end of the growing season when conditions are unfavourable for crop growth. Oospores are incorporated in the soil together with crop debris providing a continuous inoculum source in areas where no crop rotation is practised. During the growing season, secondary infection is possible through asexually produced sporangia when relative humidity exceeds 95%. In the drier regions of the Sahel, oospores are probably not only the initial but also the main source of infection.

The role of oospores in the epidemiology of pearl millet downy mildew was studied in the sahelian parts of southwest Niger. Possibilities for downy mildew control, that may be implemented in low-input agriculture, were also studied.

2. Materials and Methods

2.1. EPIDEMIOLOGY

2.1.1. *Oospore Production*
Production of oospores was studied in different parts of pearl millet plants at two periods in time (Gilijamse, 1997). One partially resistant (HKP) and two susceptible (NHB3 and

P.T.N. Spencer-Phillips et al. (eds.), Advances in Downy Mildew Research, 189–193.
© 2002 *Kluwer Academic Publishers. Printed in the Netherlands.*

7042) cultivars were studied. Inoculation took place by mixing the seed with ground infected pearl millet leaves containing approximately 5 x 10^5 oospores per gram dry weight. At 60 and 90 days after sowing (d.a.s.) plants were harvested, dried and different plant parts were ground. The number of oospores was determined in the roots, stems, leaves and ears of the three cultivars.

2.1.2. *Role of Oospores in Downy Mildew Infection*

In two regions in the southwest of Niger around the villages Dosso and Gaya downy mildew incidence and severity were monitored during the growing season following the method of Williams (1984) modified by Mbaye (1994). In each region 10 fields were visited at 30 and 60 d.a.s. and at crop maturity (90 d.a.s.). At the beginning of the rainy season soil samples were taken from the upper layer of each field and analysed for several soil characteristics. Oospore density was determined in each soil sample through extraction by centrifugation in a sucrose suspension (Van der Gaag and Frinking, 1997).

2.2. CONTROL

Several control methods have been investigated for low-input pearl millet cultivation. Crop rotation, deep tillage, changing planting date and fertilization, however, are all of limited use in the drier parts of the Sahel. No or limited resources are available for deep tillage and fertilization whereas crop rotation and changing the planting date are not feasible because of the risk of harvest failure. Traditional control methods using herbal extracts are practised sometimes (Gilijamse *et al.*, 1997). The most common control method is roguing of diseased plants (Gilijamse *et al.*, 1997). According to Mbaye (1994) this method results in the reduction of asexual spores of *S. graminicola* when it is done within the first month after sowing. No reports exist on the long-term effects. Roguing is also rather time-consuming and therefore not practised sufficiently often to control the disease adequately.

The use of chemical fungicides as a seed treatment might be an option for control of pearl millet downy mildew. The amount of fungicide applied is low and a limited input of labour is necessary, making it a rather inexpensive control method. In the southwest of Niger, downy mildew control by seed treatment was investigated using Apron Plus 50 DS (Ciba Geigy) which consists of two fungicides (metalaxyl and carboxine) and one insecticide (furathiocarb). Apron Plus was applied at the recommended dose of 10g per kg of seed. A local pearl millet variety was used in an experiment in five replications. Seedling emergence, disease incidence, severity and yield were assessed.

3. Results and Discussion

3.1. EPIDEMIOLOGY

3.1.1. *Oospore Production*

At 60 d.a.s. the number of oospores ranged from about 1 x 10^3 oospores per gram dry weight in roots and stems of the two susceptible cultivars to about 1 x 10^4 oospores in ears and leaves. No oospores were found in any part of the partially resistant cultivar. At

90 d.a.s. oospores were found in all plant parts. There were no differences between the three cultivars. The number of oospores in roots and stems differed significantly from that in leaves and ears, ranging from 1×10^3 oospores per gram in roots and stems to 2.3×10^4 in leaves and ears. In ears collected from farmers' fields, up to 5×10^5 oospores per gram were found.

These data largely correspond with those of others although Michelmore *et al.* (1982) found oospores in seedlings as early as 13 days after sowing. Since *S. graminicola* is heterothallic, the co-existence of two mating types in the same zone of host tissue is the primary determinant for oospore production. This is one explanation for why oospores are not always or only at later times present in diseased host tissue. Williams (1984) suggested that oospore production is related to environmental factors and physiological changes in the maturing plant. In the sahelian area with a dry, crop-free period, the formation of thick-walled oospores at the end of the growing season is the only strategy for the pathogen to survive.

In pearl millet cultivars with a certain level of (partial) resistance, oospore production might be restricted. It was found that fungal structures are encased by thin membranous structures which deprive the fungus of essential nutrients (Sharada *et al.*, 1995). However in our experiment there were no differences in oospore production between the three cultivars at the end of the growing season.

3.1.2. *Role of Oospores in Downy Mildew Infection*

In the Dosso region the oospore density ranged from 26 to 133 oospores per gram soil whereas around Gaya it ranged from 74 to 768 oospores per gram soil. Downy mildew incidence (percentage of diseased plants) ranged from 0 to 23.2% in Dosso and from 7.6 to 42.0% in Gaya. Disease severity (degree of infection of individual plants systemically infected) showed a similar pattern ranging from 0 to 12.9% in Dosso and from 4.3 to 24.8% in Gaya. Oospore density in the soil was highly correlated ($r = 0.83$, 18 df) with disease incidence at 90 d.a.s. During the crop-free period no alternative hosts are present for *S. graminicola* to survive on and therefore no sporangia are present to infect pearl millet at the beginning of the cropping season, thus strenghtening the view that oospores are responsible for initial infection by *S. graminicola*. Since *S. graminicola* is the only downy mildew pathogen occurring on pearl millet in the Sahel, it was assumed that the oospores found were from *S. graminicola*.

Soil characteristics were found to play a role in disease development. Organic C content, $pH(H_2O)$ and fraction of loam were correlated significantly with disease incidence at 90 d.a.s. ($r = 0.70$, 0.47 and 0.46 respectively, 18 df). No significant correlation was found between disease incidence at 90 d.a.s. and the fraction clay and sand ($r = 0.39$ and 0.13 respectively, 18 df).

This is the first time a relationship has been shown between oospore density in the soil and pearl millet downy mildew incidence. For sorghum downy mildew, Schuh *et al.* (1988) found that the spatial pattern of oospores in soil was comparable to the pattern of infected sorghum plants. However, Pratt and Janke (1978) did not find a significant correlation between oospore density of this pathogen in the soil and disease incidence ($r = 0.33$, 16 df). In both studies the number of oospores per gram soil was low (0.7-56 and 8-95 respectively) compared to this survey (26-768). Possibly with such low numbers, disease incidence is much more influenced by secondary infection due to sporangia and the percentage of viable oospores in the soil (Shetty, 1987).

It is probable that soil characteristics determine to a large extent the percentage viability of oospores and hence indirectly disease incidence. Other studies indicated the relation between soil characteristics and disease incidence although results are sometimes contradictory. Disease incidence of sorghum downy mildew on sorghum was higher on sandy soils than on clay soils according to studies by Pratt and Janke (1978) and Schuh *et al.* (1987). However, Van der Westhuizen (1977) found the opposite for sorghum downy mildew on maize. None of the authors mentioned other soil characteristics such as pH and organic C content. The relationship between soil characteristics and epidemiology of downy mildew of pearl millet and sorghum certainly needs further investigation in the development of new control strategies.

3.2. CONTROL

Disease incidence and severity ranged from 11.1 to 25% and from 6.9 to 14.6% respectively for treated seeds, and from 5.6 to 27.8% and from 4.2 to 17.4% respectively for non-treated control seeds. Statistical analysis showed no significant differences between the treatment and the control. No differences were found for seedling emergence and yield. These results largely correspond with preliminary data collected during the two previous years of the study. Under the conditions of the experiment it seems that seed treatment using the chemical fungicide Apron Plus 50 DS does not result in an efficient control of pearl millet downy mildew.

Metalaxyl however, has been shown to be effective for seed treatment followed by foliar application with metalaxyl (Shishupala *et al.*, 1990). Combined applications for control of downy mildew is probably more important for pearl millet compared to sorghum and maize, since pearl millet produces tillers that keep the crop susceptible for up to 60 days. However, foliar application of metalaxyl is again too expensive in low-input agriculture. Additionally, the fungicide might easily be washed off the leaves during heavy showers which occur regularly in the rainy season in the Sahel.

The most effective strategy to control or at least reduce pearl millet downy mildew in the Sahel is probably through the production of resistant cultivars that meet the characteristics (i.e. taste, drought resistance and low fertiliser need) of the traditional varieties, but these are not currently available.

4. References

Gilijamse, E. (1997) Downy mildew on pearl millet in south Niger, Joint report of WAU Department of Phytopathology, IAH Larenstein and DFPV Niamey/Niger.

Gilijamse, E., Frinking, H.D. and Jeger, M.J. (1997) Occurrence and epidemiology of pearl millet downy mildew, *Sclerospora graminicola*, in southwest Niger, *International Journal of Pest Management* **43**, 279-283.

Jeger, M.J., Gilijamse, E., Bock, C.H. and Frinking, H.D. (1998) The epidemiology, variability and control of the downy mildews of pearl millet and sorghum, with particular reference to Africa, *Plant Pathology* **47**, 544-569.

Mbaye, D.F. (1992) Les maladies du mil au Sahel: Etat des connaissances et propositions de lutte, in Institut du Sahel (ed), *La lutte intégrée contre les ennemis des cultures vivrières dans le Sahel*, CILLS/INSAH, Bamako, Mali, pp. 42-63.

Mbaye, D.F. (1994) Une étude du pathosysteme *Pennisetum glaucum-Sclerospora graminicola*, Application à la gestion du mildiou du mil au Senegal, Montpellier, France: Ecole nationale supérieure agronomique de Montpellier, PhD thesis.

Michelmore, R.W., Pawar, M.N. and Williams, R.J. (1982) Heterothallism in *Sclerospora graminicola*, *Phytopathology* **72**, 1368-1372.

Pratt, R.G. and Janke, G.D. (1978) Oospores of *Sclerospora sorghi* in soils of south Texas and their relationships to the incidence of downy mildew in grain sorghum, *Phytopathology* **68**, 1600-1605.

Schuh, W., Jeger, M.J. and Frederiksen, R.A. (1987) The influence of soil temperature, soil moisture, soil texture and inoculum density on the incidence of sorghum downy mildew, *Phytopathology* **77**, 125-128.

Schuh, W., Jeger, M.J. and Frederiksen, R.A. (1988) Comparisons of spatial patterns of oospores of *Peronosclerospora sorghi* in the soil and of sorghum plants with systemic downy mildew, *Phytopathology* **78**, 432-434.

Sharada, M.S., Shetty, S.A. and Shetty, H.S. (1995) Infection processes of *Sclerospora graminicola* on *Pennisetum glaucum* lines resistant and susceptible to downy mildew, *Mycological Research* **99**, 317-322.

Shetty, H.S. (1987) Biology and epidemiology of downy mildew of pearl millet, in ICRISAT (ed), *Proceedings of the International Pearl Millet Workshop*, Patancheru, India, pp. 147-160.

Shishupala, S., Lokesha, S. and Shetty, H.S. (1990) Efficacy of metalaxyl and oxadixyl formulations against pearl millet downy mildew, *Advances in Plant Sciences* **3**, 28-33.

Van der Gaag, D.J. and Frinking, H.D. (1997) Extraction of oospores of *Peronospora viciae* from soil, *Plant Pathology* **46**, 675-679.

Van der Westhuizen, G.C.A. (1977) Downy mildew fungi of maize and sorghum in South Africa, *Phytophylactica* **9**, 83-89.

Werder, J. and Manzo, S.K. (1992) Pearl millet diseases in Western Africa, in W.A.J. de Milliano, R.A. Frederiksen and G.D. Bengston (eds), *Sorghum and Millets Diseases:A second World Review*, ICRISAT, Patancheru, India, pp. 109-114.

Williams, R.J. (1984) Downy mildew of tropical cereals, *Advances in Plant Pathology* **3**, 1-103.

EFFECT OF AZOXYSTROBIN ON THE OOSPORES
OF *PLASMOPARA VITICOLA*

A. VERCESI[1], A. VAVASSORI [1], F. FAORO[2] AND M. BISIACH[1]

[1]*Istituto di Patologia Vegetale, Università di Milano and* [2]*Centro Miglioramento Sanitario Colture Agrarie, Consiglio Nazionale delle Ricerche, Milano, Italy*

1. Introduction

Azoxystrobin, a diffusant fungicide belonging to the β-methoxyacrylate group, showing a broad spectrum of activity (Godwin *et al.*, 1992; Gisi, this volume) exerts its inhibitory activity against many stages of the asexual life cycle of *Plasmopara viticola* (Berk. & M. A. Curtis) Berl. & De Toni, the causal agent of grapevine downy mildew (Godwin *et al.*, 1997). No information is available on the effect of azoxystrobin on the sexual structures of the pathogen. In temperate climates oospore differentiation, occurring from July until the leaf fall, provides the overwintering inoculum of the pathogen (Zachos, 1959) and their germination is an essential prerequisite for the occurrence of primary infections (Vercesi, 1995). Thus a possible reduction in the oospore number and germinability would affect both the date of appearance and the frequency of primary foci, and consequently the epidemic build-up.

The aim of the present work was to evaluate the effect of azoxystrobin on the production and germinability of oospores of *P. viticola*.

2. Materials and methods

2.1 SOURCE OF *P. VITICOLA* OOSPORES

A vineyard of Riesling Italico located at Vescovera (PV), Northern Italy, was divided into two plots of 0.5 ha each. In 1996 and 1997, two treatments with azoxystrobin (25 g ai/hl) were applied to the first plot respectively at the end of August and at the middle of September. No treatment was applied to the second plot.

Infected areas of leaves randomly collected in both plots were examined by light microscopy at a magnification of 40-80x and the zones showing the presence of the oospores carefully excised. Two series of samples, the first one containing leaf fragments rich in oospores excised from the treated leaves (series 1) and the second one analogous leaf areas obtained from the untreated ones (series 2), stored in nylon bags (pore size = 100 μm) were overwintered at the soil surface of the vineyard of the Institute of Plant Pathology in Milan. The area of the 20 leaf fragments contained in each sample was assessed by image analysis (Global Lab. Data Translation, USA).

P.T.N. Spencer-Phillips et al. (eds.), Advances in Downy Mildew Research, 195–199.
© 2002 *Kluwer Academic Publishers. Printed in the Netherlands.*

2.2 ASSESSMENT OF OOSPORE DENSITY AND GERMINABILITY

Oospore density and germinability were evaluated at the end of April 1997 on ten samples for each of the two series and at 20-15 day intervals from October 1997 to June 1998, on a sample from both the treated and untreated leaves for every experimental assay. The oospore suspensions were prepared according to the method described by Vercesi and coworkers (1999). The oospore density of each suspension was assessed for five replicates and the number of oospores/cm^2 of leaf fragments calculated. Oospore germination was assessed at 20°C for 400-500 oospores/sample, distributed in five Petri dishes, containing sterilized water agar (Agar Noble, Difco, 1% w/v). Germination percentages (G) of each sample have been calculated as the average value of G observed in each of the five Petri dishes. Analysis of variance was carried out on both oospore density and germination percentages.

2.3 ELECTRON MICROSCOPY

Each month from October to May, oospores of *P. viticola* in leaf samples from both series of samples were processed for Transmission Electron Microscopy (TEM), following the procedure of Vercesi and coworkers (1999).

3. Results and Discussion

3.1 OOSPORE DENSITY AND GERMINABILITY

Data on average oospore density and germination (G, seen as sporangium formation) assessed in 1997 and 1998 on the samples of the two series, show a significant decrease of both parameters in samples obtained from the treated leaves in comparison with the untreated ones (Table 1). Inoculum potential (IP), expressed as the number of oospores able to germinate/cm^2, is reduced in the samples derived from the leaves treated with AZ by 91.88% in 1997 and 95.41% in 1998.

TABLE 1: Density, germination (%, G) and inoculum potential (IP) of *Plasmopara viticola* oospores from leaves untreated and treated with azoxystrobin- 1997 and 1998

	1997			1998		
	treated with azoxystrobin	untreated	% reduction	treated with azoxystrobin	untreated	% reduction
oospore density	40.74B	175.04A	76.73	63.17B	308.5A	79.54
G	1.32B	3.73A	64.62	1.94B	8.62A	77.49
IP	0.53B	6.52A	91.88	1.22B	26.61A	95.41

Oospore density and germination assessed from October 1997 to June 1998 for the two series of samples obtained are reported in Figure 1. Oospore density was always higher in samples obtained from untreated leaves in comparison with the treated ones. In both series of samples, no germination occurred until December, and then G was statistically similar for the following three germination assays.

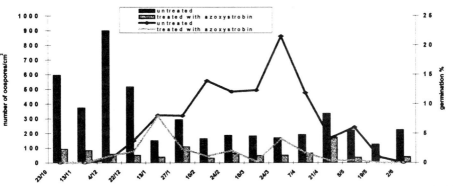

Figure 1 : Density (bars) and germination (line graph) of oospores from leaves untreated and treated with azoxystrobin in 1998

Significant differences have been found between the G obtained for the samples collected from February onwards. Germination dynamics of *P. viticola* differed in the two series of samples examined. In the untreated samples, after a progressive increase observed from December, G reaches its highest value at the end of March. Afterwards, except from a slight increase at the beginning of May, G is progressively decreasing. No sporangium formation was observed in June. In the treated samples the increase in G was detected until the middle of January, when G reaches its highest value. Afterwards, a very low G characterized the oospores differentiated in leaves treated with azoxystrobin. In these samples oospore germination did not occur from the middle of May onwards.

The detection of the first sporangium was often delayed in oospores obtained from treated leaves in comparison with the untreated ones, particularly at the end of April and in May (data not shown).

3.2 ELECTRON MICROSCOPY

TEM observations of all the samples of *P. viticola* oospores from the untreated leaves show the typical ultrastructure already described by Vercesi and coworkers (1999) (Fig. 2), characterized by a thin outer wall (OOW), a very thick inner wall (IOW) and a central ooplast (O) surrounded by many lipid droplets (L), intermingled with dehydrated mitochondria. Furthermore, in the oospores processed in May, the inner surface of the IOW sometimes showed different degrees of lysis, indicating incipient germination. The great majority of *P. viticola* oospores from treated samples collected from October to April showed a different organization of the ooplast, which appeared fragmented and localized at the cell periphery (Fig. 3). Lipid globules were also altered, often coalescing to form large lipid aggregates. The IOW was scarcely affected by azoxystrobin treatments though its thickness was usually less than untreated oospores, particularly in samples collected in October.

In the treated leaf fragments collected in May, the very low number of apparently unaltered oospores indicates that those affected by the azoxystrobin treatment have already been broken down.

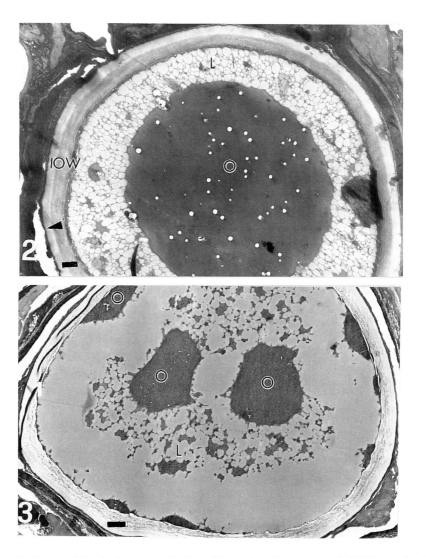

Figure 2 : Untreated *P. viticola* oospore showing a thin electron dense outer wall (OOW, arrowhead), a thick inner wall (IOW), a central ooplast (O) and lipid droplets (L); bar = 1 µm.

Figure 3 : Azoxystrobin-treated oospore with a fragmented ooplast (O) and coalescing lipid droplets (L); bar = 1 µm.

3.3 DISCUSSION

Azoxystrobin affects both the density and the germinability of *P. viticola* oospores. The lower oospore number detected in treated leaves probably depends on the detrimental effects of azoxystrobin on *P. viticola* development. Moreover the oospores differentiated in treated leaves show abnormal ultrastructure: azoxystrobin seems to affect all the oospore structures with accumulation of storage substances in lipid droplets and in the ooplast altered substantially, including in some cases a complete degeneration of the cytoplasm. These effects and the reduced wall thickness could be due to a decreased efficacy of the energy producing system, as mitochondria are the site of action of azoxystrobin. A similar reason is given for its effects on zoospore motility (Young *et al.*, this volume).

The decrease in germinability of oospores differentiated in treated leaves is probably related to their abnormal structure and particularly to the abnormal status of the storage substances. Cell wall, lipid globules and ooplast completely disappear during the germination process, indicating that all their components are utilized for sporangium formation (Vercesi *et al.*, 1999). The alterations due to the azoxystrobin treatment could induce either a degeneration of the abnormal oospores or alternatively a disrupted utilization of the storage compounds. Ultrastructural observations and the oospore density detected in the leaf fragments seems to indicate that oospore breakdown is a phenomenon occurring late in the season, from the end of April onwards, while both the decrease in oospore germinability and the delay in the sporangium formation suggest a slower utilization of storage compounds.

The large decrease in the inoculum potential of *P. viticola* due to azoxystrobin has important epidemiological consequences. In fact the presence of numerous primary foci in vineyards early in the season has often hampered the control of downy mildew epidemics (Vercesi, 1995). A decreased inoculum potential, represented by a low number of slowly germinating oospores, could efficiently contribute to control downy mildew epidemics, by reducing the consistency of primary foci and delaying their occurrence in the vineyard. Field trials are needed in order to evaluate in natural conditions the detrimental effect of azoxystrobin treatment carried out late in the previous season on the early stages and subsequent progress of these downy mildew epidemics.

4. Acknowledgements

The authors thank dr Federico Zerbetto for the statistical analysis and Solplant - Zeneca group for financial support.

5. References

Godwin, J.R., Anthony, V.M., Clough J.M. and Godfrey C.R.A. (1992) ICIA5504: a novel, broad spectrum, systemic β-methoxyacrylate fungicide, *Proceedings of Brighton Crop Conference-Pests and Diseases* 1, 435-442.

Godwin, J.R., Young, J.E., Woodward, D.J. and Hart, C.A. (1997) Azoxystrobin: effect on the development of grapevine downy mildew (*Plasmopara viticola*), *A.N.P.P. - Fifth International Conference on Plant Diseases, Tours 3-4-5 December 1997.*

Vercesi, A. (1995) Strumenti innovativi per la gestione della difesa contro la peronospora della vite, *Informatore fitopatologico* 45, 12-19.

Vercesi, A., Tornaghi, R., Sant, S., Burruano, S. and Faoro, F. (1998) A cytological and ultrastructural study on the maturation and germination of oospores of *Plasmopara viticola* from overwintering vine leaves, *Mycological Research*, **103**, 193 -202.

Zachos, D.G. (1959) Recherches sur la biologie et l'épidémiologie du Mildiou de la vigne en Grèce. Bases de prévisions et d'avertissements, *Annales de l'Institut Phytopathologique Benaki* 2, 193-335.

EFFECTS OF AZOXYSTROBIN ON INFECTION DEVELOPMENT OF *PLASMOPARA VITICOLA*

J.E. YOUNG, J.A. SAUNDERS, C.A. HART AND J.R. GODWIN
Syngenta Crop Protection, Jealott's Hill International Research Centre, Bracknell, Berkshire, RG42 6ET, UK

1. Introduction

Azoxystrobin is a novel, highly active strobilurin fungicide with an outstanding breadth of spectrum (Godwin *et al.,* 1992; Gisi, this volume). Azoxystrobin is unique in that it gives excellent control of both downy mildew (*Plasmopara viticola*) and powdery mildew (*Uncinula necator*) of grapevine (Cohadon *et al.,* 1994). The biochemical mode of action of azoxystrobin is the inhibition of mitochondrial electron transfer between cytochrome b and cytochrome c_1 in fungi. *In planta* and *in vitro* light and cryoscanning electron microscopy studies were undertaken to examine the preventative and curative effects of azoxystrobin on infection development of *P. viticola*. Additionally, effects on fungal development due to redistribution via the xylem were examined. A study of this nature is important because if the stage in the fungal life cycle at which a compound is active can be determined, this can help to define the most effective application timing for disease control. The effects of azoxystrobin were compared to the non-systemic, protectant dithiocarbamate fungicide, mancozeb and the systemic phenylamide fungicide, metalaxyl.

2. Materials and Methods

2.1. PREVENTATIVE AND CURATIVE EFFECTS

Grapevine seedlings, variety Ohanez, were grown from vernalised seed in a glasshouse (day/night temperature 24/17 °C, day length 16 hours). Fungicide treatments were applied to maximum retention on the lower surface of the youngest leaf of 4 week old seedlings using a handheld pressurised sprayer. Applications were made either 2 hours before inoculation (preventative) or 2 days after inoculation (curative). The seedlings were inoculated by spraying a suspension (40,000 sporangia/ml) of *P. viticola* sporangia (phenylamide sensitive isolate) on to the lower surface of the treated leaf. Inoculated plants were incubated in a high humidity chamber for 24 hours in the dark and then placed in a controlled environment growth room (day/night temperature 24/18°C, day/night relative humidity 60/95%, day length 16 hours, light intensity 6000 lux) until assessment.

201

P.T.N. Spencer-Phillips et al. (eds.), Advances in Downy Mildew Research, 201–205.

To assess the preventative effect of each fungicide on sporangium germination and zoospore encystment the treated, inoculated leaves were excised 48 h after inoculation (HAI) and mounted in 0.1% w/v of the fungal cell wall stain Uvitex BHT, and examined using a Leitz Ortholux II fluorescence microscope fitted with filter block 'A' (BP 360, LP 430) and viewed at a magnification of x 160. The number of full sporangia, empty sporangia and encysted zoospores were counted in five fields of view for each treatment. Three replicates per treatment were assessed. A parallel disease development test was set up by treating and inoculating the plants in the same manner as described above and assessing the percentage cover of sporulating mildew on the lower leaf surface 7 days after inoculation (DAI). To aid sporulation the plants were placed in a high humidity chamber for 24 h prior to assessment.

Low temperature scanning electron microscopy (LTSEM) was used to examine the curative effect of the three fungicides on internal mycelium. A segment of each inoculated, treated leaf was rapidly frozen in sub-cooled liquid nitrogen and subsequently freeze fractured before transfer to the nitrogen gas-cooled specimen stage of an Oxford Instruments' CT500 preparation unit fitted to a JEOL JSM6300 scanning electron microscope (SEM). The specimen surface was partially freeze dried at -90°C before gold sputter-coating and subsequent study on the SEM cryo-stage at below -130°C and with a beam accelerating voltage of 4.9 KV. A parallel disease development test was set up to record the percentage cover of sporulating mildew on the lower leaf surface of each treatment 7 days after inoculation. To aid sporulation the plants were placed in a high humidity chamber for 24 h prior to assessment.

2. 2. EFFECTS ON ZOOSPORE MOTILITY

To examine the effect of the three fungicides on zoospore motility, 0.5 ml of each fungicide treatment, at twice the intended concentration of active ingredient, was put into separate small glass vials. Each fungicide rate was replicated twice. A suspension of sporangia was made up in mineral water (150,000 sporangia/ml) and examined at intervals until zoospores had been released. A 0.5 ml volume of the suspension containing motile zoospores was added to each of the fungicide treatments. Immediately the zoospore suspension was added to the fungicide, the vial was shaken and a drop of the resulting suspension examined using a light microscope to determine zoospore motility at 10, 30, 60, 120, 300 and 600 s following their addition.

2.3. EFFECTS VIA XYLEM MOBILITY

A 1 cm wide band was marked, using a permanent waterproof ink pen, just above the petiole on the lower surface of the youngest fully expanded leaf of 6 week old grapevine seedlings (grown in the same manner as described in section 2.1.). One hundred 5 µl droplets of each fungicide preparation were evenly applied within the marked band using a Burkhard automatic spotter, to give four replicates per treatment. Forty-eight h after treatment the whole of the lower surface of each treated leaf was inoculated with a suspension of sporangia (40,000 sporangia/ml) of *P. viticola*. Two days after inoculation the number of encysted zoospores in the marked area of application and in a 1 cm untreated band 0.5 cm above the zone of application were assessed by mounting the leaves in Uvitex BHT and examining them by fluorescence microscopy. A 0.5 cm wide gap between the treated and untreated bands was not assessed in order that any movement of fungicide due to local diffusion did not influence the results of this test designed to examine transport of fungicide in the leaf via the xylem. A parallel disease development test was set up to examine the amount of disease control achieved by each treatment in the band of application and in the untreated band.

3. Results

3.1. PREVENTATIVE AND CURATIVE EFFECTS

When applied preventatively, azoxystrobin reduced the frequency with which sporangia germinated and totally inhibited zoospore encystment at rates of 0.04 mg ai/l and above (Table 1). This effect led to excellent disease control. Treatment with mancozeb in the same studies gave no consistent effect on the frequency of sporangium germination but the frequency of zoospore encystment was reduced and disease controlled, albeit at higher rates than for azoxystrobin. Metalaxyl gave similar results to mancozeb, but at even higher rates.

TABLE 1. Effect of preventative treatments of azoxystrobin, metalaxyl, and mancozeb on *Plasmopara viticola* sporangium germination (48 HAI), zoospore encystment (48 HAI) and disease development (7 DAI)

Treatment	Rate (mg ai/l)	% germination of sporangia	Number zoospores encysted	% leaf area with sporulating disease
Untreated	0	76	29	46
Azoxystrobin	5	20	0	0
	1	19	0	0
	0.2	32	0	0
	0.04	27	0	6
	0.008	54	11	40
	0.0016	61	32	33
Mancozeb	125	48	0	0
	25	77	0	0
	5	44	0	16
	1	71	22	45
	0.2	76	39	33
Metalaxyl	625	44	0	1
	125	56	15	17
	25	76	31	23
	5	69	41	55

Examination by LTSEM revealed that curative application of azoxystrobin (250 mg ai/l) 2 days after inoculation, i.e. at 29% of the latent period, caused established mycelium of *P. viticola* to collapse. In the parallel disease development test, 80% disease control was achieved with azoxystrobin. Metalaxyl (200 mg ai/l) also caused established mycelium to collapse giving 100% disease control when applied as a 2 day curative treatment. Mancozeb (2600 mg ai/l) demonstrated no curative effect.

3.2. EFFECTS ON ZOOSPORE MOTILITY

Azoxystrobin prevented zoospore motility at rates as low as 0.008 mg ai/l (Table 2). Metalaxyl and mancozeb also affected zoospore motility, but at much higher rates.

TABLE 2. Effects of *in vitro* treatments of azoxystrobin, mancozeb and metalaxyl on *P. viticola* zoospore motility

Treatment	Rate (mg ai/l)	Zoospore motility (seconds after treatment)					
		10	30	60	120	300	600
Untreated	0	✓✓	✓✓	✓✓	✓✓	✓✓	✓✓
Azoxystrobin	1	✓✓	✓✓	✓✓	✗✗	✗✗	✗✗
	0.2	✓✓	✓✓	✓✓	✓✗	✗✗	✗✗
	0.04	✓✓	✓✓	✓✓	✓✓	✗✗	✗✗
	0.008	✓✓	✓✓	✓✓	✓✓	✓✓	✗✗
	0.0016	✓✓	✓✓	✓✓	✓✓	✓✓	✓✓
Mancozeb	125	✓✓	✓✓	✓✓	✗✗	✗✗	✗✗
	25	✓✓	✓✓	✓✓	✗✗	✗✗	✗✗
	5	✓✓	✓✓	✓✓	✓✗	✗✗	✗✗
	1	✓✓	✓✓	✓✓	✓✓	✓✗	✗✗
	0.2	✓✓	✓✓	✓✓	✓✓	✓✓	✓✓
Metalaxyl	625	✗✗	✗✗	✗✗	✗✗	✗✗	✗✗
	125	✓✓	✓✓	✓✓	✓✓	✓✓	✗✗
	25	✓✓	✓✓	✓✓	✓✓	✓✓	✓✓

✓ = motile,   ✗ = immotile; 2 replicate experiments for each timing

3.3. EFFECTS VIA XYLEM MOBILITY

Azoxystrobin gave good disease control in the treated zone and significantly reduced the level of disease in the untreated zone (Table 3). Microscopic examination of the grapevine leaves revealed that the number of encysted zoospores present in the treated and untreated zones were significantly less on the azoxystrobin treated leaves than in the corresponding zones on the water treated leaves. Mancozeb gave good disease control and prevented zoospore encystment in the zone of application, but gave no significant disease control nor reduction in zoospore encystment in the untreated zone.

TABLE 3. Effects of azoxystrobin and mancozeb on *P. viticola* development due to redistribution via the xylem

Treatment	Rate (mg ai/l)	Treated zone		Untreated zone	
		% disease (7 DAI)	Total no. encysted zoospores (48 HAI)	% disease (7 DAI)	Total no. encysted zoospores (48 HAI)
Water	---	90 A	132 A	93 A	138 A
Azoxystrobin	250	3 B	8 B	42 B	29 B
Mancozeb	2800	13 B	0 *	84 A	108 A

Treatments with no letter in common were significantly different at 5% LSD
* Treatment omitted from statistical analysis as all replicates were zero; inclusion would have led to an under-estimation of the random variation.

4. Discussion and Conclusions

The results of these studies show that azoxystrobin is effective against many stages of the asexual life cycle of *P. viticola*. However, zoospore motility appears to be the stage most sensitive to azoxystrobin, presumably because zoospore motility is a high energy demanding process and azoxystrobin disrupts energy production within the fungus. A similar reason is given for its effect on oospore viability (Vercesi *et* al., this volume). Azoxystrobin gives optimum effect when applied preventatively, undoubtedly because of its potent inhibition of zoospore motility, thereby preventing zoospores from locating a stoma and encysting on the grapevine leaf surface. However, unlike classic protectant fungicides such as mancozeb, azoxystrobin also demonstrates some curative activity (expressed as mycelial collapse and effective at up to 29% of the latent period in this study). This gives azoxystrobin a degree of flexibility in application timing against *P. viticola*, but it should not be used as a specialist curative fungicide.

Following preventative application, azoxystrobin reduced the number of zoospores encysting in an untreated area towards the tip of the leaf, most likely by inhibiting zoospore motility. In order to do this, azoxystrobin, which is not vapour active, must have been taken up into the leaf, moved towards the leaf tip in the xylem and diffused onto the leaf surface. Radiochemical studies have confirmed that azoxystrobin moving in the xylem diffuses onto the leaf surface. Inhibition of zoospore motility via xylem mobility has never previously been reported for a downy mildewicide.

Metalaxyl, a systemic, curative phenylamide fungicide inhibits ribosomal RNA synthesis, thus interfering with protein synthesis. As in previous studies (Staub *et al.*, 1980), metalaxyl was most effective against the later stages of *P. viticola* infection, causing the collapse of mycelium within the leaf. Interestingly, the current study did show some effect of metalaxyl on zoospore motility, whereas the study of Staub *et al.*, showed no such effect. This difference probably reflects different methodologies between the two studies.

Mancozeb, a dithiocarbamate fungicide was effective only as a preventative treatment inhibiting zoospore motility and reducing zoospore encystment. It showed neither an effect on mycelial growth nor redistribution in the xylem of grapevine leaves.

5. References

Cohadon, P., Roques, J.F., Godwin, J.R. and Heaney, S.P. (1994) Le ICIA5504 : Un fongicide à large spectre doté d'un nouveau mode d'action pour lutter contre les maladies des céréales et de la vigne, *Proceedings ANPP Quatrième Conférence Internationale sur les Maladies des Plantes, Bordeaux, 6-8 Decembre 1994* **3**, 931-938.

Godwin, J.R., Anthony, V.M., Clough, J.M. and Godfrey, C.R.A. (1992) ICIA5504: a novel, broad spectrum, systemic β-methoxyacrylate fungicide, *Proceedings 1992 Brighton Crop Protection Conference - Pests and Diseases* **1**, 435-442.

Staub, T.H., Dahmen, H. and Schwinn, F.J. (1980) Effects of Ridomil® on the development of *Plasmopara viticola* and *Phytophthora infestans* on their host plants, *Journal of Plant Diseases and Protection* **87**, 83-91.

LOCAL AND SYSTEMIC ACTIVITY OF BABA (DL-3-AMINOBUTYRIC ACID) AGAINST *PLASMOPARA VITICOLA* IN GRAPEVINES

Y. COHEN[1], M. REUVENI[2] AND A. BAIDER[1]
[1]*Dept. of Life Sciences, Bar-Ilan University, Ramat Gan 52900, Israel*
[2]*Haifa University, Golan Research Institute, Katzrin 12900, Israel*

1. Introduction

A recent technology for plant disease control is based on the activation of the plant's own defence system with the aid of low molecular weight synthetic molecules. Compounds such as salicylic acid (SA), 2,6-dichloroisonicotic acid (INA) and benzo[1,2,3]thiadiazole-7-carbothionic acid-S-methyl ester (BTH) are able to induce systemic acquired resistance (SAR) in a variety of plants against a wide range of microbial pathogens without possessing direct antimicrobial activity *in vitro*. SA, INA and BTH are structurally related thus sharing the activation of similar genes (coding for PR-protein synthesis) in the plant. Both INA and BTH act independently of the presence of SA. They are active in monocot as well as dicot plants and normally require about 2 days for gene activation (Kessmann et al., 1994; Ryals et al., 1996; Sticher et al., 1997). They failed to induce SAR in *nim* mutants (non-inducible) of Arabidopsis (Delaney, 1997), thus proving again that they work only through the plants. One major hurdle with commercial application of INA-like compounds is their intolerance by certain crops (Sticher et al., 1997).

Aminobutyric acids are a new class of SAR compounds recently reported by Cohen and coworkers. A recent review by Sticher et al. (Sticher et al., 1997) states that "local treatments with DL-3-aminobutyric acid (BABA) protect tomato, potato, and tobacco, systemically, against *Phytophthora infestans* and *Peronospora tabacina*, respectively. BABA has no detectable antifungal activity *in vitro* or in planta and has a curative effect, which is surprising for an SAR inducer. A hypothesis was formulated whereby the action of BABA is based on covalent cell-wall modification, since after [14C]-BABA treatment, covalently bound label is found in the cell wall that can only be released with cell-wall degrading enzymes (P. Schweizer and Y. Cohen, unpublished data).

SAR was extensively studied with at least 20 plant species (see Table 1 in Sticher *et al.*, 1997); none of them was a woody, perennial species. In the present study, we show that BABA provides a durable, local and systemic protection of grape plants against the downy mildew disease incited by the fungus *Plasmopara viticola* when applied pre- or post-infectionally (Cohen *et al.*, 1999). This widely distributed and highly destructive disease of grapevines (Pearson; Goheen, 1988) is currently controlled by fungicides whose efficacy is often hampered by resistant fungal mutants (LeRoux;

P.T.N. Spencer-Phillips et al. (eds.), Advances in Downy Mildew Research, 207–224.

Clerjeau, 1985). The discovery of BABA as an SAR molecule may offer an important opportunity to develop a new technology to control downy mildew in grapes. The high tolerance of BABA by grape plants adds to its value in agricultural practice.

2.   Materials and Methods

2.1. CHEMICALS

DL-2-amino-n-butanoic acid (AABA), L-2-amino-n-butanoic acid, DL-3-amino-n-butanoic acid (BABA), 2-amino *iso*butanoic acid, DL-3-amino *iso*butanoic acid and 4-amino-n-butanoic acid (GABA) were purchased from Sigma. The R and S enantiomers of BABA and the ^{14}C-BABA [CH_3-$^{14}CH(NH_2)$-CH_2-COOH] with a specific activity of 9.68 μCi per mg (1 m Ci per m mol) and a purity of >98% were kindly supplied by Novartis (Sandoz) Pharma, Basel. INA 25% WP and CGA 245704 (BTH) formulated as 50% WG were a gift of Novartis AgroResearch, Basel.

2.2. PLANTS

Plants were raised from either cuttings or seeds. Unless stated otherwise, the cultivar used was Emerald Riesling. Plants were grown in 1 liter pots containing a mix of (1:1:1 v/v) sand, vermiculite and peat under glasshouse conditions (15-25°C). Plants were used when they had 4-10 expanded leaves. For leaf disc assays, leaf discs were punched from the second and third leaves from the top of a plant using a 10 or 12 mm diameter cork borer. Leaf discs were floated, lower surface uppermost, on chemical solutions in 24-well titer plates (10 mm discs) or in 9 cm Petri dishes (ten 12 mm discs per dish). For whole plant assays, plants were sprayed first (see Results) and thereafter inoculated on their lower leaf surfaces.

2.3. THE FUNGUS

Experiments were carried out using an isolate of *Plasmopara viticola* collected in 1996 in the coastal plain of Israel. Fungus was maintained on detached grape leaves or whole potted plants by repeated inoculations at 20°C.

2.4. INOCULATION

Sporangia of *P. viticola* were harvested from freshly sporulating leaves into ice-cold double distilled water, using a camel hair brush. Sporangial concentration was adjusted to 1×10^4 per ml using a haemocytometer unless otherwise stated. To inoculate whole plants, the sporangial suspension was sprayed onto the lower leaf surfaces with the aid of a glass sprayer. To inoculate leaf discs, one (or two) 10 μl inoculum droplet(s) was applied to each disc using an Eppendorf pipette dispenser. Inoculated plants were immediately transferred to a dew chamber (18°C, darkness) for 20 h and then maintained in a growth chamber at 20°C (12 h photoperiod, 100 μE $m^{-2}.s^{-1}$ light intensity and RH of 60-70%) until symptoms developed or sporulation was induced. Petri dishes and titer plates with inoculated leaf discs were placed in a 20°C growth chamber. To induce fungal sporulation infected plants were returned to the dew chamber for 24 h.

2.5. ASSESSING DISEASE DEVELOPMENT AND FUNGAL SPORULATION

Disease intensity in leaf discs was visually assessed using a 0-4 scale, in which: 0 = no symptoms visible; 1 = up to 25% of the leaf disc area chlorotic or necrotic; 2 = 25-50%; 3 = 50-75% and 4 = >75% of the leaf disc area chlorotic or necrotic. Fungal sporulation in whole plants was assessed visually using a 0-4 scale in which 0 = no sporulation visible and 4 = all of abaxial leaf surface covered with sporangia and sporangiophores. For quantitative determination of sporulation in leaf discs, infected leaf discs were placed in 5 ml per disc of 50% ethanol, shaken for 5 min and the number of sporangia was counted with the aid of a haemocytometer.

2.6. MICROSCOPICAL EXAMINATIONS

2.6.1. Autofluorescence
Inoculated leaf discs were placed on a glass slide, a few water droplets were pipetted to each disc, covered with a cover slip and examined with a Zeiss UV epifluorescent microscope equipped with filter combination 390-420/FT 425/LP450 (excitation 390-420 nm, transmission > 450 nm). Under such conditions normal host cells fluoresced red (chlorophyll), necrotic host cells looked dark and phenolics fluoresced blue. Fungal structures were not visible (Cohen et al., 1990).

2.6.2. Fungal Structures
Inoculated leaf discs were treated with a few droplets of aqueous Calcofluor (0.01%, Sigma) and similarly examined. Sporangiophores and sporangia fluoresced strong blue. Fungal structures inside the leaf could not be seen.

2.6.3. Callose Accumulation
Inoculated leaf discs were first clarified in boiling ethanol and then treated with basic (pH 9.8) aniline blue in the manner described before (Cohen et al., 1990). Callose deposits in the host or the fungus fluoresced strong yellow.

2.6.4. Lignin Deposition
Inoculated leaf discs were clarified with ethanol and then treated with phloroglucinol (Sigma) followed by 25% HCl in the manner described before (Cohen et al., 1990). Leaf discs were examined microscopically under bright field illumination. Lignin deposits in host cells stained red.

2.7. SYSTEMIC TRANSLOCATION OF ^{14}C-BABA

Plants were produced from 10 cm cuttings and when they had 7 leaves they were treated with ^{14}C-BABA on a single leaf. Fifty 10 µl droplets of the radioactive compound were placed onto the upper leaf surface of either leaf number (from stem base) 1, 4 or 6 (total of 1 µCi per leaf) and plants were kept in the glasshouse (15-25 °C) for 5 days until assayed for radioactivity.

Two types of analyses were made:
(a) measuring the radioactivity extractable from the plants;
(b) autoradiography;

For (a), leaves were detached, washed with excessive water, blotted dry and extracted in MCW (methanol:chloroform:water, 12:5:3, v/v) as described before (Cohen; Gisi, 1994a; Bieleski; Turner, 1996). An aliquot of 0.1 ml was taken for measuring radioactivity (Cohen; Gisi, 1994a).

For (b), plants were carefully uprooted, washed copiously with water, the woody stem cutting removed, blotted dry, and pressed between thick papers for a week, and then exposed to X-ray film for 50h (Cohen; Gisi, 1994a).

2.8. SPRAY APPLICATION OF INTACT PLANTS

Plants of cv. Superior, made from cuttings and carrying 4 expanded leaves, were sprayed on both leaf surfaces with aqueous BABA solutions of various concentrations, placed in humid plastic tents in growth chambers at 20°C for 1 day to facilitate BABA uptake, and inoculated two days afterwards by spraying the lower leaf surfaces with sporangial suspension (1×10^4/ml). Seven days after inoculation plants were placed in a dew chamber (18°C darkness) for 24 h to induce fungal sporulation.

2.9. PROTECTION BY APPLYING BABA TO THE ROOT SYSTEM

Two experiments were conducted to evaluate the protection of grape plants from downy mildew attack by root applications of BABA. In the first, 4-leaf plants of cv. Superior were produced from cuttings were used. In the second, 9-leaf plants of cv. Emerald Reisling, produced from seeds, were used. In both experiments, plants were carefully uprooted, their root system washed with water, and placed in vials containing 50 ml of aqueous BABA solution of various concentrations. Plants were left in the solutions for either 3 days (cv. Superior) or 1 day (cv. Emerald Reisling), the volume of the solution taken up was measured, plants were transferred to new vials filled with water, and inoculated with sporangia of *P. viticola*. A week after inoculation sporulation was induced and the extent of sporulation on lower leaf surfaces was estimated visually.

3. Results

3.1. LEAF DISC ASSAYS

Six isomers of aminobutyric acid were tested for their efficacy to suppress downy mildew development in grape leaf discs. They were: DL-2-amino, L-2-amino, DL-3-

tubes and mixed with aqueous solutions of aminobutyric acids to various final concentrations, so that sporangial concentration (5000 per ml) remained constant. Twenty μl of the mixtures were pipetted onto depressions of microscope slides and incubated at 14°C for 20 h in the dark. Microscopical examination revealed that in water controls most of the sporangia were empty, many zoospores were motile and only a few cystospores germinated. A similar picture was observed with DL-2-aminobutyric acid, DL-3-aminobutyric acid and 4-aminobutyric acid at concentrations ranging from 250-2000 μg/ml. Based on our experience on the enhancement of cystospore germination with sucrose, a 10 μl droplet of 50 mM sucrose was added to each depression. After an additional 20 h, germination of approximately 100 cystospores per depression as noticed in the water-sucrose controls, with a germ tube of 200-600 μm in length. With DL-3-aminobutyric (BABA) and 4-aminobutyric (GABA) (+ sucrose) of up to 2000 μg/ml, results were similar to the control. However, in DL-2-aminobutyric acid (AABA) of 1000-2000 μg/ml, germ tubes were much longer, reaching a length of 1000-2000 μm, and branched once or twice. Lowering the concentration of this compound to 250 μg/ml diminished this phenomenon.

3.3. WHOLE PLANT ASSAYS

3.3.1. Uptake and Systemic Translocation of ^{14}C-BABA

Because of the scarcity of ^{14}C-BABA tests were conducted with only foliar applications. Five days after such application, most of the ^{14}C-label was recovered from the leaf surface with water. TLC analysis (Cohen; Gisi, 1994a) revealed that this label on the leaf surface is due to ^{14}C-BABA only.

The data on the uptake into the treated leaves and translocation to other organs are presented in Table 3. They show that out of the 1 μCi of ^{14}C-BABA applied to leaf 1 or 4, a total of only 5.02% and 3.49% of the label applied could be re-extracted with MCW from the various organs of the plant (leaves, green stem and roots). In contrast, a total of 29.99% could be recovered with MCW when the ^{14}C-BABA was applied to leaf 6 (youngest treated) indicating that uptake is readily occurring through the upper surface of young rather than old leaves. Most of the penetrating label was recovered from the treated leaf itself. The recovery of ^{14}C-BABA from the remote younger leaves indicates on acropetal transport. Label was recovered also from the root system, especially when leaf 1 was treated, indicating also on basipetal translocation (Fig. 5). Other experiments, conducted with detached leaves in Petri dishes proved that uptake after 1 week ranged between 56.3-92.7% of the applied ^{14}C-BABA, suggesting that the low uptake in intact plants resulted from the hydrophilic nature of BABA and the quick drying of the label droplets. It should be mentioned that MCW extracts of ^{14}C-BABA-treated grape leaves contained ^{14}C-BABA, as proved by thin layer chromatography. This solvent mixture extracted 96% of the label present inside the tissue. The remaining 4% were bound to the cell walls, in a similar manner described in tomato (Cohen; Gisi, 1994a).

TABLE 3.   Uptake and distribution of ^{14}C-3-aminobutyric acid in grape plants after foliar application to either leaf 1, 4 or 6, while leaf 1 is nearest the stem base[a].

Organ	Percentage of recovered radioactivity		
	1	4	6
leaf 1	26.0 [b]	1.1	0.2
leaf 2	2.4	1.1	0.2
leaf 3	2.2	1.1	0.3
leaf 4	2.8	31.2 [b]	0.3
leaf 5	5.4	3.7	0.4
leaf 6	18.6	12.0	92.6 [b]
leaf 7	18.8	21.8	3.2
leaf 8	-	20.3	-
Stem	18.1	6.0	1.9
Root	5.7	1.7	0.9
% of applied[c]	5.02	3.49	29.99

a   Fifty 10 µl droplets of ^{14}C-BABA, containing a total of 1 µCi label, were applied to the upper surface of either leaf 1, 4 or 6 of 7-leaf grape plants. Five days later, plants were washed with water and organs extracted to determine radioactivity inside the tissue.
b   Treated leaf.
c   The rest of the ^{14}C-BABA applied (1 µCi) was recovered from the leaf surface by water.
-   Leaf not present.

3.4. SPRAY APPLICATION OF INTACT PLANTS

Percentage leaf area (abaxial surfaces) covered with sporangiophores and sporangia was found to decrease significantly as the concentration of BABA applied increased. Thus, plants sprayed with 0, 250, 500, 1000 and 2000 µg/ml solution of BABA sporulated on 90, 70, 35, 5 and 0% (mean values, n = 3) of their leaf total area, respectively.

To evaluate the protection associated with systemic translocation of BABA, the compound was formulated as a 25% wettable powder (WP) and applied (2000 µg/ml a.i.) to either the whole plant or the 3 bottom leaves of 8-leaf plants grown from seeds. Two days after spraying, plants were inoculated with *P. viticola* and 7 days later were placed in moist plastic tents in a growth chamber for 2 days to induce fungal sporulation. Results shown in Table 4 indicate that a total spray application with 25% WP BABA was highly effective in suppressing fungal sporulation. A similar treatment with a control (WP minus BABA) formulation resulted in a limited inhibition of fungal sporulation as compared to water-treated inoculated plants. When the formulated BABA was applied to the 3 bottom leaves, protection was observed in the treated leaves as well as in the upper two youngest leaves, but not in the middle leaves. The later result accords with the preferable acropetal translocation of ^{14}C-BABA (see Fig. 5).

amino, 4-amino, 2-amino *iso* and DL-3-amino *iso* butyric acids. Test solutions contained either 6.25, 12.5, 25, 50 or 100 µg/ml of tested compound (~0.06-1 mM). BTH was also used, for comparison purposes, at 0.3, 1.25, 5, 20 and 40 µg/ml (~0.001-0.2 mM). Nine days after inoculation, all leaf discs, except those treated with DL-3-aminobutyric acid (BABA) or BTH, showed profuse sporulation of *Plasmopara viticola*. Sporangial counts for DL-2-aminobutyric acid (AABA), BABA, and 4-aminobutyric acid (GABA) are given in Fig. 1. Sporangial counts for L-2-aminobutyric acid, 2-amino *iso* butyric acid and DL-3-amino *iso* butyric acid were similar to the water control (data not shown). Untreated-inoculated (control) leaf discs yielded mean of a 120×10^3 sporangia/disc (Fig. 1). BABA at 6.25 µg/ml decreased this yield by 11%, while 12.5, 25, 50 and 100 µg/ml reduced this yield by 85, 96, 99 and 99%, respectively (Fig. 1). AABA at 100 µg/ml significantly reduced sporulation by 37% whereas GABA at all concentrations enhanced it, although insignificantly. BTH at 1.25, 5 and 20 µg/ml reduced sporulation by 27, 75 and 99%, respectively. At 40 µg/ml, it was phytotoxic.

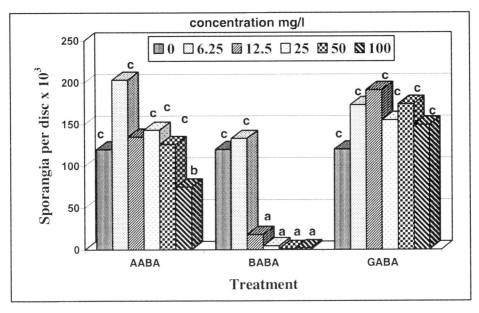

Figure 1. Downy mildew (*Plasmopara viticola*) development assessed as number of sporangia on grape leaf discs (2.27 cm^2) floating on aminobutyric acids of various concentrations. AABA = DL-2-aminobutyric; BABA = DL-3-aminobutyric; GABA = 4-aminobutyric acid. Data collected at 10 days post-inoculation. Bars with different letters are significantly different at 55 level according to Duncan's Multiple Range Test.

Inoculated leaf discs treated with BABA of ≥12.5 µg/ml showed pale-brownish lesions beneath the inoculum droplets. No such lesions were seen in water-treated inoculated discs, profuse sporulation is seen), nor in BABA-treated leaf discs mock-inoculated with water. Bright-field microscopical examination of non-stained leaf discs revealed a hypersensitive-like response (HR) in the BABA-treated inoculated tissue as against normally-looking tissue bearing sporangia in water-treated inoculated tissue. BTH at a concentration of 20 μg/ml also produced HR. No further studies were devoted to BTH.

In UV micrographs of calcofluor-stained water-treated and BABA-treated inoculated leaf discs, the following observations were made: compared to water, BABA strongly suppressed both steps of sporulation: the emergence of sporangiophores from stomata and the sporangial production. In water-treated tissue, sporangiophores emerged in groups of 3-5/stoma, they were branched and bear abundant sporangia, whereas in BABA-treated tissue, only a few non-branched sporangiophores, 1/stoma, with only a few sporangia, were developed. The numerical data collected from this experiment are given in Table 1.

TABLE 1. Sporulation of *Plasmopara viticola* on grape leaf discs as affected by BABA (DL-3-aminobutyric acid). A UV-microscopical analysis with Calcofluor.

BABA, mg/L	Sporangia/mm^2		% protection	sporangiophores/mm^2		% protection
0	508 ± 102	a	-	72 ± 7	a	-
12.5	28 ± 11	b	94	11 ± 6	bc	85
25	19 ± 5	b	96	10 ± 2	bc	86
50	14 ± 14	b	97	10 ± 1	bc	86
100	7 ± 7	b	99	4 ± 4	c	94

Ten mm diameter leaf discs were floated on 1 ml of the test solution, lower side uppermost, in 24-well titer plates. Two 10 μl droplets of 2×10^4 sporangia/ml of *P. viticola* were applied to each disc one day after floating. Fourteen days post-inoculation, discs were placed on glass slides and 0.2 ml calcofluor (0.01% in water) was pipetted on each disc and covered with a cover slip. Sporulation was determined with the aid of a UV epifluorescence microscope.
Figures followed by different letters are significantly different at 5% level (n=4, Duncan's multiple range test).

Other microscopical examinations were aimed to ascertain whether the HR response of inoculated BABA treated leaf discs was associated with autofluorescence of the tissue, callose formation or lignification.. At 10 days after inoculation, no autofluorescence was found in either water control-inoculated or BABA-treated inoculated leaf discs, when mounted in water and examined under UV light using an epifluorescence microscope. Similarly, when first clarified with ethanol and then placed for 48 h in basic (pH 9.8) aniline-blue solution and examined under UV light no yellow fluorescence was visible in the affected plant tissue, indicating that resistance acquired by BABA treatment was not associated with callose accumulation. With basic aniline-blue staining, however, a strong yellow emission of fluorescence from the sporangiophores was observed, indicating the presence of β-1,3-glucans in walls and cytoplasm of the sporangiophore.

Phloroglucinol-HCl staining for lignin was negative for control-infected leaf discs but positive for BABA-treated inoculated discs. While no red-stained cells were observed in the control infected tissue, many such cells were seen in the mesophyll (not in the epidermis – probably because fungal penetration occurs via stomata) of BABA-treated infected tissue. The red staining, characteristic for lignin accumulation, was mostly confined to the cells surrounding the inoculated site. The walls and cytoplasm of such cells were stained intensively.

The effect of BABA on mildew development was examined in leaf discs of 3 other cultivars. The cultivars used were: Sultanina, Emerald Reisling and Chardonnay. No significant differences were found between cultivars in their response to BABA, namely, visible sporulation was suppressed totally at ≥25 µg/ml. At 12.5 µg/ml, Chardonnay, Sultanina and Emerald Reisling showed sporulation in about 25% of the inoculated leaf discs.

To estimate the effect of inoculum density on the efficacy of BABA, leaf discs in Petri dishes were placed on BABA solutions and inoculated with either 500 or 2000 sporangia (in 50 µl) each. Data in Fig. 2 show that at the lowest BABA concentration tested, 10 µg/ml, 83 and 46% reduction in sporulation occurred in discs inoculated with 500 and 2000 sporangia/disc, respectively. At the highest concentration of BABA, 100 µg/ml, complete inhibition was seen with the low inoculum dose, as against 98.7% inhibition with the high inoculum dose.

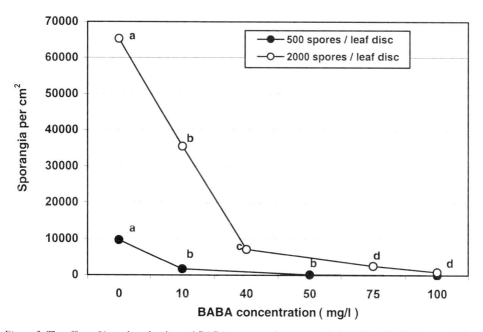

Figure 2. The effect of inoculum density and BABA concentration on sporulation of *P. viticola* in grape leaf discs.

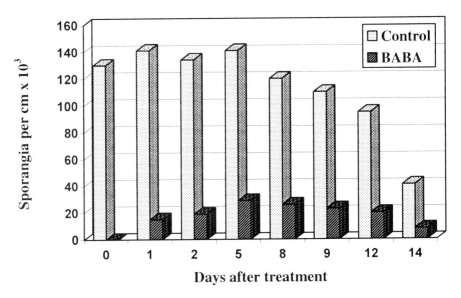

Figure 3. Persistence of BABA activity against *P. viticola* in grape leaf discs. Leaf discs were pulse-loaded with 100 μg/ml BABA for one day and then transferred to water and inoculated at 1-14 day intervals. Leaf discs in the 0 treatment were kept continuously on BABA. Figures in BABA treatments are significantly different at 5% level from the controls.

To evaluate for how long BABA activity can persist in grape leaf tissue, the following experiment was conducted: leaf discs in Petri dishes were placed on BABA solution of 100 μg/ml, or water, for one day, then removed, blotted dry between paper towels and placed on water. Inoculation took place 1-14 days after such pulse-loading with BABA (or water). Twelve days after inoculation sporangial counts were taken from the leaf discs to determine the magnitude of protection. Results (Fig. 3) show that at 1, 2, 5, 8, 9, 12 and 14 days after treatment, sporangial production in BABA-treated leaf discs were significantly reduced by 89, 86, 79, 78, 79, 79 and 80%, respectively. The lower counts at 14 days seemed to result from tissue senescence. It should be noted that leaf discs which were inoculated 0 days after floating, were maintained continuously on BABA (Fig. 3).

The post-infectional activity against grape downy mildew was examined in leaf discs inoculated with a 50 μl droplet/disc containing 1500 sporangia of *P. viticola*. At 0, 24, 48, 72 and 96 hours after inoculation discs were transferred to Petri dishes containing 200 μg/ml BABA solution on filter paper. Nine days after inoculation disease intensity was estimated visually and sporangial yield in discs was determined using a haemocytometer. The data presented in Fig. 4 show that BABA possessed a strong post-infectional activity against *P. viticola*. When applied at time 0, sporulation was completely inhibited but the tissue beneath the inoculated site showed an HR. When BABA was applied 24 h after inoculation the HR symptoms remained but a few sporangia were produced. At 48 h, the lesions expanded slightly, but sporulation was still almost completely (98.7%) suppressed. Inhibition was only partially removed at 72 h. At 96 h, symptoms appeared almost normal but sporulation was still inhibited by 66% (Fig. 4). Other experiments revealed that BABA had no effect on sporulation when applied to leaf discs bearing chlorotic lesions (8 days after inoculation).

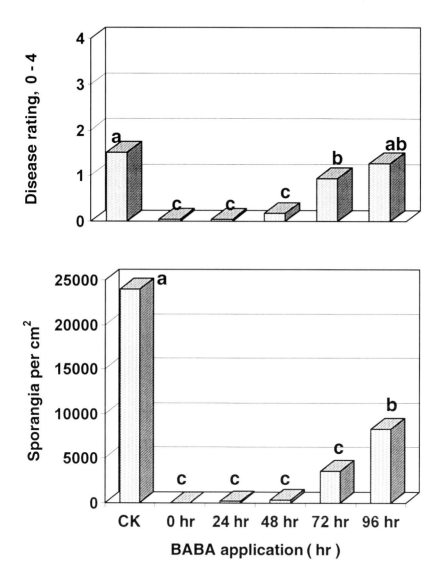

Figure 4. The post-infectional activity of BABA (200 μg/ml) against *P. viticola* in grape leaf discs. All leaf discs were inoculated at time 0. At various time intervals, discs were transferred from water to BABA. Upper panel – disease rating. Lower panel – number of sporangia produced.

To compare the efficacy of the S- and the R- enantiomers of BABA on grape downy mildew development, leaf discs (cultivar Superior) were floated in titer plates on aqueous solutions of the racemate or each of the enantiomers. Twelve days after inoculation, clear differences between the compounds were observed (Table 2). The S-enantiomer was totally ineffective, as sporulation occurred on all leaf discs at all concentrations. The racemate behaved as previously described, namely HR was observed at 16 and 32 µg/ml. The R-enantiomer was highly effective in suppressing fungal development: at ≥4 μg/ml it induced an HR and prevented visible sporulation (Table 2).

TABLE 2. The effect of the racemate, the S- and the R- enantiomers of 3-aminobutyric acid (BABA) on the development of *P. viticola* in grape leaf discs[a].

	Percentage leaf discs with visible sporulation at concentration, µg/ml					
BABA	0	2	4	8	16	32
Racemate	100	100	100	67	0[b]	0[b]
S-enantiomer	100	100	100	100	100	100
R-enantiomer	100	67	0[b]	0[b]	0[b]	0[b]

a   Ten mm diameter leaf discs of cultivar Superior were floated on 1 ml solutions of a 3-aminobutyric acid, lower surface uppermost, and inoculated with two 10 µl droplets of sporangial suspension of *P. viticola* containing 1000 sporangia each. Sporulation was assessed 12 days after inoculation. Data represent means from 3 experiments with 4 replicates per treatment (n = 12).
b   Hypersensitive response (HR).

In another experiment sporangial formation was quantitated in Sultanina leaf discs floating on water or R-BABA of 10 µg/ml for 15 days. Control-inoculated discs produced (as counted by a haemocytometer) a mean of $112 \pm 21.4 \ 10^3$ sporangia/disc as against $1.5 \pm 2.5 \ 10^3$ in R-BABA treated discs. Calcofluor staining of such discs followed by UV microscopical examination revealed the emergence of 600-900 sporangia from 65-100 sporangiophores from 6.15 mm^2 tissue (microscope field) in control leaves, against none in R-BABA-treated discs. In the later discs, the old sporangial inoculum was still visible after staining but was disregarded. Other discs from the same experiment showed, after phloroglucinol-HCl staining, a massive accumulation of lignin in mesophyll cells in R-BABA treated discs, but none was detected in control-inoculated discs. Here again, no callose formation was visible with basic aniline-blue staining.

3.2. ACTIVITY OF AMINOBUTYRIC ACID AGAINST P. VITICOLA *IN VITRO*

Due to its obligate biotrophic nature, germination of sporangia was the only phase in the life cycle of *P. viticola,* against which sensitivity to aminobutyric acids could be tested *in vitro*. To assess such an activity, sporangial suspensions were kept on ice in 1.5 ml tubes and mixed with aqueous solutions of aminobutyric acids to various final concentrations, so that sporangial concentration (5000 per ml) remained constant. Twenty μl of the mixtures were pipetted onto depressions of microscope slides and incubated at 14°C for 20 h in the dark. Microscopical examination revealed that in water controls most of the sporangia were empty, many zoospores were motile and only a few cystospores germinated. A similar picture was observed with DL-2-aminobutyric acid, DL-3-aminobutyric acid and 4-aminobutyric acid at concentrations ranging from 250-2000 μg/ml. Based on our experience on the enhancement of cystospore germination with sucrose, a 10 μl droplet of 50 mM sucrose was added to each depression. After an additional 20 h, germination of approximately 100 cystospores per depression as noticed in the water-sucrose controls, with a germ tube of 200-600 μm in length. With DL-3-aminobutyric (BABA) and 4-aminobutyric (GABA) (+ sucrose) of up to 2000 μg/ml, results were similar to the control. However, in DL-2-aminobutyric acid (AABA) of 1000-2000 μg/ml, germ tubes were much longer, reaching a length of 1000-2000 μm, and branched once or twice. Lowering the concentration of this compound to 250 μg/ml diminished this phenomenon.

3.3. WHOLE PLANT ASSAYS

3.3.1. *Uptake and Systemic Translocation of ^{14}C-BABA*
Because of the scarcity of ^{14}C-BABA tests were conducted with only foliar applications. Five days after such application, most of the ^{14}C-label was recovered from the leaf surface with water. TLC analysis (Cohen; Gisi, 1994a) revealed that this label on the leaf surface is due to ^{14}C-BABA only.

The data on the uptake into the treated leaves and translocation to other organs are presented in Table 3. They show that out of the 1 μCi of ^{14}C-BABA applied to leaf 1 or 4, a total of only 5.02% and 3.49% of the label applied could be re-extracted with MCW from the various organs of the plant (leaves, green stem and roots). In contrast, a total of 29.99% could be recovered with MCW when the ^{14}C-BABA was applied to leaf 6 (youngest treated) indicating that uptake is readily occurring through the upper surface of young rather than old leaves. Most of the penetrating label was recovered from the treated leaf itself. The recovery of ^{14}C-BABA from the remote younger leaves indicates on acropetal transport. Label was recovered also from the root system, especially when leaf 1 was treated, indicating also on basipetal translocation (Fig. 5). Other experiments, conducted with detached leaves in Petri dishes proved that uptake after 1 week ranged between 56.3-92.7% of the applied ^{14}C-BABA, suggesting that the low uptake in intact plants resulted from the hydrophilic nature of BABA and the quick drying of the label droplets. It should be mentioned that MCW extracts of ^{14}C-BABA-treated grape leaves contained ^{14}C-BABA, as proved by thin layer chromatography. This solvent mixture extracted 96% of the label present inside the tissue. The remaining 4% were bound to the cell walls, in a similar manner described in tomato (Cohen; Gisi, 1994a).

TABLE 3.   Uptake and distribution of ^{14}C-3-aminobutyric acid in grape plants after foliar application to either leaf 1, 4 or 6, while leaf 1 is nearest the stem base[a].

Organ	Percentage of recovered radioactivity		
	1	4	6
leaf 1	26.0 [b]	1.1	0.2
leaf 2	2.4	1.1	0.2
leaf 3	2.2	1.1	0.3
leaf 4	2.8	31.2 [b]	0.3
leaf 5	5.4	3.7	0.4
leaf 6	18.6	12.0	92.6 [b]
leaf 7	18.8	21.8	3.2
leaf 8	-	20.3	-
Stem	18.1	6.0	1.9
Root	5.7	1.7	0.9
% of applied[c]	5.02	3.49	29.99

a   Fifty 10 µl droplets of ^{14}C-BABA, containing a total of 1 µCi label, were applied to the upper surface of either leaf 1, 4 or 6 of 7-leaf grape plants. Five days later, plants were washed with water and organs extracted to determine radioactivity inside the tissue.
b   Treated leaf.
c   The rest of the ^{14}C-BABA applied (1 µCi) was recovered from the leaf surface by water.
-   Leaf not present.

3.4. SPRAY APPLICATION OF INTACT PLANTS

Percentage leaf area (abaxial surfaces) covered with sporangiophores and sporangia was found to decrease significantly as the concentration of BABA applied increased. Thus, plants sprayed with 0, 250, 500, 1000 and 2000 µg/ml solution of BABA sporulated on 90, 70, 35, 5 and 0% (mean values, n = 3) of their leaf total area, respectively.

To evaluate the protection associated with systemic translocation of BABA, the compound was formulated as a 25% wettable powder (WP) and applied (2000 µg/ml a.i.) to either the whole plant or the 3 bottom leaves of 8-leaf plants grown from seeds. Two days after spraying, plants were inoculated with *P. viticola* and 7 days later were placed in moist plastic tents in a growth chamber for 2 days to induce fungal sporulation. Results shown in Table 4 indicate that a total spray application with 25% WP BABA was highly effective in suppressing fungal sporulation. A similar treatment with a control (WP minus BABA) formulation resulted in a limited inhibition of fungal sporulation as compared to water-treated inoculated plants. When the formulated BABA was applied to the 3 bottom leaves, protection was observed in the treated leaves as well as in the upper two youngest leaves, but not in the middle leaves. The later result accords with the preferable acropetal translocation of ^{14}C-BABA (see Fig. 5).

Figure 5. Autoradiograph of a grape plant treated with 1 μ Ci of ^{14}C-BABA (2 mg per ml). Fifty 10 μl droplets of the radioactive compound (total 2.17×10^6 dpm) were placed on leaf 4 (where leaf 1 is nearest the stem base) of an 8-leaf plant. Plant was uprooted 5 days after treatment, washed with water, the 3-node woody stem removed, blotted dry and pressed for 2 weeks between thick papers. Plants were then exposed to X-ray film for 50 h. Heavy label is seen in the treated leaf, the youngest leaf and root apices (compared with Table 4). Label is not translocated to leaves 1, 2, 3, 5 and 6.

TABLE 4. Systemic activity of BABA against *P. viticola* in grape plants[a].

No. of leaf from stem base	Percentage leaf area showing sporulation			
	Water control	Blind formulation	25% WP BABA	25% WP BABA
	Treated leaves			
	1-8	1-8	1-8	1-3
1	100 a	50 b	5 c	0 c
2	100 a	60 a	7.5 b	15 b
3	100 a	50 b	0 c	0 c
4	100 a	50 b	0 c	95 a
5	100 a	60 b	2.5 c	100 a
6	100 a	80 a	2.5 b	80 a
7	100 a	80 a	5 b	0 b
8	100 a	30 b	0 c	0 c
mean	100.0	57.5	2.8	36.3

a   Eight-leaf plants grown from seed (cv. Emerald Reisling) were sprayed with either 0.8% of a 25% WP (wettable powder) BABA or with 0.6% WP blind formulation (containing no BABA) on all leaves (1-8) or the bottom 3 leaves only (1-3). Spray was applied to both leaf surfaces. Two days after spraying, plants were inoculated with *P. viticola*. Seven days after inoculation the plants were kept for 2 days in moist plastic tents and leaf lower surface area showing sporulation of the fungus was estimated visually. Figures in rows followed by the same letter are not significantly different (n = 3. Duncan's Multiple Range Test).

3.5. PROTECTION BY APPLYING BABA TO THE ROOT SYSTEM

Results (Fig. 6) show that exposure of the root system to BABA protected the foliage from the mildew indicating the uptake and translocation of the compound in the vascular system of the plant. The extent of protection was directly related to the quantity of BABA taken up: 7.5-10 mg per plant sufficed for 50% protection and 15-22 mg/plant totally abolished fungal sporulation (Fig. 6).

4. Discussion

The data presented in this paper demonstrate that DL-3-amino-n-butanoic acid (BABA) protects grape plants from infection by the obligate downy mildew pathogen *P. viticola*. A similar capacity was performed by BTH. Five other isomers of aminobutyric acid, namely L-2-amino, DL-2-amino, DL-2-isoamino, DL-3-isoamino and 4-aminobutyric acids were totally ineffective, suggesting that the amino group has to be in a 3-position to the terminal carboxylic group for activity. The 3-amino acids are not incorporated into proteins in nature (Rosenthal, 1982) except in one reported case (Helms et al., 1988). A single report on the occurrence of BABA in nature was as a secretion from tomato roots following soil solarization (Gamliel; Katan, 1992). The DL racemate of BABA is composed of the S- and R-enantiomers. When tested separately the R-enantiomer was highly active whereas the S-enantiomer was totally inactive, indicating a specific stereo structure affinity of the molecule to a putative ligand in the host or in the fungus.

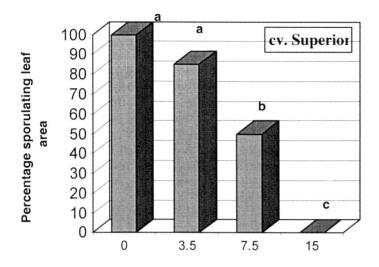

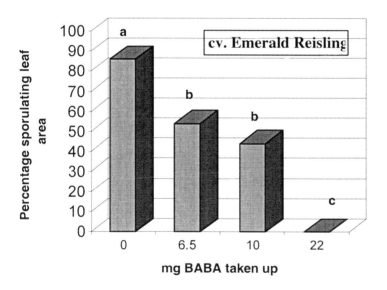

Figure 6. Protection of grape plants against *Plasmopara viticola* by BABA applied to the root system. Four-leaf plants of cv. Superior produced from cuttings were carefully uprooted and their root system was dipped in 50 ml of BABA solutions of 500, 1000 and 2000 μg/ml for 3 days. Roots were then washed, placed in water, and plants were inoculated (n = 3). Nine leaf plants of cv. Emerald Reisling produced from seeds were treated as above except that exposure of the root systems to BABA lasted for 24 h (n = 3). Figures in columns, for each cultivar, followed by different letters, are significantly different at 5% level (Duncan's Multiple Range Test).

Similar differential activity of the enantiomers were reported in tobacco (Cohen, 1994a). Because BABA (as well as R-BABA) had no adverse effect on fungal growth *in vitro*, we postulate that it acts as an inducer of systemic acquired resistance (SAR). The compound also did not affect fungal penetration into the host, as judged from the hypersensitive response associated with the inoculum droplets placed on treated leaf discs. Lignification of BABA-treated leaf discs occurred only upon inoculation, thus supporting the case for induced resistance. Unlike INA and BTH, BABA produced no phytotoxic symptoms when administered to either leaf discs, whole plants or the root system of grape plants (data not shown).

One-day feeding of leaf discs with BABA was enough to protect them from disease for at least 14 days. Enhanced senescence of such discs prevented further pursuing the persistence of activity. We assume that persistence in intact growing plants will decline with time due to dilution and translocation.

SAR normally requires a lapse period between treatment and inoculation (Kessmann et al., 1994; Ryals et al., 1996; Sticher et al., 1997). This allows for gene activation and signal transduction to take place. This is the case with SAR induction by either biological agents (e.g. TMV in N-tobacco) or chemical activators (e.g. SA, INA and BTH). Interestingly, BABA protected grape leaves from the mildew when applied post-infectionally. Even when applied at 48 h after inoculation protection achieved was 98.7%, based on number of sporangia produced relative to the control. This means that BABA activates host defence in such a quick and effective manner that it can suppress further development of established infections. It may be speculated that BABA deteriorates the fungus-penetrated host cells so that translocation of nutrients into the haustoria is blocked, thus prohibiting further mycelial growth and sporangial production. More histological data are required to prove such an hypothesis. Another hypothesis is that BABA quickly binds to host cell walls in a covalent manner, and modifies them so that further fungal ingress is blocked (Schwizer and Cohen, unpublished data).

BABA, as other amino acids, is translocated from source to sink in the plant via the phloem (Peoples; Gifford, 1990). When applied to leaves of intact plants it reaches the youngest leaves, as could be judged from either the acropetal translocation of ^{14}C-BABA or from the protection observed in top leaves following treatment of the bottom leaves with unlabelled BABA. Exposing the root system of bare-root intact plants to BABA also protected the leaves from the mildew indicating xylem transportation from the roots. Xylem transportation of ^{14}C-BABA from roots was also demonstrated in tomato (Cohen; Gisi, 1994a). Xylem transport and phloem reallocation of amino acids is well-known in plants (Peoples; Gifford, 1990).

BABA was reported to protect tomato against *Phytophthora infestans* (Cohen, 1994b), tobacco against *Peronospora tabacina* (Cohen, 1994a) and peppers against *Phytophthora capsici* (Sunwoo et al., 1996). Other reports demonstrated activity of BABA against soil-borne pathogens such as *Aphanomyces euiches* in peas (Papavizas, 1964), *Fusarium lycopersici* f.sp. *solani* in tomato (Li et al., 1996; Kalix et al., 1996), *Verticilium dahliae* in cotton (Li et al., 1996; Kalix et al., 1996), *Fusarium oxysporium* f.sp. *melonis* in melons (Cohen, 1996) and *Fusarium oxysporium* f. sp. *niveum* in watermelon (Y. Cohen, unpublished). Activity was also observed against *Plasmopara hatstedii* in sunflower attacking via either the roots or the foliage (Y.

Cohen, unpublished), and *Alternaria brassicicola* in broccoli (Cohen, 1996). It thus appears that BABA has a wide range of activity against fungal pathogens belonging to various taxonomic groups.

How BABA confers its activity against fungal disease is partially understood. In tomato and tobacco, foliar spray with BABA induced the accumulation of the pathogenesis-related (PR) proteins PR-1, chitinase and β-1, 3-glucanase (Cohen, 1994a; Cohen et al., 1994b). Surprisingly, no PR-proteins were deteced in tobacco stem-injected with BABA (Cohen, 1994a), although plants were highly protected, indicating that PR-protein accumulation is not the sole mechanism of resistance. In melons, BABA induced peroxidase and lignification in roots upon inoculation with *Fusarium* (R. Bitton and Y. Cohen, unpublished). In grape plants, no data are available as yet on gene activation or biochemical changes, if any, associated with the application of BABA. In preliminary experiments, we detected PR-1 but failed to detect resveratrol in BABA-treated grape leaf discs. Induction of PR proteins by SA or *Botrytis cinerea* in grape leaves was demonstrated but no data on systemic resistance against foliar pathogens are given (Renault et al., 1996). The induction by BABA of lignin accumulation in HR-responding mesophyll cells hints at the possible involvement of enzymes like phenylalanine-ammonia lyase (PAL) and peroxidase. In fact, BABA was reported (Newton et al., 1997) to increase PAL activity and induce resistance against late blight in potato leaf discs.

The role of salicylic acid in BABA-treated plants was investigated (A. Ovadia and Y. Cohen, unpublished) in the transgenic NahG tobacco plants which constitutively expresses salicylic acid hydroxylase (Delaney, 1997). BABA was found to be equally effective in NahG as well as the wild type Xanthi nc against blue mold suggesting that BABA may trigger down-stream SA-dependent event(s).

Induced resistance in general and systemic-induced resistance in particular were mostly studied with annual, herbaceous plants such as cucumber, tobacco, tomato and Arabidospis (Sticher et al., 1997). Very few studies were done with perennial, woody plants such as avocado (Dolan et al., 1986) and none, that we are aware of, on grape plants.

The data presented in this paper, together with our findings (Reuveni *et al.*, 2001) on the efficacy of BABA in vineyards against natural attack by downy mildew, makes BABA an attractive compound for practical agronomic use against *P. viticola* in grapes.

5. References

Bieleski, R.L. and Turner, N.A. (1996) Separation and estimation of amino acids in crude plant extracts by thin-layer electrophoresis and chromatography, *Anal.ytical Biochemistry* **17**, 278-293.

Cohen, Y. (1994a) 3-Aminobutyric acid induces systemic resistance against *Peronospore tabacina*, *Physiological and Molecular Plant Pathology* **44**, 273-288.

Cohen, Y. (1994b) Local and systemic control of *Phytophthora infestans* in tomato plants by DL-3-amino-n-butanoic acids, *Phytopathology* **84**, 55-59.

Cohen, Y. (1996) Induced resistance against fungal diseases by aminobutyric acids, in H. Lyr, P.E. Russel and H.D. Sisler (eds), *Modern Fungicides and Antifungal Compounds*, Intercept, Andover, UK, pp. 461-466.

Cohen, Y., Eyal, H. and Hanania, J. (1990) Ultrastructure, autofluorescence, callose deposition and lignification in susceptible and resistant muskmelon leaves infected with the powdery mildew fungus *Sphaerotheca fuliginea*, *Physiological and Molecular Plant Pathology* **36**, 191-204.

Cohen, Y. and Gisi, U. (1994a) Systemic translocation of ^{14}C-DL-3-aminobutyric acid in tomato plants in relation to induced resistance against *Phytophthora infestans*, *Physiological and Molecular Plant Pathology* **45**, 441-456.

Cohen, Y., Niderman, T., Mosinger, E. and Fluhr, R. (1994b) b-aminobutyric acid induces the accumulation of pathogenesis- related proteins in tomato (*Lycopersicon escullentum* Mill.) and resistance to late blight infections caused by *Phytophthora infestans*, *Plant Physiology* **104**, 59-66.

Cohen, Y., Reuveni, M. and Baider, A. (1999) Local and systemic activity of BABA (DL-3-aminobutyric acid) against *Plasmopara viticola* in grapevines, *European Journal of Plant Pathology* **105**, 351-361.

Delaney, T.P. (1997) Genetic dissection of acquired resistance to disease, *Plant Physiology* **113**, 5-12.

Dolan, T.E., Cohen, Y., and Coffey, M.D. (1986) Protection of *Persea* species against *Phytophthora cinnamomi* and *P. citricola* by prior inoculation with a citrus isolate of *P. parasitica*, *Phytopathology* **76**, 194-198.

Gamliel, A. and Katan, J. (1992) Influence of seed and root exudates on fluorescent pseudomonas and fungi in solarized soil, *Phytopathology* **82**, 320-327.

Helms, G.L., Moore, R.E., Niemczura, W.P., Patterson, M.L., Tomer, K.B. and Gross, M.L. (1988) Scytonemin A, a novel calcium antagonist from blue green alga, *Journal of Organic Chemistry* **53**, 1298-1307.

Kalix, S., Anfoka, G., Fi, Y., Stadnik, M. and Buchenauer, H. (1996) Induced resistance in some selected crops: prospects and limitations, in H. Lyr, P.E. Russell and H.D. Sisler (eds), *Modern Fungicides and Antifungal Compounds*, Intercept, Andover, UK, pp. 451-460.

Kessmann, H., Staub, T., Hofmann, C., Maetzke, T., Herzog, J., Ward, E., Uknes, S. and Ryals, J. (1994) Induction of systemic acquired resistance in plants by chemicals, *Annual Review of Phytopathology* **32**, 439-459.

LeRoux, P. and Clerjeau, M. (1985) Resistance of *Botrytis cinerea* Pers. and *Plasmopara viticola* (Berk. & Curt.) Berl. and de Toni to fungicides in the French vineyards, *Crop Protection* **4**, 137-160.

Li, J., Zingen-Sell, I., and Buchenauer, H. (1996) Induction of resistance of cotton plants to *Verticillium* wilt and of tomato plants to *Fusarium* wilt by 3-aminobutyric acid and methyl jasmonate, *Journal of Plant Disease Protection* **103**, 288-299.

Newton, A.C., Miller, S.K., Lyon, G.D. and Reglinski, T. (1997) Resistance elicitors as crop protectants, *Phytopathology* **87** (Suppl), S 69 (Abstract).

Papavizas, F.G. (1964) Greenhouse control of *Aphanaomyces* root rot of peas with aminobutyric acid and methylaspartic acid, *Plant Disease Reporter* **48**, 537-541.

Pearson, R.C. and Goheen, A.C. (1988) *Compendium of Grape Diseases*, APS Press, St. Paul, MN, USA.

Peoples, M.B. and Gifford, R.M. (1990) Long-distance transport of nitrogen and carbon from sources to sinks in higher plants, in D.T. Dennis and D.M. Turnip (eds), *Plant Physiology, Biochemistry and Molecular Biology*, Longman Scientific & Technical, UK, pp. 434-497.

Renault, A.S., Deloire, A. and Bierne, J. (1996) Pathogenesis-related proteins in grapevines induced by salicylic acid and *Botrytis cinerea*, *Vitis* **35**, 49-52.

Reuveni, M., Zahavi, T. and Cohen, Y. (2001) Controlling downy mildew (*Plasmopara viticola*) in field-grown grapevine by β-aminobutyric acid (BABA), *Phytoparasitica* **29**, 125-133.

Rosenthal, G.A. (1982) *Plant Nonprotein Amino and Imino Acids: Biological, Biochemical and Toxicological Properties*, Academic Press, New York.

Ryals, J.A., Neuenschwander, U.M., Willitis, M.G., Molina, A., Steiner, H. and Hunt, M.O. (1996) Systemic acquired resistance, *Plant Cell* **8**, 1809-1819.

Sticher, L., Mauch-Mani, B. and Metraux, J.P. (1997) Systemic acquired resistance, *Annual Review of Phytopathology* **35**, 235-270.

Sunwoo, J.Y., Lee, Y.K. and Hwang, B.K. (1996) Induced resistance against *Phytophthora capsici* in pepper plants in response to DL-b-amino-n-butyric acid, *European Journal of Plant Pathology* **102**, 663-670.

BINOMIALS IN THE PERONOSPORALES, SCLEROSPORALES AND PYTHIALES

M. W. DICK
Centre for Plant Diversity and Systematics, Department of Botany,
School of Plant Sciences, University of Reading,
READING RG6 6AU, U.K.

1. Introduction

The following lists of genera, species and authorities for downy mildews (DMs) and related pathogens have been extracted from the complete lists for straminipilous fungi (Dick, 2001*b*). The genera are listed in alphabetic order under families; the species are listed in alphabetic order under each genus, with the type species identified, in Tables 2-7. The lists are concordant with the summary classification of the orders in the sub-classes Peronosporomycetidae and Saprolegniomycetidae (Dick, 1995, 2001*a, b*; Dick *et al.*, 1989) and given in Table 1. The validity of the sub-classes has now been substantiated by 18S rDNA data (Dick *et al.*, 1999), but the position of the Sclerosporaceae has yet to be ascertained by molecular data.

For uniformity, it is desirable that the recommendations of Brummitt and Powell (1992) or Kirk and Ansell (1992) for authority abbreviations are followed, but it should be noted that some authors do not fully accept these recommendations. Unfortunately, the most recent monographs on *Peronospora* (Kochman and Majewski, 1970; Constantinescu, 1991*a*); earlier entries in the *Index of Fungi*; the check-lists for *Pythium* (Dick, 1990*b*), and the most recent key to *Phytophthora* (Stamps *et al.*, 1990), did not use these forms of author abbreviations consistently. Erwin and Ribeiro (1996) adopted the potentially misleading style of citing the authors of the publication, rather than the authority for the binomial, after the masthead binomial.

Note that the spelling and use of accents in 'authority' notation may differ from the forms used in published articles, for example: Itô for Ito; Plaäts-Nit. for Plaats-Niterink; Skvortzov for Skvortzow; Sorokīn for Sorokin/Sorokine). Korf (1996) has been followed in that '*ex*' combinations have been shortened and '*et al.*' is used for authorities of three or more names (though these details are given in square brackets after the entry). For more recent taxa (and for most binomials in *Plasmopara*) the page reference to the *Index of Fungi* is given, with part numbers for the incomplete volume 6.

Many intra-specific taxa (e.g., Skalický, 1964; Skidmore and Ingram, 1985) are without justification in the sense that they have not been named in accordance with the

P.T.N. Spencer-Phillips et al. (eds.), Advances in Downy Mildew Research, 225–265.

TABLE 1. Partial classification of the Peronosporomycotina (with references to Table numbers following)

PERONOSPOROMYCOTINA

    PERONOSPOROMYCETES

        PERONOSPOROMYCETIDAE
           PERONOSPORALES
                PERONOSPORACEAE (Table 2)
                ALBUGINACEAE (Table 3)

           PYTHIALES
                PYTHIACEAE (Table 4)
                PYTHIOGETONACEAE (Table 5)

        SAPROLEGNIOMYCETIDAE
           SCLEROSPORALES
                SCLEROSPORACEAE (Table 6)
                VERRUCALVACEAE (Table 7)

ICBN rules; but because they may have some value for plant pathologists where they are supported by herbarium specimens (Constantinescu, 1991*a*), they have been included. Authors of non-taxonomic articles should note that the use of the binomial alone, where varietal names have been published, automatically includes all varieties and this may not be the intention! For example, the use of *Pythium ultimum* alone will include both *P. ultimum* var. *ultimum* and *P. ultimum* var. *sporangiiferum*. It should be remembered that while subspecies, varieties and forms (*formae*) have formal taxonomic status, with recommendations governing the use of one or other of these intraspecific categories, *formae speciales* are not governed by *ICBN* rules and recommendations. The list for *Plasmopara*, but not that for *Peronospora*, includes all of these *formae* and *formae speciales* names for the purposes of retrieval, but without taxonomic reappraisal.

    Check-lists for the principal genera of the dicotyledonous downy mildews are not readily available (but see below) although taxonomic reviews of some of the genera of the Peronosporales can be found in Barreto and Dick (1991, 2001*b*), Biga (1955), Constantinescu (1979, 1989, 1991*b*, 1992, 1996*a*, *b*, 1998), de Bary (1863), Gäumann (1923), Gustavsson (1959*a*, *b*), Kochman and Majewski (1970), Săvulescu (1962), Săvulescu and Săvulescu (1952), Shaw (1978, 1981), Skalický (1966) and Waterhouse and Brothers (1981). Hall (1996) has reviewed species concepts in the downy mildews. For brevity, excluded taxa and accepted synonyms are not given here, except where they have been used in the CABI database. The treatment of the Peronosporaceae here is not consistent from genus to genus because of the disparate numbers of species in genera, particularly *Peronospora* and *Plasmopara*; more detailed information on the treatment adopted is given under the genus name. The most complete, annotated, strictly alphabetic list of *Peronospora* names is in Constantinescu (1991*a*). There is no recent compilation for names in *Plasmopara*. *Albugo* has not been reviewed since the account by Biga (1955). In Table 2, to facilitate comprehension of relationships, binomials are listed alphabetically under the sub-heads of the supra-ordinal groupings of the host angiosperm orders according to the most recent molecular phylogeny of the angiosperms (Soltis, Soltis, Chase *et al.*, 1998*a*, *b*; Soltis, Soltis and Chase, 1999) and the Angiosperm Phylogeny Group (APG, 1998). Note that the Urticaceae and related families are classified with the Rosids and that the asclepiads are included within the Apocynaceae. See Dick, this volume, for a summary table of host ranges according to this classification and Dick (2001*b*) for an integrated table of host relationships for all genera of the DMs and *Albugo*.

Waterhouse (1964) provided an earlier summary of the graminicolous downy mildews. Since then, Dick *et al.* (1984, 1989) and Dick (1988, 1990*a*, 1995) have proposed a reclassification of these fungi into the Sclerosporales.

New check-lists for accepted species, and taxa of debatable synonymy, for *Phytophthora* and *Pythium* are included because it seems increasingly probable that reference will need to be made to various species of *Phytophthora* (see Stamps *et al.*, 1990; Erwin and Ribeiro, 1996; Cooke *et al.* this volume; Dick, this volume) and *Pythium* (see Waterhouse, 1968; Plaats-Niterink, 1981; Dick, 1990*b*) when discussing the phylogeny and systematics of the downy mildews. Minor genera related to *Phytophthora* are also included.

A debate is being vigorously developed regarding the value of hierarchical versus hierarchy-less systematics (e.g., Hibbett and Donoghue, 1998; Dick, 2001*b* and this volume). The argument stems from the natural relationships and cladistic branching (phylogenies) presumed from molecular biological investigations. However, it is unlikely that all the known or named taxa will be characterized by molecular means in the foreseeable future. This is not because of technical ability, but because of the unavailablity of material or finance to carry out this work. Thus it is important that hierarchical systematics should continue to be developed alongside molecular biology. Nomenclatural changes based on molecular data can only be made when the type species have been characterized by these techniques. It is also pertinent to restate that these phylogenies are only probabilities (even if very highly significant probabilities) and that they are based on the sequences for very few genes (in many cases restricted solely to parts of the rDNA gene - 18S rDNA and the ITS and 5.8S regions). There is no indication that these genes are functionally important in determining the biodiversity, including host preferences, of the downy mildews. Evaluation of the possible relationships between incipient speciation and intra-specific variation will also need a molecular biodiversity input.

TABLE 2. Peronosporales: Peronosporaceae

BASIDIOPHORA Roze & Cornu

Basidiophora entospora Roze & Cornu **[type species]**   [Saccardo: *Sylloge Fungorum* **6**: 239]
  Basidiophora kellermanii (Sacc.) G. W. Wilson [Swingle *ex* Sacc.] [≡ *Benua kellermanii*]
Basidiophora montana R. W. Barreto   [*Index of Fungi* **6(4)**: 196]

BENUA Constant.

Benua kellermanii (Sacc.) Constant. **[type species; monotypic]**
                                        [Swingle *ex* Sacc.] [*Index of Fungi* **6(17)**: 900]

BREMIA Regel

    [NOTE: Some authorities would regard *Bremia sensu stricto* as monotypic. Infra-specific
                 taxa may be of dubious standing. Synonymy has not been evaluated.]

Bremia betae H. C. Bai & X. Y. Cheng   [*Index of Fungi* **6(8)**: 422]
Bremia centaurea Syd.
Bremia cicerbitae C. J. Li & Z. Q. Yuan   [*Index of Fungi* **6(20)**: 1085]
Bremia cirsii (Uljan.) J. F. Tao & Y. N. Yu   [Jacz. *ex* Uljan.] [*Index of Fungi* **6(5)**: 260]
Bremia elliptica Sawada
Bremia gemminata (Unger) Kochman & T. Majewski   [*Index of Fungi* **4**: 4]
Bremia graminicola Naumov var. *graminicola*
Bremia graminicola Naumov var. *indica* M. K. Patel   [*Index of Fungi* **1**: 277]
Bremia lactucae Regel var. *lactucae* **[type species]**
Bremia lactucae Regel var. *arctii* Uljan.   [*Index of Fungi* **4**: 4]
Bremia lactucae Regel var. *cardui* Uljan.   [*Index of Fungi* **4**: 4]
Bremia lactucae Regel var. *hedypnoidis* Uljan.   [*Index of Fungi* **4**: 4]
Bremia lactucae Regel var. *pterothecae* Uljan.   [*Index of Fungi* **4**: 4]
Bremia lactucae Regel var. *willemetiae* Uljan.   [*Index of Fungi* **4**: 4]
Bremia lactucae Regel var. *xeranthemi* Uljan.   [*Index of Fungi* **4**: 4]
Bremia lactucae Regel forma *mulgedii* C. B. Benua   [*Index of Fungi* **4**: 303]
Bremia lactucae Regel forma *taraxaci* L. Ling & F. L. Tai   [*Index of Fungi* **1**: 226]
Bremia lactucae Regel forma *taraxaci* C. B. Benua [later homonym] [*Index of Fungi* **4**: 303]
Bremia lactucae Regel forma specialis *carthami* M. O. Milovtzova
Bremia lactucae Regel forma specialis *centaureae* D. I. Skidmore & D. S. Ingram [*Index of Fungi* **5**: 470]
Bremia lactucae Regel forma specialis *chinensis* L. Ling & M. C. Tai
Bremia lactucae Regel forma specialis *cirsii* D. I. Skidmore & D. S. Ingram   [*Index of Fungi* **5**: 470]
Bremia lactucae Regel forma specialis *crepidis* D. I. Skidmore & D. S. Ingram [*Index of Fungi* **5**: 470]
Bremia lactucae Regel forma specialis *hieracii* D. I. Skidmore & D. S. Ingram [*Index of Fungi* **5**: 470]
Bremia lactucae Regel forma specialis *lapsanae* D. I. Skidmore & D. S. Ingram [*Index of Fungi* **5**: 470]
Bremia lactucae Regel forma specialis *leontodi* D. I. Skidmore & D. S. Ingram [*Index of Fungi* **5**: 470]
Bremia lactucae Regel forma specialis *picridis* D. I. Skidmore & D. S. Ingram [*Index of Fungi* **5**: 470]
Bremia lactucae Regel forma specialis *ovata* L. Ling & M. C. Tai
Bremia lactucae Regel forma specialis *senecionis* D. I. Skidmore & D. S. Ingram [*Index of Fungi* **5**: 470]
Bremia lactucae Regel forma specialis *sonchi* D. I. Skidmore & D. S. Ingram   [*Index of Fungi* **5**: 470]
Bremia lactucae Regel forma specialis *sonchicola* L. Ling & M. C. Tai
Bremia lactucae Regel forma specialis *taraxaci* L. Ling & M. C. Tai
Bremia lactucae Regel forma specialis *taraxaci* D. I. Skidmore & D. S. Ingram [*Index of Fungi* **5**: 470]

Table 2, continued.

Bremia lagoceridis Y. N. Yu & J. F. Tao   [*Index of Fungi* 6(5): 260]
Bremia lapsanae Syd. [as '*lampsanae*']
Bremia leibnitziae J. F. Tao & Y. Qin   [*Index of Fungi* 5: 280]
Bremia moehringiae T. R. Liu & C. K. Pai   [*Index of Fungi* 5: 421]
Bremia microspora Sawada
Bremia ovata Sawada
Bremia picridis S. Ito & Tokun.
Bremia picridis-hieracioidis Savinceva   [*Index of Fungi* 4: 303]
Bremia saussurea Sawada
Bremia sonchi Sawada
Bremia stellata (Desm.) Kochman & T. Majewski   [*Index of Fungi* 4: 4]
Bremia taraxaci S. Ito & Tokun.
Bremia tulasnei (Hoffm.) Syd.

BREMIELLA G. W. Wilson

   [NOTE: See Constantinescu (1979; 1991*b*).]

Bremiella baudysii (Skalický) Constant. & Negrean   [*Index of Fungi* 2: 197; **4**: 589]
Bremiella megasperma (Berl.) G. W. Wilson **[type species]**
      Bremiella oenanthes J. F. Tao & Y. Qin [= *B. baudysii*]   [*Index of Fungi* 5: 421]
Bremiella sphaerosperma Constant.

PARAPERONOSPORA Constant.

   [NOTE: All hosts in Asteraceae. See Constantinescu (1989; 1996*a*).]

Paraperonospora apiculata Constant.   [*Index of Fungi* 6 (14): 768]
Paraperonospora artemisiae-annuae (L. Ling & M. C. Tai) Constant.
                                          [*Index of Fungi* 1: 168; **3**: 415; **5**: 1065]
Paraperonospora artemisiae-biennis (Gäum.) Constant.
                           [Saccardo: *Sylloge Fungorum* **26**: 31;*Index of Fungi* 3: 415; **5**: 1065]
Paraperonospora chrysanthemi-coronarii (Sawada) Constant.   [Petrak's Lists **6**]
Paraperonospora leptosperma (de Bary) Constant. **[type species]**   [*Index of Fungi* 3: 415; **5**: 1065]
Paraperonospora minor (Săvul. & Rayss) Constant.   [*Index of Fungi* 5: 1065]
Paraperonospora multiformis (J. F. Tao & Y. Qin) Constant.   [*Index of Fungi* 5: 1065]
Paraperonospora sulphurea (Gäum.) Constant.   [*Index of Fungi* 3: 415; **5**: 1065]
Paraperonospora tanaceti (Gäum.) Constant.   [*Index of Fungi* 3: 415; **5**: 1065]

Table 2, continued.

PERONOSPORA Corda

Peronospora rumicis Corda **[type species]**

[NOTES: Perhaps *ca* 75 species, 800 binomials and infra-specific taxa proposed; for the full list of
binomials see Constantinescu (1991*a*) and *Index of Fungi* therafter. *Index of Fungi* references
are mostly excluded because full information is available in Constantinescu (1991*a*).   Dr
Constantinescu writes (*in litt.*, 10 Jul 96): "I do not have a check-list of taxonomically correct
or accepted *Peronospora* species. Such a list can only be made after a taxonomic study of the
fungi involved. There is no such study covering the whole genus *Peronospora*".

To facilitate relationships in the following long list, species are entered alphabetically according
to the phylogenetic arrangement of their hosts, using the Soltis, Soltis, Chase *et al.*, 1998*a*, *b*;
Soltis, Soltis and Chase, 1999 and the Angiosperm Phylogeny Group (APG, 1998) system.

Names for *Peronospora* as used in Kochman & Majewski (1970) (i.e., including
Pseudoperonospora: are referenced by [page number]-[entry number] to descriptions);
valid [bold-face] binomials for *Peronospora sensu stricto*, as moderated by Constantinescu
(1991*a*) are justified left; updated from *Index of Fungi* to Vol 6 (16);
indented entries are those generally considered conspecific, either ≡ (nomenclatural synonym)
or = (taxonomic synonym) [with relevant binomial with priority];
synonyms used in the CABI database have been included;
F is used to indicate information available in Farr *et al.* (1989);
valid, infra-specific taxa listed in Constantinescu (1991*a*) and Farr *et al.* (1989) are omitted.]

EUDICOTS; no supra-ordinal phylogeny

RANUNCULALES: Berberidaceae

Peronospora achlydis S. Ito & Tokun.

RANUNCULALES: Papaveraceae with Fumariaceae

Peronospora affinis Rossmann	[140-58]
Peronospora arborescens (Berk.) Casp.	[143-61; **F**]
Peronospora argemones Gaüm.	[142-60]
Peronospora bocconiae Syd.	
Peronospora bulbocapni Beck	[138-55]
Peronospora chelidonii Jacz. & P. A. Jacz.	[T. Miyabe ex Jacz & P. A. Jacz.]
Peronospora corydalis de Bary	[139-56; **F**]
Peronospora corydalis-intermediae Gaüm.	[140-57]
Peronospora cristata Tranzschel	
Peronospora dicentrae Syd.	
Peronospora gaeumannii Mundk. [= *P. indica* Gaum.]	
Peronospora glaucii Lobik	[141-59]
Peronospora hylomeconis S. Ito & Tokun.	
Peronospora hypecoi Bremer [*nom. illegit.*]	[Bremer in Petrak] [*Index of Fungi* **2**: 150]
Peronospora hypecoi Jacz. & P. A. Jacz. [as '*hypecoumis*']	
Peronospora meconopsidis Mayor [replaces *Peronospora mayorii* Gaüm.]	[214-149; **F**]

Table 2, continued.

Peronospora papaveris Tul. [*nom. invalid.*]   [Saccardo: *Sylloge Fungorum* **7**: 251]
Peronospora papaveris-pilosi Vienn.-Bourg. [*nom. invalid.*]
Peronospora roemeriae Zaprom.

RANUNCULALES: Ranunculaceae

Peronospora aconiti Y. N. Yu
Peronospora alpicola Gaüm.   [133-49]
    Peronospora alpina Johans. [≡ *Plasmopara alpina*]   [Saccardo: *Sylloge Fungorum* **9**: 343]
Peronospora anemonones Tramier [*nom. invalid.*]
Peronospora apiospora G. Poirault
Peronospora consolidae Jacz. & P. A. Jacz.   [Lagerh. ex Jacz & P . A. Jacz.] [131-45]
    Peronospora eranthidis-hyemalis (Pass.) A. Fisch. [as 'eranthidis'] [= *P. myosuri*]
          [Saccardo: *Sylloge Fungorum* **17**: 521]
Peronospora ficariae de Bary   [Tul. ex de Bary] [137-54]
Peronospora gigantea Gaüm.   [135-51]
    Peronospora glacialis (A. Blytt) Gaüm. [=*P. ficariae*]   [135-50]
Peronospora hellebori-purpurascentis Sāvul. & Rayss
    Peronospora hepaticae de Bary [= *Plasmopara pygmaea*]   [Saccardo: *Sylloge Fungorum* **7**: 253]
Peronospora hiemalis Gaüm.   [137-53; **F**]
Peronospora illyrica Gaüm.
Peronospora iwatensis S. Ito & Muray.
Peronospora leptopyri C. K. Pai
    Peronospora macrocarpa Corda [= *Plasmopara pygmaea*]   [Saccardo: *Sylloge Fungorum* **7**: 240]
Peronospora myosuri Fuckel   [133-48]
Peronospora parvula Jacz. & P. A. Jacz. [≡ *Plasmopara parvula*]
          [W. G. Schneid. *ex* Jacz. & P. A. Jacz.] [132-47]
Peronospora pennsylvanica Gaüm.   [**F**]
Peronospora pulveracea Fuckel   [131-46]
    Peronospora pygmaea Unger [≡ *Plasmopara pygmaea*]   [Saccardo: *Sylloge Fungorum* **7**: 240]
Peronospora ranunculi Gaüm.   [136-52; **F**]
Peronospora ranunculi-carpatici Sāvul. & Rayss
    Peronospora ranunculi-flabellati Vienn.-Bourg. [*nom. invalid.*]
Peronospora ranunculi-oxyspermi Jacz. & Sergeeva
Peronospora ranunculi-peduncularis Roiv.
Peronospora ranunculi-sardoi Sāvul. & Rayss
Peronospora ranunculi-steveni Sāvul. & Rayss
Peronospora yamadana Togashi

CORE EUDICOTS; no supra-ordinal phylogeny

<u>Vitaceae</u>

Peronospora viticola (Berk. & M. A. Curtis) de Bary [≡ *Plasmopara viticola*]
          [Saccardo: *Sylloge Fungorum* **3**: 111]

CARYOPHYLLALES: <u>Aizoaceae</u>

Peronospora mesembryanthemi Verwoerd

Table 2, continued.

CARYOPHYLLALES: Amaranthaceae [see also Chenopodiaceae]

Peronospora amaranthi Gaüm.   [130-44]

CARYOPHYLLALES: Cactaceae

Peronospora cactorum Lebert & Cohn [as 'E. Cohn & Lebert'] [≡ *Phytophthora cactorum*]
[Saccardo: *Sylloge Fungorum* 7: 238]

CARYOPHYLLALES: Caryophyllaceae [see also Illebraceae]

Peronospora agrostemmatis Gaüm.	[102-11]
Peronospora alsinearum Casp.	[118-32; **F**]
Peronospora arenariae (Berk.) Tul.	[112-25]
Peronospora atlantica Gaüm.	
Peronospora campestris Gaüm.	[103-12]
Peronospora cerastii-anomali Săvul. & Rayss	
Peronospora cerastii-brachypetali Săvul. & Rayss	[107-17]
Peronospora cerastii-glandulosi S. Ito & Tokun.	
Peronospora cucubali S. Ito & Tokun.	
Peronospora dianthi de Bary	[113-26]

    Peronospora dianthicola Barthelet [*nom. invalid.*]   [107-18; **F**]

Peronospora fontana A. Gustavsson	
Peronospora gypsophilae Jacz. & P. A. Jacz.	[108-19]
Peronospora helvetica Gaüm.	
Peronospora holostei de Bary	[Casp. *ex* de Bary] [109-21]

    Peronospora honckenyae (Syd. & P. Syd.) Syd. [= *P. alsinearum* var. *honckeniae*]   [110-22]
    Peronospora jaczewskii Săvul. & Rayss [= *P. gypsophilae*]

Peronospora lepigoni Fuckel	[117-31]
Peronospora lychnitis Gaüm.	[110-23]
Peronospora media Gaüm.	
Peronospora melandrii Gaüm.	[111-24]
Peronospora melandrii-noctiflori Săvul. & Rayss	
Peronospora moenchiae C. Camara & Oliviera	
Peronospora obovata Bonord.	[115-29; **F**]
Peronospora parva Gaüm.	[119-33]
Peronospora paula A. Gustavsson	[106-16]
Peronospora polycarpi Mayor & Vienn.-Bourg. [as '*polycarponis*']	
Peronospora pseudostellariae G. Y. Yin & Z. S. Yang	[*Index of Fungi* **6(11)**: 621]
Peronospora septentrionalis Gaüm.	[104-13]
Peronospora silenes G. W. Wilson [≡ *P. arenariae* var. *macrospora*]	[114-28; **F**]

    Peronospora stellariae de Bary [*nom. invalid.*]

Peronospora stellariae-aquaticae Sawada	
Peronospora stellariae-radiantis S. Ito & Tokun.	
Peronospora stellariae-uliginosae Sawada	
Peronospora tomentosa Fuckel [= *P alsinearum*]	[105-15]
Peronospora tornensis Gaüm.	
Peronospora trivialis Gäum.	
Peronospora uralensis Jacz. & P. A. Jacz.	
Peronospora vernalis Gaüm.	[116-30]

    Peronospora vexans Gäum. [*nom. invalid.*]

Table 2, continued.

CARYOPHYLLALES: Chenopodiaceae [now subsumed in Amaranthaceae]

Peronospora atriplicis-halimi Săvul. & Rayss
Peronospora atriplicis-hastatae Săvul. & Rayss
Peronospora atriplicis-hortensis Săvul. & Rayss
Peronospora atriplicis-tataricae Oescu & Rădul.
Peronospora axyridis C. Benois
    Peronospora betae J. G. Kühn [= *P. farinosa*]
Peronospora bohemica Gaüm.
Peronospora boni-henrici Gaüm.   [125-39]
    Peronospora ceratocarpi Kalymb. [*nom. invalid.*]   [*Index of Fungi* **3**: 193]
Peronospora chenopodii Schltdl.   [123-37]
    Peronospora chenopodii-ambrosioides Golinia [*nom. invalid.*]
Peronospora chenopodii-ficifolii Sawada
Peronospora chenopodii-glauci Gaüm.   [125-38]
Peronospora chenopodii-opulifolii Săvul. & Rayss
Peronospora chenopodii-polyspermi Gaüm.   [126-40]
Peronospora chenopodii-rubri Gaüm.
Peronospora chenopodii-urbici Săvul. & Rayss
Peronospora chenopodii-vulvariae Săvul. & Rayss
Peronospora daturae Hulea
Peronospora echinopsili Tunkina
Peronospora effusa (Grev.) Tul.   [129-43]
    Peronospora epiphylla (Pers.) Pat. & Lagerh. [= *P. farinosa*]
Peronospora eurotiae Kalymb.
Peronospora farinosa (Fr.) Fr. *sensu lato*   [(Fr. : Fr.) Fr.] **[F]**
Peronospora farinosa (Fr.) Fr. *sensu stricto*   [(Fr. : Fr.) Fr.]
Peronospora iolotanica Kolosch.
Peronospora kochiae Gaüm.
    Peronospora kochiae-prostratae Sandu & Iacob [= *P. kochiae-scopariae*]   [*Index of Fungi* **4**: 315]
Peronospora kochiae-scopariae Kochman & T. Majewski   [127-41]
Peronospora litoralis Gaüm.   [120-34]
    Peronospora minor (Casp.) Gaüm. [= *P. effusa* var. *minor*]   [120-35]
    Peronospora minor (Casp.) Gäum. [≡ *P. farinosa*]
Peronospora monolepidis Gaüm.
Peronospora muralis Gaüm.
Peronospora nitens Oescu & Rădul.
Peronospora obionis-verruciferae Săvul. & Rayss
Peronospora salicorniae Jenkina
    Peronospora schachtii Fuckel [≡ *P. farinosa*]   [122-36]
    Peronospora spinaciae Laubert [≡ *P. farinosa*]
Peronospora tatarica Săvul. & Rayss
Peronospora teloxydis Jacz. & P. A. Jacz.
Peronospora ussuriensis Jacz. & P. A. Jacz. [new name for *P. effusa* var. *manshurica* Naumov]
Peronospora variabilis Gaüm.
Peronospora vistulensis Wróbl.   128-42]

CARYOPHYLLALES: Illecebraceae [now subsumed in Caryophyllaceae]

Peronospora herniariae de Bary   [108-20]
Peronospora scleranthi Rabenh.   [Rabenh. ex J. Schröt. in Cohn] [113-27]

Table 2, continued.

CARYOPHYLLALES: Nyctaginaceae

Peronospora oxybaphi Ellis & Kellerm.   [F]

CARYOPHYLLALES: Plumbaginaceae

Peronospora constantineanui Săvul. & Rayss
Peronospora limonii Simonyan
Peronospora statices Lobik

CARYOPHYLLALES: Polygonaceae

Peronospora americana Gaüm.   [F]
Peronospora ducometii Siemaszko & Jank.   [98-6]
Peronospora eriogoni H. Solheim & Gilb.   [*Index of Fungi* 4: 221; F]
Peronospora fagopyri Jacz. & P. A. Jacz.
Peronospora jaapiana Magnus   [100-9]
Peronospora polygoni (Thüm.) A. Fisch. [≡ *P. effusa* var. *polygoni*]   [99-7]
Peronospora polygoni-convolvuli A. Gustavsson   [99-8]
Peronospora rumicis Corda [**type species**]   [101-10]
Peronospora rumicis-rosei Rayss
Peronospora sinensis D. Z. Tang

CARYOPHYLLALES: Portulacaceae

Peronospora calindriniae Speg.
Peronospora claytoniae Farl.   [F]

SANTALALES: Santalaceae

Peronospora thesii Lagerh.   [97-5]

SAXIFRAGALES: Crassulaceae

Peronospora sempervivi Schenk. [= *Phytophthora cactorum*]   [Saccardo: *Sylloge Fungorum* 7: 238]

SAXIFRAGALES: Grossulariaceae

Peronospora ribicola J. Schröt. [≡ *Plasmopara ribicola*]   [Saccardo: *Sylloge Fungorum* 7: 243]

SAXIFRAGALES: Saxifragaceae

Peronospora chrysosplenii Fuckel   [181-108]
Peronospora minima G. W. Wilson   [182-109]
Peronospora saxifragae Bubák   [182-110; F]
Peronospora whippleae Ellis & Everh.   [F]

Table 2, continued.

CORE EUDICOTS: ROSIDS: sister groups to Eurosids I and II

Zygophyllaceae

Peronospora tribulina Pass.

GERANIALES: Geraniaceae

    Peronospora beccarii Pass. [= *P. conglomerata*]
Peronospora conglomerata Fuckel   **[F]**
Peronospora effusa-ciconia Becc.
Peronospora erodii Fuckel   [215-151; **F**]
    Peronospora geranii Peck [≡ *Plasmopara geranii*]   [Saccardo: *Sylloge Fungorum* 7: 242]

CORE EUDICOTS: ROSIDS: EUROSIDS I

CUCURBITALES: Cucurbitaceae

    Peronospora actinostemmatis (Sawada) Skalický [= *Pseudoperonospora cubensis*]
                                                 [*Index of Fungi* 3: 413]
    Peronospora australis Speg. [≡ *Plasmopara australis*]   [Saccardo: *Sylloge Fungorum* 7: 260-261]
    Peronospora cucumeris (Sawada) Skalický [= *Pseudoperonospora cubensis*]
                                                 [*Index of Fungi* 3: 413]
    Peronospora luffae (Sawada) Skalický [= *Pseudoperonospora cubensis*]   [*Index of Fungi* 3: 413]
    Peronospora momordicae (Sawada) Skalický [= *Pseudoperonospora cubensis*]
                                                 [*Index of Fungi* 3: 413]
    Peronospora sicyicola Peck [= *Plasmopara australis*]   [Saccardo: *Sylloge Fungorum* 7: 260]

FABALES: Fabaceae

Peronospora aestivalis Syd.   [201-134]
Peronospora astragali Syd.
Peronospora astragali-purpurei Mayor & Vienn.-Bourg.
Peronospora astragalina Syd.   [191-121]
Peronospora cilicia Bremer & Gäum.
Peronospora coronillae Gäum.   [193-123]
    Peronospora coronillae-minimae Vienn.-Bourg. [*nom. invalid.*]
Peronospora coronillicola C. Camara & Oliviera
Peronospora cytisi Rostr.   [195-126]
Peronospora desmodii S. Ito & Tokun.
Peronospora dipeltae Jacz. & P. A. Jacz.
Peronospora dorycnii Uljan.
Peronospora ervi A. Gustavsson [= *P. viciae sensu lato*]   [214-150]
Peronospora esperaussensis Mayor
Peronospora fabae Jacz. & Sergeeva   [213-148]
Peronospora fulva Syd.   [198-130]
Peronospora galegae Săvul. & Rayss   [193-124]
Peronospora lagerheimii Gäum.   [192-122]
Peronospora lathyri-aphacae Săvul. & Rayss
Peronospora lathyri-hirsuti Săvul. & Rayss
Peronospora lathyri-humilis C. Benois

Table 2, continued.

Peronospora lathyri-maritimi Jermal.
Peronospora lathyri-palustris Gaüm.   [F]
Peronospora lathyri-pisiformis M. I. Nikol.
Peronospora lathyri-rosei Osipian
Peronospora lathyri-verni A. Gustavsson   [197-128]
Peronospora lathyri-versicolaris Săvul. & Rayss
Peronospora lathyrina Vienn.-Bourg.
Peronospora lentis Gaüm.   [199-131]
Peronospora lotorum Syd.   [200-133; F]
Peronospora manshurica (Naumov) Syd. [≡ *Peronospora trifoliorum* var. *manshurica*]   [194-125; F]
Peronospora mayorii Gaüm.   [214-149; F]
Peronospora medicaginis-minimae Gapon.
Peronospora medicaginis-orbicularis Rayss
Peronospora medicaginis-tianschanicae Gapon.
Peronospora meliloti Syd.   [203-136]
Peronospora melissiti Byzova & Dejeva
Peronospora moreaui Rayss
Peronospora narbonensis Gaüm.   [F]
Peronospora ononidis G. W. Wilson   [204-138]
Peronospora ornithopi Gaüm.   [205-139]
Peronospora orobi Gaüm.   [198-129]
Peronospora oxytropidis Gaüm.   [206-140]
Peronospora phacae Gaüm.
Peronospora pisi Syd. [≡ *P. viciae*]   [206-141; F]
Peronospora pratensis Syd.
Peronospora romanica Săvul. & Rayss [≡ *P. aestivalis* f. *medicaginis-falcatae*]
    [= *P. trifoliorum sensu lato*]   [202-135]
Peronospora ruegeriae Gaüm.   [204-137]
Peronospora savulescui Rayss
Peronospora senneniana Gonz. Frag. & Sacc.   [195-127]
    Peronospora sepium Gaüm. [≡ *P. viciae*]   [211-146]
    Peronospora sojae Lehman & F. A. Wolf [≡ *P. manschurica*]
Peronospora tetragonolobi Gaüm.   [200-132]
Peronospora tetragonolobi-palestini Rayss
Peronospora trifolii-alpestris Gäum.
Peronospora trifolii-arvensis Syd.   [210-145]
    Peronospora trifolii-cherleri Rayss [*nom. invalid.*]   [*Index of Fungi* **3**: 120]
    Peronospora trifolii-clypeati Rayss [*nom. invalid.*]   [*Index of Fungi* **3**: 120]
Peronospora trifolii-formosi Rayss
Peronospora trifolii-hybridi Gaüm.   [209-144]
Peronospora trifolii-minoris Gaüm.
Peronospora trifolii-pratensis A. Gustavsson
    Peronospora trifolii-pilularis Rayss [*nom. invalid.*]   [*Index of Fungi* **3**: 120]
    Peronospora trifolii-purpurei Rayss [*nom. invalid.*]   [*Index of Fungi* **3**: 120]
Peronospora trifolii-repentis Syd.
Peronospora trifoliorum de Bary *sensu lato*   [208-143; F]
Peronospora trifoliorum de Bary *sensu stricto*
Peronospora trigonellae Gäum.
Peronospora viciae (Berk.) Casp. *sensu lato*   [212-147; F]
Peronospora viciae (Berk.) Casp. *sensu stricto*
    Peronospora viciae-sativae Gaüm. [≡ *P. viciae*]
Peronospora viciae-venosae Jacz. & P. A. Jacz.

Table 2, continued.

Peronospora viciicola Campbell [as *'vicicola'*] [*nom. invalid.*]   **[F]**
Peronospora viennotii Mayor [as *'viennoti'*]   [207-142]

FAGALES: Fagaceae

Peronospora fagi R. Hartig [≡ *Phytophthora fagi* (? = *P. quercina*)]
                                                                    [Saccardo: *Sylloge Fungorum* **7**: 238]

MALPIGHIALES: Euphorbiaceae

Peronospora andina Speg.
    Peronospora chamaesycis G. W. Wilson [= *P. euphorbiae*]
Peronospora cyparissiae de Bary   [218-154]
Peronospora embergeri Mayor & Vienn.-Bourg.
Peronospora esulae Gäum.
Peronospora euphorbiae Fuckel   [219-155; **F**]
Peronospora euphorbii-glyptospermae Gäum.
Peronospora euphorbii-thymifoliae Sawada
Peronospora favargeri Mayor & Vienn.-Bourg.
Peronospora hypericifoliae Sinha & Mathur
Peronospora valesiaca Gäum.

MALPIGHIALES: Linaceae

Peronospora lini J. Schröt.   [217-153; **F**]

MALPIGHIALES: Violaceae

Peronospora megasperma Berl. [≡ *Bremiella megasperma*]   [Saccardo: *Sylloge Fungorum* **14**: 458]
Peronospora violae J. Schröt.   [de Bary *ex* J. Schröt.] [220-156]

OXALIDALES: Oxalidaceae

Peronospora oxalidis Verwoerd & du Plessis

ROSALES: Rosaceae

    Peronospora agrimoniae Syd. [= *P. sparsa*]   [183-111]
    Peronospora alchemillae G. H. Otth [=*P. sparsa*]   [184-112]
    Peronospora fragariae Rozc & Cornu [= *P. sparsa*]   **[F]**
    Peronospora gei Syd. [=*P. sparsa*]   [184-113]
    Peronospora ibarakii S. Ito & Muray. [=*P. sparsa*]
Peronospora potentillae de Bary   [185-114; **F**]
    Peronospora potentillae-americanae Gäum. [= *P. sparsa* (*P. potentillae-anserinae*)]
                                                                             [Petrak's Lists **3**: [542]]
    Peronospora potentillae-anserinae Gäum. [= *P. sparsa*]   [188-117]
    Peronospora potentillae-reptantis Gäum. [=*P. sparsa*]   [187-116]
    Peronospora potentillae-sterilis Gäum. [=*P. sparsa*]   [186-115]
    Peronospora rosae-gallicae Săvul. & Rayss [≡ *P. sparsa*]
    Peronospora rubi Rabenh. [=*P. sparsa*]   [190-119; **F**]
    Peronospora sanguisorbae Gäum. [=*P. sparsa*]   [191-120]

Table 2, continued.

Peronospora sparsa Berk. *sensu lato*   [188-118; **F**]
Peronospora sparsa Berk. *sensu stricto*

ROSALES: Ulmaceae

    Peronospora celtidis Waite [≡ *Pseudoperonospora celtidis*]   [Saccardo: *Sylloge Fungorum* **11**: 243]

ROSALES: Urticaceae

    Peronospora cannabina G. H. Otth [≡ *Pseudoperonospora cannabina*]   [93-1]
Peronospora boehmeriae G. Y. Yin & Z. S. Yang   [*Index of Fungi* **6(11)**: 621]
Peronospora debaryi E. S. Salmon & Ware   [97-4; **F**]
Peronospora parietariae Vanev & E. G. Dimitrova
    Peronospora humuli (Miyabe & M. Takah.) Skalický [≡ *Pseudoperonospora humuli*]   [94-2]
    Peronospora illinoensis Farl. [= *Plasmopara obducens*]   [Saccardo: *Sylloge Fungorum* **7**: 261]
    Peronospora urticae (Lib.) Casp. [≡ *Pseudoperonospora urticae*]   [96-3]

CORE EUDICOTS: ROSIDS: EUROSIDS II

BRASSICALES: Brassicaceae [see also Capparaceae]

    Peronospora aethionomatis Simonyan [*nom. invalid.*]
Peronospora alliariae-wasabi Gaüm.   [133-49]
Peronospora alyssi-calycini Gaüm.   [145-53]
Peronospora alyssi-incani Gaüm.
Peronospora alyssi-maritimi Kochman
Peronospora arabidis-alpinae Gaüm.   [147-65]
Peronospora arabidis-glabrae Gaüm.
Peronospora arabidis-hirsutae Gaüm.   [148-67]
Peronospora arabidis-oxyphyllae Gaüm.
Peronospora arabidis-strictae Jacz. & P. A. Jacz.
Peronospora arabidis-turritae Gaüm.
Peronospora arabidopsidis Gaüm.   [149-68]
Peronospora aubrietae Mayor
Peronospora barbareae Gaüm.   [151-70]
Peronospora berteroae Gaüm.   [152-71]
Peronospora biscutellae Gaüm.   [152-72]
Peronospora brassicae Gaüm. [≡ *P. parasitica brassicae*]   [153-73]
Peronospora buniadis Gaüm.   [155-74]
Peronospora cakiles Savile
Peronospora calepinae Gaüm.
Peronospora camelinae Gaüm.   [156-75]
Peronospora cardamines-laciniatae A. Gaüm.
Peronospora cardaminopsis A. Gustavsson   [160-79]
Peronospora cardariae-repentis N. P. Golovina
Peronospora carpoceratis Byzova
Peronospora chartomatis Golvin & Kalymb.
Peronospora cheiranthi Gaüm.   [161-80]
Peronospora chorisporae Gäum.
Peronospora cochleariae Gaüm.   [149-69; **F**]

Table 2, continued.

Peronospora coringiae Gäum.
Peronospora coronopi Gaüm.   [162-82]
   Peronospora coronopi-procumbentis Vienn.-Bourg. [*nom. invalid.*]
Peronospora crambes Jacz. & P. A. Jacz.
Peronospora cryptosporae Annal.
Peronospora dentariae Rabenh.   [159-78]
Peronospora dentariae-macrophyllae Gaüm.   [158-77]
Peronospora desertorum Jacz. & P. A. Jacz.
Peronospora diplotaxidis Gaüm.   [163-84]
Peronospora diptychocarpi Kalymb.
Peronospora drabae Gaüm.   [164-85]
Peronospora drabae-majusculae Lindtner
Peronospora eigii Rayss
Peronospora erophilae Gaüm.   [165-86]
Peronospora erucastri Gaüm.   [166-87]
Peronospora erysimi Gaüm.   [166-88]
Peronospora euclidii Săvul. & Rayss   [167-89]
Peronospora gaeumanniana Jaap
Peronospora galligena S. Blumer   [146-64]
Peronospora goldbachiae Jacz. & P. A. Jacz.
Peronospora golovinii Kalymb.
Peronospora heliophilae E. Müll. & Poelt
Peronospora hesperidis Gaüm.   [168-90]
Peronospora hornungiae A. Gustavsson
Peronospora hymenolobi Annal.
Peronospora iberidis Gaüm.   [Gaüm. *ex* Gaüm.] [169-91]
Peronospora iranica Petr. & Esfand.
Peronospora isatidis Gaüm.   [169-92]
Peronospora jordanovii R. Krusch.
   Peronospora lepidii (McAlpine) G. W. Wilson [≡ *Peronospora parasitica* var. *lepidii*]   [170-93]
Peronospora lepidii-perfoliati Săvul. & Rayss   [171-94]
Peronospora lepidii-sativi Gaüm.
Peronospora lepidii-virginici Gaüm.
Peronospora leptalei Kolosch.
Peronospora litwinowiae Kalymb.
Peronospora lobulariae Ubrizsy & Vörös   [171-95]
Peronospora lunulariae Gaüm.   [172-96]
Peronospora malcolmiae Lobik
Peronospora malyi Lindtner
Peronospora matthiolae Gaüm.   [173-97]
Peronospora maublancii Săvul. & Rayss
Peronospora menioci Kolosch.
   Peronospora minuta Vienn.-Bourg. [*nom. invalid.*]
Peronospora myagri Mayor
Peronospora nasturtii-aquatici Gaüm.   [174-98]
Peronospora nasturtii-montani Gaüm.
Peronospora nasturtii-palustris S. Ito & Tokun.
Peronospora nesliae Gaüm.   [174-99]
Peronospora niessliana A. Berl.   [144-62]
Peronospora norvegica Gaüm. [as 'norwegica']
   Peronospora ochracea Ces. [*nom. invalid.*]   [147-66]
   Peronospora ochroleuca Ces. [= *Peronospora ochracea* Ces.] [*nom. invalid.*]   [147-66]

Table 2, continued.

Peronospora pachyphragmatis-macrophylli Savinceva
Peronospora parasitica (Fr.) Fr.   [(Pers. *ex* Fr.) Fr.] [156-76]
Peronospora rapistri Jacz. & Sergeeva [as '*rapistrae*']
Peronospora rhaetica Gaüm.   [176-102]
Peronospora rorippae-islandicae Gaüm.   [176-101]
    Peronospora savulescui Sandu [*nom. invalid.*]
    Peronospora senecionis Fuckel [= *P. parasitica*]
Peronospora simonianii Osipian
Peronospora sisymbrii-intermedii Gaüm.
Peronospora sisymbrii-loeselii Gaüm.
Peronospora sisymbrii-officinalis Gaüm.   [177-103]
Peronospora sisymbrii-orientalis Gaüm.
Peronospora sisymbrii-sophiae Gaüm.   [163-83]
Peronospora sophiae-pinnatae Gaüm.
Peronospora streptolomatis Golovin & Kalymb. [as '*streptolomatae*']
    Peronospora syreniae Jenkina [*nom. invalid.*]
Peronospora syreniae-cuspidatae Oescu & Rădul.
Peronospora takahashii S. Ito
Peronospora taurica Jacz. & P. A. Jacz.
Peronospora tauscheriae Kalymb.
Peronospora teesdaliae Gaüm. [as '*teesdalae*']   [178-104]
Peronospora thlaspeos-alpestris Gaüm.   [179-106]
Peronospora thlaspeos-arvensis Gaüm.   [179-105]
Peronospora thlaspeos-perfoliati Gaüm.
    Peronospora torulariae Domashova [*nom. invalid.*]   [*Index of Fungi* **3**: 333]
Peronospora turritidis Gäum.
Peronospora vvedenskyi Golovin

BRASSICALES: Capparaceae [now subsumed in <u>Brassicaceae</u>]

    Peronospora capparidis Sawada [*nom. invalid.*]

BRASSICALES: <u>Limnanthaceae</u>

Peronospora floerkeae Kellerm.   **[F]**

BRASSICALES: <u>Resedaceae</u>

Peronospora crispula Fuckel   [180-107]

MALVALES: <u>Cistaceae</u>

Peronospora alpestris Gaüm.   [133-49]
Peronospora leptoclada Sacc.   [221-157]

MYRTALES: <u>Onagraceae</u>

Peronospora arthurii Farl.   [222-159; **F**]
    Peronospora epilobii G. H. Otth [≡ *Plasmopara epilobii*]   [Saccardo: *Sylloge Fungorum* **11**: xxiii]
    Peronospora epilobii Rabenh. [≡ *Plasmopara epilobii*]   [Saccardo: *Sylloge Fungorum* **7**: 243]

Table 2, continued.

SAPINDALES: Meliaceae

Peronospora portoricensis (Lamkey) Skalický [≡ *Plasmopara portoricensis*]
                                                                    [*Index of Fungi* **3**: 413]

CORE EUDICOTS: ASTERIDS: sister groups to ASTERIDS I and II

ERICALES: Balsaminaceae

Peronospora impatientis Ellis & Everh. [= *Plasmopara obducens*]
                                              [Saccardo: *Sylloge Fungorum* **9**: 243]
Peronospora obducens J. Schröt. [≡ *Plasmopara obducens*]   [Saccardo: *Sylloge Fungorum* **7**: 242]

ERICALES: Polemoniaceae

Peronospora phlogina Dietel & Holw.   [**F**]
    Peronospora giliae Ellis & Everh. [= *P. phlogina*]

ERICALES: Primulaceae

Peronospora agrorum Gaüm.   [224-162]
Peronospora anagallidis J. Schröt.   [Saccardo: *Sylloge Fungorum* **7**: 248; **F**]
Peronospora androsaces Niessl   [223-161]
Peronospora candida Fuckel   [223-160; **F**]
Peronospora cortusae Gaüm. & S. Blumer   [224-163]
Peronospora gregoriae S. Blumer
Peronospora mirabilis Jacz. & P. A. Jacz.
Peronospora oerteliana J. G. Kühn   [225-164]

ASTERIDS: EUASTERIDS I

ASTERIDS: EUASTERIDS I: families without ordinal grouping

Boraginaceae [see also Hydrophyllaceae]

Peronospora alkannae Vienn.-Bourg.
Peronospora anchusae Ziling
Peronospora arnebiae Golovin
Peronospora asperuginis J. Schröt.   [231-172]
Peronospora bothriospermi Sawada
Peronospora cerinthes Uljan. [as '*cerinthe*']   [232-173]
Peronospora cynoglossi Swingle   [Burrill *ex* Swingle] [233-174; **F**]
    Peronospora echii (Krieger) Jacz. & P. A. Jacz. [= *P. myosotidis*]
    Peronospora echinospermi (Swingle) Swingle [= *P. cynoglossi*]   [233-175]
Peronospora eritrichii S. Ito & Tokun.
Peronospora lithospermi Gaüm.   [234-176]
Peronospora microulae Y. R. Meng & G. Y. Yin
Peronospora myosotidis de Bary   [235-177; **F**]

Table 2, continued.

Peronospora noneae Jacz. & Sergeeva [as '*nonneae*']	[236-178]
Peronospora omphalodis Gaüm.	[236-179]
Peronospora pulmonariae Gaüm. [? = *P. myosotidis* f. *pulmonariae*]	[237-180]
Peronospora rocheliae Kalymb.	
Peronospora rugosa Jacz. & P. A. Jacz.	
Peronospora solenanthi Byzova	
Peronospora symphyti Gaüm.	[237-181]
Peronospora thyrocarpi L. Ling & M. C. Tai [as '*thyrocarpii*']	
Peronospora trigonotidis S.Ito & Tokun.	
Peronospora uljanishchevii Tunkina	

Hydrophyllaceae [now subsumed in Boraginaceae]

Peronospora hydrophylli Waite	**[F]**
Peronospora nemophilae C. G. Shaw	**[F]**

ASTERIDS: EUASTERIDS I: orders and families

GENTIANALES: Apocynaceae

Peronospora vincae J. Schröt.   [227-167]

GENTIANALES: Asclepiadiaceae [now subsumed in Apocynaceae]

   Peronospora gonolobii Lagerh. [≡ *Plasmopara gonolobi*]   [Saccardo: *Sylloge Fungorum* **11**: 243]

GENTIANALES: Gentianaceae

Peronospora canscorina Thite & M. S. Patil	
Peronospora carnicola Gaüm. [*nom. invalid.*]	
Peronospora chlorae de Bary	
Peronospora erythraeae Gaüm.	[J. G. Kühn *ex* Gaüm.] [226-165]
Peronospora gentianiae Rostr.	[226-166]

GENTIANALES: Rubiaceae

Peronospora aparines (de Bary) Gaüm. [≡ *P. calotheca* var. *aparines*]	[228-169]
Peronospora borealis Gaüm.	
Peronospora borreriae Lagerh. [≡ *Plasmopara borreriae*]	[Saccardo: *Sylloge Fungorum* **11**: 243]
Peronospora calotheca de Bary	[227-168]
Peronospora crucianellae Maire	
Peronospora galii Fuckel [= *P. calotheca*]	[229-170]
Peronospora galii-anglici Uljan.	
Peronospora galii-pedemontani Săvul. & Rayss	
Peronospora galii-rubioides Săvul. & Rayss	
Peronospora galii-trifidii S. Ito & Tokun.	
Peronospora galii-veri Gäum.	
Peronospora hiratsukae S. Ito & Tokun.	
Peronospora hommae S. Ito & Tokun.	
Peronospora insubrica Gaüm.	

Table 2, continued.

Peronospora rubiae Gäum.
Peronospora sakamotoi S. Ito & Tokun.
Peronospora seymourii Burrill in Underwood   [F]
Peronospora sherardiae Fuckel   [230-171]
Peronospora silvatica Gäum.

LAMIALES: Buddlejaceae [formerly Loganaceae *pro parte*]

Peronospora hariotii Gäum.   [246-193]

LAMIALES: Lamiaceae

Peronospora amethysteae Jacz. & P. A. Jacz.
Peronospora calaminthae Fuckel   [238-182]
Peronospora chamaesphaci Kalymb.
Peronospora clinopodii Terui
Peronospora dracocephali C. J. Li & Zhen Y. Zhao   [*Index of Fungi* **6(19)**: 1048]
Peronospora davisii C. G. Shaw   [F]
Peronospora elsholtziae T. R. Liu & C. K. Pai
Peronospora galeopsidis Lobik   [238-183]
Peronospora glechomae Oescu & Rădul.
    Peronospora glechomatis (Willi Krieg.) T. Majewski [≡ *P. lamii* var. *glechomae*]   [239-184]
Peronospora hedeomatis Kellerm. & Swingle   [F]
Peronospora ibrahimovii T. M. Achundov
Peronospora lallemantiae Golovin & Kalymb.
Peronospora lamii A. Braun   [240-185; F]
    Peronospora lamii var. *glechomae* Willi Krieg. [as '*glechomatis*'] [≡ *P. glechomae*]
                                              [Saccardo's Omissions: 48]
Peronospora leonuri T. R. Liu & C. K. Pai
Peronospora lophanthi Farl.   [F]
Peronospora menthae X. Y. Cheng & H. C. Bai
Peronospora perillae Miyabe in S. Ito & Tokun.
Peronospora rossica Gäum.
Peronospora saturejae-hortensis Osipian
Peronospora sideritidis Byzova
Peronospora scutellariae Bejlin   [242-188]
Peronospora stachydis Syd.   [243-189]
Peronospora stigmaticola Raunk.   [241-186; F]
    Peronospora swinglei Ellis & Kellerm.   [241-187; F]
Peronospora teucrii Gäum.   [243-190]
    Peronospora thymi Syd [*nom. invalid.*]
Peronospora ziziphorae Byzova

LAMIALES: Plantaginaceae

Peronospora akatsukae S. Ito & Muray.
Peronospora alta Fuckel   [262-213; F]
Peronospora canescens C. Benois
Peronospora plantaginis Underw.   [F]
Peronospora plantaginis Burrill   [264-21]
    Peronospora lanceolatae Gapon. [*nom. invalid.*]   [*Index of Fungi* **4**: 352]

Table 2, continued.

LAMIALES: Scrophulariaceae

Peronospora agrestis Gaüm.	[260-210
Peronospora andicola Speg.	
Peronospora antirrhini J. Schröt.	[246-194; **F**]
Peronospora aquatica Gaüm.	[257-207]
Peronospora arvensis Gaüm.	[259-209]
Peronospora canadensis Gaüm.	
Peronospora celsiae Syd. & P. Syd.	
Peronospora densa Rabenh. [≡ *Plasmopara densa*]	[Saccardo: *Sylloge Fungorum* **7**: 243]
Peronospora digitalis Gaüm.	[247-195]
Peronospora erini Vienn.-Bourg. [*nom. invalid.*]	
Peronospora flava Gaüm.	[249-198; **F**]
Peronospora grisea (Unger) Unger	[257-208; **F**]
Peronospora indica Syd. & P. Syd.	
Peronospora jacksonii C. G. Shaw	[252-202; **F**]
Peronospora lapponica Lagerh.	[Saccardo: *Sylloge Fungorum* **9**: 344; 248-196]
Peronospora linariae Fuckel	[250-199; **F**]
Peronospora linariae-genistifoliae Săvul. & Rayss	[249-197]
Peronospora melampyri (Bucholz) Davis	[251-200]
Peronospora melampyri-cristati Săvul. & Rayss	
Peronospora orontii J. Schröt. [*nom. invalid.*]	
Peronospora palustris Gaüm.	
Peronospora pedicularis Palm	
Peronospora pocutica T. Majewski	[253-203]
Peronospora satarensis M. S. Patil	
Peronospora saxatilis Gaüm.	
Peronospora silvestris Gaüm.	
Peronospora sordida Berk. & Broome	[253-204; **F**]
Peronospora tozziae S. Blumer	[254-253]
Peronospora tranzscheliana Bakhtin	[252-201]
Peronospora verbasci Gaüm.	[256-206]
Peronospora verna Gaüm.	
Peronospora veronicae-cymbalariae Rayss	

SOLANALES: Convolvulaceae

Peronospora convolvuli J. C. Lindq.
Peronospora fritzii J. Schröt.

SOLANALES: Solanaceae

Peronospora capsici J. F. Tao & T. J. Li	
Peronospora devastatrix Casp. [= *Phytophthora infestans*]	[Saccardo: *Sylloge Fungorum* **7**: 237]
Peronospora dubia A. Berl.	
Peronospora fintelmannii Casp. [= *Phytophthora infestans*]	[Saccardo: *Sylloge Fungorum* **15**: 241]
Peronospora hyoscyami de Bary	[**F**]
Peronospora infestans (Mont.) de Bary [as 'de Bary'] [≡ *Phytophthora infestans*]	
	[Saccardo: *Sylloge Fungorum* **7**: 237]
Peronospora infestans (Mont.) Casp. [as 'Casp.'] [≡ *Phytophthora infestans*]	
	[Saccardo: *Sylloge Fungorum* **1**: 224]
Peronospora lycii L. Ling & M. C. Tai	

Table 2, continued.

Peronospora nicotianae Speg.
    Peronospora tabacina D. B. Adam [≡ *P. hyoscyami*]   [245-192; **F**]
    Peronospora trifurcata Unger [= *Phytophthora infestans*]   [Saccardo: *Sylloge Fungorum* **7**: 237]

ASTERIDS: EUASTERIDS II: orders and families

APIALES: Apiaceae

    Peronospora conii Tul. [*nom. invalid.*; ? = *Plasmopara umbelliferarum sensu lato*]
                            [Saccardo: *Sylloge Fungorum* **7**: 241]
  Peronospora macrocarpa Corda *sensu* Rabenh. [= *Plasmopara umbelliferarum*]
                            [Saccardo: *Sylloge Fungorum* **7**: 240]
    Peronospora macrospora (Unger) Unger [= *Plasmopara umbelliferarum*]
                            [Saccardo: *Sylloge Fungorum* **7**: 241]
    Peronospora podagrariae G. H. Otth [≡ *Plasmopara podagrariae*]
                            [Saccardo: *Sylloge Fungorum* **11**: xxiii]

APIALES: Araliaceae

    Peronospora panacis Bunkina [*nomen nudum*] [≡ *Plasmopara panacis* Bondartsev & Bunkina]
                            [*Index of Fungi* **3**: 56]

ASTERALES: Asteraceae

    Peronospora achilleae Săvul. & Vánky [= *Paraperonospora leptosperma*]   [*Index of Fungi* **2**: 515]
    Peronospora anthemidis Gäum. [= *Paraperonospora leptosperma*]
                            [Saccardo: *Sylloge Fungorum* **26**: 30]
    Peronospora artemisiae-annuae L. Ling & M. C. Tai [≡ *Paraperonospora artemisiae-annuae*]
    Peronospora artemisiae-biennis Gäum. [≡ *Paraperonospora artemisiae-biennis*]
    Peronospora brachycomes Enkina [= *Paraperonospora leptosperma*]   [*Index of Fungi* **3**: 561]
    Peronospora buhrii Săvul. & Vánky [= *Paraperonospora leptosperma*]   [*Index of Fungi* **2**: 515]
    Peronospora crossostephii Sawada [= *Paraperonospora multiformis*]   [*Index of Fungi* **2**: 395]
  Peronospora danica Gäum.
    Peronospora dimorphothecae Săvul. & Vánky [= *Paraperonospora leptosperma*]
                            [*Index of Fungi* **2**: 515]
    Peronospora ganglioniformis Berk.) de Bary [as 'ganglïformis'] [= *Bremia lactucae*]
                            [Saccardo: *Sylloge Fungorum* **7**: 244]
    Peronospora ganglioniformis (Berk.) Tul. [as 'Tul.'] [= *Bremia lactucae*]
                            [Saccardo: *Sylloge Fungorum* **7**: 244]
    Peronospora halstedii Farl. [≡ *Plasmopara halstedii*]   [Saccardo: *Sylloge Fungorum* **7**: 242]
    Peronospora helianthi Rostr. [=]   [Saccardo: *Sylloge Fungorum* **26**: 39]
    Peronospora helichrysi Togashi & Egami [= *Paraperonospora sulphurea*]   [Petrak's Lists 7]
    Peronospora kellermanii Fuckel [as 'kellermannii'] [≡ *Benua kellermanii*]
                            [Saccardo: *Sylloge Fungorum* **7**: 263]
    Peronospora leptosperma de Bary [≡ *Paraperonospora leptosperma*]   [*Index of Fungi* **5**: 445] [**F**]
    Peronospora pospelovii Gaponenko [probably an *Aspergillus*]   [*Index of Fungi* **4**: 353]
  Peronospora radii de Bary   [269-221]
    Peronospora simplex Peck [= *Basidiophora entospora*]   Sacc. *Syllog. Fung.***7**: 239]
    Peronospora sonchi Gapon. [a doubtful taxon - ? *Bremia*, see Constantinescu, 1991*a*]
                            [*Index of Fungi* **2**: 442]

Table 2, continued.

Peronospora sulphurea Gäum. [≡ *Paraperonospora sulphurea*]   Sacc. *Syllog. Fung.***26**: 55]
Peronospora tanaceti Gäum. [≡ *Paraperonospora tanaceti*]   [Saccardo: *Sylloge Fungorum* **26**: 56]
Peronospora ursiniae Săvul. & Vánky [= *Paraperonospora leptosperma*]   [*Index of Fungi* **2**: 515]

ASTERALES: Campanulaceae

  Peronospora campanulae G. Nicholas & Aggéry [*nom. invalid.*]
Peronospora corollae Tranzschel   [267-218]
Peronospora erinicola Durrieu
Peronospora phyteumatis Fuckel   [268-219]
Peronospora speculariae Gäum.   [268-220]

DIPSACALES: Dipsacaceae

Peronospora cephalariae Vincens
Peronospora cephalariae-laevigatae Săvul. & Rayss
Peronospora dipsaci de Bary   [Tul. *ex* de Bary] [264-215; **F**]
Peronospora karelii Bremer & Gäum.
Peronospora knautiae J. Schröt.   [Fuckel *ex* J. Schröt.] [265-216]
Peronospora violacea Berk. [variously cited as 'Berk. *ex* Cooke' or Berk. in Cooke (preferred)] [266-217]

DIPSACALES: Valerianaceae

  Peronospora centranthi Massenot [*nom. invalid.*]
Peronospora patriniae Kalymb.
Peronospora valerianae Trail   [261-211; **F**]
Peronospora valerianellae Fuckel   [261-212]

MONOCOTS

ASPARAGALES: Alliaceae

  Peronospora alliorum Fuckel [= *P. destructor*]   [Saccardo: *Sylloge Fungorum* **7**: 257]
Peronospora destructor (Berk.) Casp.   [270-222; **F**]
Peronospora fugitai S. Ito & Tokun.
  Peronospora schleidenii Unger [= *P. destructor*]

LILIALES: Liliaceae

  Peronospora lilii Stenina [very dubious taxon]

ALISMATALES: Araceae

  Peronospora trichotoma Massee [unidentifiable, but cross-reference to *Phytophthora colocasiae*]
                                                   [Saccardo: *Sylloge Fungorum* **9**: 344]

Table 2, continued.

COMMELINOIDS

POALES: Cyperaceae

Peronospora cyperi Ideta [*non P. cyperi* Miyabe & Ideta ≡ *Phytophthora cyperi*]   [F]

POALES: Poaceae

Peronospora diplachnis Milovtz. [? = *Sclerospora graminicola*]
Peronospora graminicola (Sacc.) Sacc. [≡ *Sclerospora graminicola*]
[Saccardo: *Sylloge Fungorum* **7**: 238]
Peronospora maydis Racib. [≡ *Peronosclerospora maydis*]   [Saccardo: *Sylloge Fungorum* **14**: 460]
Peronospora setariae Pass. [= *Sclerospora graminicola*]   [Saccardo: *Sylloge Fungorum* **7**: 238]

Peronospora names in the CABI-data base, but excluded from *Peronospora*

Peronospora barcinonae Ferrán [as 'Ferraris', = *P. ferrani* Gimeno]
[name used in connection with the agent for cholera] [Saccardo: *Sylloge Fungorum* **15**: 240]
Peronospora dilachnis [see *P. diplachnis* on Poaceae]   [Petrak's Lists **8**]
Peronospora exigua Wm. G. Sm.
[a hyphomycete, = *Ovularia sphaeroidea*[Saccardo: *Sylloge Fungorum* **11**: 598]
Peronospora gossypina Averna-Saccá
[a hyphomycete, *Olpitrichum tenellum*] [Petrak's Lists **1**: [157/233]]
Peronospora interstitialis Berk. & Broome
[a hyphomycete, *Ramularia primulae*] [Saccardo: *Sylloge Fungorum* **7**: 259]
Peronospora lutea Carmona y Valle   [a bacterium] [Saccardo: *Sylloge Fungorum* **8**: 1054]
Peronospora muscorum Sorokīn [very doubtful taxon, on moss]
[Saccardo: *Sylloge Fungorum* **9**: 343]
Peronospora nivea (Unger) Unger [as 'Unger']
[excluded (*nomen confusum*): see Costantinescu, 1991*a*] [Saccardo: *Sylloge Fungorum* **7**: 244]
Peronospora nivea de Barỵnot listed in Costantinescu, 1991*a*] [Saccardo: *Sylloge Fungorum* **7**: 240]
Peronospora obliqua Cookựa hyphomycete, *Ovularia obliqua*] [Saccardo: *Sylloge Fungorum* **4**: 145]
Peronospora rufibasis Berk. & Broome
[a hyphomycete, *Ramularia rufibasis*] [Saccardo: *Sylloge Fungorum* **7**: 261]

PLASMOPARA J. Schröt.

[NOTE: All names from *Sylloge Fungorum*, *Petrak's Lists* and *Index of Fungi* are included
here, except for a few recombinations now listed under other genera. Infra-specific
taxa may be of dubious standing. Synonymy has not been evaluated. Hosts of
thirteen species (in as many families) may be found under *Peronospora* synonyms;
families are given below when not the type of the order; for other symbols, see note
under *Peronospora*.]

Plasmopara acalyphae (G. W. Wilson) G. W. Wilson
[Saccardo: *Sylloge Fungorum* **24**: 64; **F**] [host(s) in Malpighiales (Euphorbiaceae)]
Plasmopara achilleae (Săvul. & L. Vánky) Skalický [≡ *Paraperonospora leptosperma*]
[*Index of Fungi* **3**: 415; 70-11]

Table 2, continued.

Plasmopara achyranthis J. F. Tao & Y. Qin
　　　　　　　　[*Index of Fungi* 5: 306] [host(s) in Caryophyllales (Amaranthaceae)]
　Plasmopara aegopodii (Casp.) Trotter [= *P. umbelliferarum*]　[Saccardo: *Sylloge Fungorum* 24: 65]
Plasmopara affinis Novot. forma *affinis*　　　　　　[*Index of Fungi* 3: 255] [host(s) in Asterales]
Plasmopara affinis Novot. forma *silphii* Novot.　　　　　　　　　　　　[*Index of Fungi* 3: 255]
Plasmopara alpina (C. J. Johanson) A. Blytt　　　　　　　　　[host(s) in Ranunculales]
Plasmopara ammi Constant.　　　　　　　　　　　[*Index of Fungi* 3: 492] [host(s) in Apiales]
Plasmopara amurensis [no authority; no reference, but in the CABI database; ? ≡ *P. viticola* var. *amurensis*]
Plasmopara anemones-dichotomae Benua　　　　　　[*Index of Fungi* 4: 316] [host(s) in Ranunculales]
Plasmopara anemones-nemorosae Săvul. & O. Săvul.　　　[*Index of Fungi* 3: 364] [host(s) in Ranunculales]
Plasmopara anemones-ranunculoidis Săvul. & O. Săvul.　　[*Index of Fungi* 3: 364] [host(s) in Ranunculales]
Plasmopara anethi Jermal.　　　　　　　　　　　[*Index of Fungi* 3: 415] [host(s) in Apiales]
Plasmopara angelicae (Casp.) Trotter　　　　[Saccardo: *Sylloge Fungorum* 24: 65] [host(s) in Apiales]
Plasmopara angustiterminalis Novot. forma *angustiterminalis*
　　　　　　　　　　　　　　　　　[*Index of Fungi* 3: 255; 76-17] [host(s) in Asterales]
Plasmopara angustiterminalis Novot. forma *ambrosiae* Novot.　　　　　[*Index of Fungi* 3: 255]
Plasmopara angustiterminalis Novot. forma *bidentis* Novot.　　　　　　[*Index of Fungi* 3: 255]
　Plasmopara anthemidis (Gäum.) Skalický [≡ *Paraperonospora leptosperma*]
　　　　　　　　　　　　　　　　　　　　　　　　[*Index of Fungi* 3: 415; 72-13]
Plasmopara apii Săvul. & O. Săvul.　　　　　　　[*Index of Fungi* 3: 364] [host(s) in Apiales]
Plasmopara archangelicae Gapon.　　　　　　　　[*Index of Fungi* 4: 354] [host(s) in Apiales]
　Plasmopara artemisiae-annuae (L. Ling & M. C. Tai) Skalický
　　[≡ *Paraperonospora artemisiae-annuae*]
　Plasmopara artemisiae-biennis (Gäum.) Skalický [≡ *Paraperonospora artemisiae-biennis*]
Plasmopara asterea Novot. forma *asterea*　　　　　　[*Index of Fungi* 3: 255] [host(s) in Asterales]
　Plasmopara asterea Novot. forma *callistephi* Novot. [= *P. callistephi*]　　[*Index of Fungi* 3: 255]
　Plasmopara asterea Novot. forma *galatellae* Novot. [= *P. galatellae*]　　[*Index of Fungi* 3: 255]
　Plasmopara asterea Novot. forma *heteropappi* Novot. [= *P. heteropappi*]　　[*Index of Fungi* 3: 255]
Plasmopara asystasiae Vienn.-Bourg.　　　[*Index of Fungi* 2: 249] [host(s) in Lamiales (Acanthaceae)]
Plasmopara australis (Speg.) Swingle
　　　　　　　[Saccardo: *Sylloge Fungorum* 9: 342; 12: 575; 20: 422; F] [host(s) in Cucurbitales]
Plasmopara baicalensis Jacz. & P. A. Jacz.　　　　　　　　　　　[host(s) in Ranunculales]
　Plasmopara baudysii Skalický [≡ *Bremiella baudysii*]　　　　　　　　[*Index of Fungi* 2: 197]
Plasmopara bidentis (Novot.) Novot.　　　　　　　　　　　　　　[host(s) in Asterales]
Plasmopara borreriae (Lagerh.) Constant.
　　　　　　　　　[*Index of Fungi* 6(4): 221] [host(s) in Gentianales (Rubiaceae)]
　Plasmopara buhrii (Săvul. & L. Vánky) Skalický [≡ *Paraperonospora leptosperma*]
　　　　　　　　　　　　　　　　　　　　　　　　　　[*Index of Fungi* 3: 415]
Plasmopara calaminthae S. H. Ou　　　　　　　[*Index of Fungi* 1: 45] [host(s) in Lamiales]
Plasmopara callistephi (Novot.) Novot.　　　　　　　　　　　　[host(s) in Asterales]
Plasmopara carlottae Savile　　　　　　　　　　[*Index of Fungi* 3: 334] [host(s) in Apiales]
Plasmopara carthami Negru　　　　　　　　　　[*Index of Fungi* 3: 469] [host(s) in Asterales]
Plasmopara caucalis Săvul. & O. Săvul.　　　　　　[*Index of Fungi* 3: 364] [host(s) in Apiales]
Plasmopara cenolophii Jermal.　　　　　　　　　[*Index of Fungi* 3: 415] [host(s) in Asterales]
Plasmopara centaureae-mollis T. Majewski　　　[*Index of Fungi* 3: 515; 73-14] [host(s) in Asterales]
Plasmopara cephalophora Davis　　　　　[Saccardo: *Sylloge Fungorum* 24: 64; F] [host(s) in Lamiales]
Plasmopara cercidis C. G. Shaw　　　　　　　[*Index of Fungi* 2: 61; F] [host(s) in Fabales]
Plasmopara chaerophyllii (Casp.) Trotter　　　[Saccardo: *Sylloge Fungorum* 24: 65] [host(s) in Apiales]
Plasmopara chinensis Gorlenko　　　　　　　　[*Index of Fungi* 3: 534] [host(s) in Vitaceae)]
　Plasmopara chrysanthemi-coronarii Sawada [≡ *Paraperonospora chrysanthemi-coronarii*]
　　　　　　　　　　　　　　　　　　　　　　　　　　　　[Petrak's Lists]
Plasmopara cimicifugae S. Ito & Tokun.　　　　　　　　　　　　[host(s) in Ranunculales]

Table 2, continued.

Plasmopara cissi Vienn.-Bourg.   [*Index of Fungi* **2**: 249] [host(s) in Vitaceae]
Plasmopara conii (Casp.) Trotter [≡ *P. umbelliferarum sensu lato*]
                                                            [Saccardo: *Sylloge Fungorum* **24**: 65]
Plasmopara conii (Wart.) Cif. & C. Camera [later homonym]   [*Index of Fungi* **3**: 224]
Plasmopara cruseae C. G. Shaw & Safeeulla [= *P. borreriae*]   [*Index of Fungi* **3**: 195]
Plasmopara crustosa (Fr.) Jørst.
                [*nomen confusum*, Constantinescu (1992); excluded] [*Index of Fungi* **3**: 224; **F**]
Plasmopara cryptotaeniae J. F. Tao & Y. Qin   [*Index of Fungi* **5**: 606] [host(s) in Apiales]
Plasmopara cubensis (Berk. & M. A. Curtis) Humphrey var. *cubensis* [≡ *Pseudoperonospora cubensis*]
                                [Saccardo: *Sylloge Fungorum* **17**: 520; **20**: 422, 1288]
[The two varieties listed below do not appear either to have been transferred or reduced to synonymy.]
    Plasmopara cubensis (Berk. & M. A. Curtis) Humphrey var. *atra* Zimm.
                                        [Saccardo: *Sylloge Fungorum* **17**: 520; **20**: 422]
    Plasmopara cubensis (Berk. & M. A. Curtis) Humphrey var. *tweriensis* Rostovzev
                                        [Saccardo: *Sylloge Fungorum* **17**: 520; **20**: 422]
Plasmopara curta (Berk.) Skalický subspecies *curta* [≡ *Plasmopara pygmaea*]
                                                            [*Index of Fungi* **2**: 173]
Plasmopara curta (Berk.) Skalický subspecies *orientalis* Skalický   [*Index of Fungi* **2**: 173]
Plasmopara curta (Berk.) Skalický var. *fusca* (Peck) Skalický   [*Index of Fungi* **2**: 173]
Plasmopara curta (Berk.) Skalický forma *curta*   [*Index of Fungi* **2**: 173]
Plasmopara curta (Berk.) Skalický forma *hellebori* Săvul. & Rayss   [*Index of Fungi* **2**: 173]
Plasmopara dahurici Benua   [*Index of Fungi* **4**: 316] [host(s) in Apiales]
Plasmopara dauci Săvul. & O. Săvul.   [*Index of Fungi* **3**: 364] [host(s) in Apiales]
Plasmopara delphinii (Gapon.) Novot.   [host(s) in Ranunculales]
Plasmopara densa (Rabenh.) J. Schröt.
    [Saccardo: *Sylloge Fungorum* **7**: 243; **12**: 575; **20**: 422] [67-8] [host(s) in Lamiales]
Plasmopara elatostemmatis (Togashi & F. Onuma) S. Ito & Tokun. [as 'elatostematis']
                                                    [host(s) in Rosales (Urticaceae)]
Plasmopara elsholtziae J. F. Tao & Y. Qin [cf. *Pseudoperonospora elsholtziae*]
                                          [*Index of Fungi* **5**: 306] [host(s) in Lamiales]
    Plasmopara entospora (Roze & Cornu) J. Schröt. [≡ *Basidiophora entospora*]
                              [Saccardo: *Sylloge Fungorum* **7**: 239; **12**: 575; **20**: 423, 1288]
Plasmopara epilobii (G. H. Otth) Sacc. & P. Syd. [as '(G. Otth) Schröt.']
                                          [63-6; **F**] [host(s) in Myrtales (Onagraceae)]
Plasmopara galatellae (Novot.) Novot.   [host(s) in Asterales]
Plasmopara galinsogae L. Campb.   [host(s) in Asterales]
Plasmopara geranii (Peck) Berl. & De Toni   [**F**] [host(s) in Geraniales]
Plasmopara geranii-pratensis Săvul. & O. Săvul.   [*Index of Fungi* **3**: 364] [host(s) in Geraniales]
Plasmopara geranii-sylvatici Săvul. & O. Săvul.   [*Index of Fungi* **3**: 364] [[host(s) in Geraniales]
Plasmopara gnaphalii Novot.   [*Index of Fungi* **3**: 255] [host(s) in Asterales]
Plasmopara gonolobi (Lagerh.) Swingle   [**F**] [host(s) in Gentianales (Apocynaceae)]
Plasmopara halstedii (Farl.) Berl. & De Toni   [**F**] [host(s) in Asterales - *Verbena*!]
Plasmopara harae S. Ito & Muray.   [*Index of Fungi* **2**: 41] [host(s) in Cornales (Hydrangeaceae)]
Plasmopara helianthi Novot. forma *helianthi*   [*Index of Fungi* **3**: 364; 75-16] [host(s) in Asterales]
Plasmopara helianthi Novot. forma *patens* Novot.   [*Index of Fungi* **3**: 364]
Plasmopara helianthi Novot. forma *perennis* Novot.   [*Index of Fungi* **3**: 364]
    Plasmopara helichrysi (S. Ito & Tokun.) J. F. Tao & Y. Qin [≡ *Paraperonospora sulphurea*]
                              [Togashi & Egami *ex* S. Ito & Tokun.] [*Index of Fungi* **5**: 776]
Plasmopara heliocarpi Lagerh.   [host(s) in Malvales (Grewioideae) (formerly Tiliaceae)]
Plasmopara hellebori-purpurascentis Săvul. & O. Săvul.
                                          [*Index of Fungi* **3**: 364] [host(s) in Ranunculales]
    Plasmopara hepaticae Casp. [=*P. pygmaea*]

Table 2, continued.

Plasmopara hepaticae (Casp.) C. G. Shaw [= *P. hepaticae*]   [*Index of Fungi* 1: 267]
Plasmopara heteropappi (Novot.) Novot.   [host(s) in Asterales]
Plasmopara humuli Miyabe & Takah. [≡ *Pseudoperonospora humuli*]
                                                    [Saccardo: *Sylloge Fungorum* 21: 861]
Plasmopara illinoensis (Farl.) Davis   [F] [host(s) in Rosales (Urticaceae)]
  Plasmopara impatientis (Ellis & Everh.) Berl. [= *P. obducens*]
Plasmopara isopyri Skalický   [*Index of Fungi* 2: 173] [host(s) in Ranunculales]
Plasmopara isopyri-thalictroidis (Săvul. & Rayss) Săvul. & O. Săvul.
                                       [*Index of Fungi* 3: 364] [host(s) in Ranunculales]
  Plasmopara justiciae (Sawada) Skalický   [host(s) in Lamiales (Acanthaceae)]
    Plasmopara kellermanii (Ellis & Halst.) Swingle [as '*kellermannii*']
                                                    [Saccardo: *Sylloge Fungorum* 9: 342]
Plasmopara lactucae-radicis Stangh. & Gilb.   [*Index of Fungi* 5: 842] [host(s) in Asterales]
Plasmopara laserpitii (Wartenw.) Săvul. & Rayss   [host(s) in Apiales]
Plasmopara latifolii Savile   [*Index of Fungi* 3: 195; F] [host(s) in Myrtales (Onagraceae)]
  Plasmopara leptosperma (de Bary) Skalický [≡ *Paraperonospora leptosperma*]
                                                    [*Index of Fungi* 3: 415; 71-12]
    Plasmopara megasperma (Berl.) Berl. [≡ *Bremiella megasperma*]
    Plasmopara megasperma Săvul. [later homonym]   [*Index of Fungi* 1: 231; 3: 255]
Plasmopara mei-foeniculi Săvul. & O. Săvul.   [*Index of Fungi* 3: 364] [host(s) in Apiales]
  Plasmopara melampyri Bucholtz [≡ *Peronospora melampyri*]
Plasmopara mikaniae Vienn.-Bourg.   [*Index of Fungi* 2: 249] [host(s) in Asterales]
Plasmopara miyakeana S. Ito & Tokun.   [host(s) in Rosales (Urticaceae)]
Plasmopara myosotidis C. G. Shaw   [*Index of Fungi* 2: 61; F] [host(s) in Boraginaceae]
Plasmopara nakanoi S. Ito & Muray.   [*Index of Fungi* 2: 41] [host(s) in Ranunculales (Papaveraceae)]
  Plasmopara nivea (Unger) J. Schröt. [*nomen confusum*, see *P. umbelliferum sensu lato*]
Plasmopara obducens (J. Schröt.) J. Schröt.   [60-4; F] [host(s) in Ericales (Balsiminaceae)]
Plasmopara oenanthes J. F. Tao & Y. Qin   [*Index of Fungi* 5: 306] [host(s) in Apiales]
Plasmopara oplismeni Vienn.-Bourg.   [*Index of Fungi* 2: 517] [host(s) in Poales (Poaceae, Paniceae)]
Plasmopara palmae L. Campb.   [host(s) in Asterales]
Plasmopara panacis Bondartsev & Bunkina   [*Index of Fungi* 3: 56] [host(s) in Apiales (Araliaceae)]
  Plasmopara parvula (Jacz. & P. A. Jacz.) Skalický [as '(Schneider) Skalický'] [≡ *Peronospora parvula*]
                                                    [W. G. Schneid. *ex* Jacz. & P. A. Jacz.]
Plasmopara pastinacae Săvul. & O. Săvul.   [*Index of Fungi* 3: 364] [host(s) in Apiales]
Plasmopara paulowniae C. C. Chen   [*Index of Fungi* 4: 251] [host(s) in Lamiales (Scrophulariaceae)]
Plasmopara penniseti R. G. Kenneth & J. Kranz
                                       [*Index of Fungi* 4: 223] [host(s) in Poales (Poaceae, Paniceae)]
Plasmopara petasitidis S. Ito & Tokun.   [host(s) in Asterales]
Plasmopara petroselini Săvul. & O. Săvul.   [*Index of Fungi* 3: 364] [host(s) in Apiales]
Plasmopara peucedani Nannf.   [host(s) in Apiales]
Plasmopara phrymae S. Ito & Hara
              [*Index of Fungi* 2: 41] [host(s) in Lamiales ([Phrymaceae] Verbenaceae]
Plasmopara pileae (Gäum.) Jacz. & P. A. Jacz.
              [Petrak's Lists: Supplement] [host(s) in Rosales (Urticaceae)]
  Plasmopara pileae S. Ito & Tokun. [later homonym]   [*Index of Fungi* 4: 286]
Plasmopara pimpinellae Săvul. & O. Săvul. var. *pimpinellae*
                                       [*Index of Fungi* 3: 364] [host(s) in Apiales]
Plasmopara pimpinellae Săvul. & O. Săvul. var. *maioris* B. Wrońska   [*Index of Fungi* 5: 690]
Plasmopara plantaginicola T. R. Liu & C. K. Pai
              [*Index of Fungi* 5: 448] [host(s) in Lamiales (Plantaginaceae)]
Plasmopara plectranthi L. Ling & M. C. Tai   [*Index of Fungi* 1: 169] [host(s) in Lamiales]
  Plasmopara plectranthi A. D. Sharma & Munjal [later homonym]   [*Index of Fungi* 4: 635]

Table 2, continued.

Plasmopara podagrariae (G. H. Otth) Nannf.   [*Index of Fungi* **2**: 26] [host(s) in Apiales]

Plasmopara portoricensis (Lamkey) Waterh.   [*Index of Fungi* **5**: 92] [host(s) in Sapindales (Meliaceae)]

Plasmopara pusilla (de Bary) J. Schröt.   [59-2] [host(s) in Geraniales]

Plasmopara pygmaea (Unger) J. Schröt. forma *pygmaea* **[type species]**
                                                      [57-1; **F**] [host(s) in Ranunculales]

Plasmopara pygmaea (Unger) J. Schröt. forma *anemones* Gapon.   [*Index of Fungi* **4**: 354]

Plasmopara pygmaea (Unger) J. Schröt. forma *delphinii* Gapon.   [*Index of Fungi* **4**: 354]

Plasmopara pygmaea (Unger) J. Schröt. forma *hellebori* T. Săvul. & Rayss   [*Index of Fungi* **3**: 534]

Plasmopara pygmaea (Unger) J. Schröt. forma *isopyri-thalictroidis*   [Petrak's Lists, Supplement]

    Plasmopara pyrethri Dudka & Burdjuk. [= *Paraperonospora sulphurea*]   [*Index of Fungi* **4**: 635]

Plasmopara ribicola Davis [J. Schröt. *ex* Davis]

                              [60-3; **F**] [host(s) in Saxifragales (Grossulariaceae)]

Plasmopara sambucinae Nelen   [*Index of Fungi* **3**: 415] [host(s) in Dipsacales (Caprifoliaceae)]

Plasmopara sanguisorbae C. J. Li *et al.*   [C. J. Li, Z. Q. Yuan & Zhen Y. Zhao] [host(s) in Rosales]

Plasmopara saniculae Săvul. & O. Săvul.   [*Index of Fungi* **3**: 364] [host(s) in Apiales]

Plasmopara satarensis P. B. Chavan & U. V. Kulkarni
                                [*Index of Fungi* **4**: 418] [host(s) in Malvales (Tiliaceae)]

Plasmopara satureiae F. L. Tai & C. T. Wei   [host(s) in Lamiales]

Plasmopara saussureae Novot.   [*Index of Fungi* **3**: 255] [host(s) in Asterales]

Plasmopara savulescui Novot.   [*Index of Fungi* **3**: 255; 75-15] [host(s) in Asterales]

Plasmopara selini B. Wrońska   [*Index of Fungi* **5**: 690] [host(s) in Apiales]

    Plasmopara sigesbeckiae (Lagerh.) J. F. Tao [*nom. invalid.*] [= *Peronospora*]
                                                         [*Index of Fungi* **5**: 776]

Plasmopara sii Gapon.   [*Index of Fungi* **4**: 354] [host(s) in Apiales]

Plasmopara silai Săvul. & O. Săvul.   [*Index of Fungi* **3**: 364] [host(s) in Apiales]

Plasmopara skvortovii Miura   [host(s) in Malvales]

Plasmopara smyrnii Săvul. & M. Bechet   [*Index of Fungi* **4**: 72] [host(s) in Apiales]

Plasmopara solidaginis Novot.   [*Index of Fungi* **3**: 255; 78-19] [host(s) in Asterales]

    Plasmopara sordida [no authority, no reference in CABI database, probably *Peronospora sordida*]

Plasmopara sphaerosperma Săvul.   [*Index of Fungi* **1**: 231; 77-18] [host(s) in Asterales]

Plasmopara spilanthicola Syd.   [host(s) in Asterales]

    Plasmopara sulphurea (Gäum.) Skalický [≡ *Paraperonospora sulphurea*]
                                                   [*Index of Fungi* **3**; 415; 70-10]

    Plasmopara tanaceti (Gäum.) Skalický [≡ *Paraperonospora tanaceti*]   [*Index of Fungi* **3**: 415; 69-9]

Plasmopara triumfettae A. D. Sharma & Munjal
                [*Index of Fungi* **4**: 635] [host(s) in Malvales (Grewioideae) (formerly Tiliaceae)]

Plasmopara umbelliferanum (Casp.) Wartenw. var. *umbelliferarum*
                                [64-7] [J. Schröt. *ex* Wartenw.] [host(s) in Apiales]

[This species may include the doubtful *P. crustosa* and *P. nivea* (Constantinescu, 1992)]

Plasmopara umbelliferarum (Casp.) J. Schröt. var. *hacquetiae* Skalický   [*Index of Fungi* **2**: 197]

    Plasmopara ursiniae (Săvul. & L. Vánky) Skalický [= *Paraperonospora leptosperma*]
                                                   [*Index of Fungi* **3**: 415]

Plasmopara venezuelana Chardón   [host(s) in Brassicales]

Plasmopara vernoniae-chinensis Sawada   [host(s) in Asterales]

Plasmopara viburni Peck   [**F**] [host(s) in Dipsacales (Caprifoliaceae)]

Plasmopara vincetoxici Ellis & Everh.   [host(s) in Gentianales (Apocynaceae)]

Plasmopara viticola (Berk. & M. A. Curtis) Berl. & De Toni var. *viticola*   [62-5; **F**] [host(s) in Vitaceae]

Plasmopara viticola (Berk. & M. A. Curtis) Berl. & De Toni var. *americana* N. P. Golovina
                                                    [*Index of Fungi* **2**: 398]

Plasmopara viticola (Berk. & M. A. Curtis) Berl. & De Toni var. *amurensis* N. P. Golovina
                                                    [*Index of Fungi* **2**: 398]

Table 2, continued.

Plasmopara viticola (Berk. & M. A. Curtis) Berl. & De Toni var. *parthica* N. P. Golovina
[*Index of Fungi* 2: 398]
Plasmopara viticola (Berk. & M. A. Curtis) Berl. & De Toni forma *aestivalis-labruscae* Săvul.
[*Index of Fungi* 3: 364]
Plasmopara viticola (Berk. & M. A. Curtis) Berl. & De Toni forma *sylvestris* Săvul.
[*Index of Fungi* 3: 364]
Plasmopara viticola (Berk. & M. A. Curtis) Berl. & De Toni forma *viniferae-ampelopsidis* Săvul.
[*Index of Fungi* 3: 364]
Plasmopara wartenweileri Skalický   [*Index of Fungi* 2: 173] [host(s) in Ranunculales]
Plasmopara wildemaniana Henn. var. *wildemaniana*
[Saccardo: *Sylloge Fungorum* 21: 861; 24: 64] [host(s) in Lamiales (Acanthaceae)]
Plasmopara wildemaniana Henn. var. *macrospora* Sawada   [Saccardo: *Sylloge Fungorum* 24: 64]
Plasmopara yunnanensis J. F. Tao & Y. Qin   [*Index of Fungi* 5: 776] [host(s) in Asterales]

PSEUDOPERONOSPORA Rostovzev

[NOTE: Hosts principally in different families of the Rosales (Cannabaceae, Celtidaceae,
                 Ulmaceae, Urticaceae).]

Pseudoperonospora aethiomenatis (Simonyan) Waterh. [= *Peronospora parasi*[*Index of Fungi* 5: 93]
Pseudoperonospora elsholtziae D. Z. Tang [cf. *Plasmopara elsholtziae*]
[*Index of Fungi* 5: 403] [host(s) in Lamiales]
Pseudoperonospora cannabina (G. H. Otth) Curzi
Pseudoperonospora cassiae Waterh.   [*Index of Fungi* 5: 93]
Pseudoperonospora celtidis (Waite) G. W. Wilson
Pseudoperonospora cubensis (Berk. & M. A. Curtis) Rostovzev **[type species]**
Pseudoperonospora humuli (Miyabe & M. Takah.) G. W. Wilson
Pseudoperonospora urticae (Berk.) E. S. Salmon & Ware   [Lib. *ex* Berk.]

TABLE 3. Peronosporales: Albuginaceae; angiosperm host orders included for cross-reference to Table 2:
Peronospora and *Plasmopara*

ALBUGO (Pers.) Roussel

Albugo achyranthis (Henn.) Miyabe   [host(s) in Caryophyllales (Amaranthaceae)]
Albugo aechmantherae Z.-y. Zhang & Y.-x. Wang[*Index of Fungi* 5: 377] [in Lamiales (Scrophulariaceae)]
Albugo amaranthi (Schwein.) Kuntze   [host(s) in Caryophyllales (Amaranthaceae)]
Albugo austro-africana Syd. & P. Syd.   [host(s) in Caryophyllales (Aizoaceae)]
Albugo bliti (Biv.) Kuntze   [host(s) in Caryophyllales (Amaranthaceae)]
Albugo candida (J. F. Gmel.) Kuntze var. *candida* **[type species]**
[(J. F. Gmel.: Pers.)] [host(s) in Brassicales]
Albugo candida (J. F. Gmel.) Kuntze var. *macrospora* Togashi
[(J. F. Gmel.: Pers.)] [host(s) in Brassicales]
Albugo capparidis (de Bary) Cif.   [host(s) in Brassicales]

Table 3, continued.

Albugo caryophyllacearum (Wallr.) Cif. & Biga	[host(s) in Caryophyllales]
Albugo centaurii (Hansf.) Cif. & Biga	[host(s) in Gentianales]
Albugo chardiniae Bremer & Petr.	[host(s) in Asterales]
Albugo chardonii W. Weston	[host(s) in Brassicales]
Albugo cynoglossi (Unamuno) Cif. & Biga	[host(s) in Boraginaceae]

Albugo eomeconis Z. Y. Zhang & Ying X. Wang
[*Index of Fungi* **5**: 73] [host(s) in Ranunculales (Papaveraceae)]

Albugo eurotiae Tranzschel	[host(s) in Caryophyllales (Amaranthaceae)]
Albugo evansii Syd.	[host(s) in Lamiales (Scrophulariaceae)]
Albugo evolvulae (Damle) Cif. & Biga	[host(s) in Lamiales]
Albugo gomphrenae (Speg.) Cif. & Biga	[host(s) in Caryophyllales (Amaranthaceae)]

Albugo hyoscyami Z. Y. Zhang *et al.*
[Z. Y. Zhang, Ying X. Wang & Z. S. Fu] [*Index of Fungi* **5**: 571] [host(s) in Solanales (Convolvulaceae)]

Albugo ipomoeae-aquaticae Sawada	[host(s) in Solanales (Convolvulaceae)]
Albugo ipomoeae-hardwikii Sawada	[host(s) in Solanales (Convolvulaceae)]

Albugo ipomoeae-panduratae (Schwein.) Swingle var. *ipomoeae-panduratae*
[host(s) in Solanales (Convolvulaceae)]

Albugo ipomoeae-panduratae (Schwein.) Swingle var. *tillaceae* Cif. & Biga
[host(s) in Solanales (Convolvulaceae)]

Albugo ipomoeae-pes-caprae Cif.	[host(s) in Solanales (Convolvulaceae)]
Albugo keeneri Solheim & Gilb.	[*Index of Fungi* **4**: 531] [host(s) in Ranunculales (Papaveraceae)]
Albugo lepidii A. N. S. Rao	[*Index of Fungi* **5**: 41] [host(s) in Brassicales]

Albugo mangenotii Mayor & Vienn.-Bourg. [= *A. molluginis*]
[*Index of Fungi* **2**: 51] [host(s) in Caryophyllales]

Albugo mauginii (Parisi) Cif. & Biga	[host(s) in Fabales]
Albugo minor (Speg.) Cif.	[host(s) in Solanales (Convolvulaceae)]

Albugo molluginis S. Ito
[also cited as S. Ito & Tokun. (*in* Ito & Tokunaga)] [host(s) in Caryophyllales (Molluginaceae)]

Albugo mysorensis (Thirum. & Safeeulla) Vasudeva [*nom. invalid.*]
[*Index of Fungi* **3**: 455] [host(s) in Caryophyllales (Molluginaceae)]

Albugo occidentalis G. W. Wilson   [host(s) in Caryophyllales (Amaranthaceae)]

Albugo pes-tigridis Gharse [as '(Verma) Gharse']
[*Index of Fungi* **3**: 269] [host(s) in Solanales (Convolvulaceae)]

Albugo pileae J. F. Tao & Y. Qin	[*Index of Fungi* **5**: 277] [host(s) in Rosales (Urticaceae)]
Albugo platensis (Speg.) Swingle	[host(s) in Caryophyllales (Nyctaginaceae)]

Albugo polygoni Z. D. Jiang & P. K. Chi
[*Index of Fungi* **6(10)**: 833] [host(s) in Caryophyllales (Polygonaceae)]

Albugo portulacae (DC.) Kuntze   [host(s) in Caryophyllales (Amaranthaceae)]

Albugo portulacearum (Schltdl.) Kochman & T. Majewski
[*Index of Fungi* **4**: 2] [host(s) in Caryophyllales (Portulacaceae)]

Albugo quadrata (Kalchbr. & Cooke) Kuntze [later homonym]   [host(s) in Lamiales (Acanthaceae)]

Albugo quadrata (Wallr.) S. D. Baker [? = *Albugo bliti*]
[*Index of Fungi* **2**: 207] [host(s) in Caryophyllales]

Albugo resedae (Jacz.) Cif. & Biga	[host(s) in Brassicales (Resedaceae)]
Albugo solivae J. Schröt.	[host(s) in Asterales]
Albugo solivarum (Speg.) Herter	[*Index of Fungi* **4**: 301] [host(s) in Asterales]
Albugo spinulosus (de Bary) Herter	[*Index of Fungi* **4**: 301] [host(s) in Asterales]
Albugo swertiae (Berl. & Kom.) G. W. Wilson	[host(s) in Gentianales]
Albugo tilleae (Lagerh.) Cif. & Biga	[host(s) in Saxifragales (Crassulaceae)]
Albugo tragopogonis (Pers.) Gray var. *tragopogonis*	[host(s) in Asterales]
Albugo tragopogonis (Pers.) Gray var. *cirsii* Cif. & Biga	[host(s) in Asterales]
Albugo tragopogonis (Pers.) Gray var. *inulae* Cif. & Biga	[host(s) in Asterales]

Table 3, continued.

Albugo tragopogonis (Pers.) Gray var. *pyrethri* Cif. & Biga   [host(s) in Asterales]
Albugo tragopogonis (Pers.) Gray var. *xeranthemi-annui* Săvul. & Rayss   [host(s) in Asterales]
Albugo trianthemae G. W. Wilson   [host(s) in Caryophyllales (Aizoaceae)]
Albugo tropica Lagerh.   [host(s) in Piperales]

TABLE 4. Pythiales: Pythiaceae

[NOTE: When there has been a fully comprehensive molecular biological account of all the
              genera listed here, together with a full sampling of all the different sections of
              Pythium sensu lato and *Phytophthora sensu lato*, there will be new monophyletic
              genera rearranged among the orders PYTHIALES and PERONOSPORALES.]

CYSTOSIPHON Roze & Cornu

[NOTE: = *Pythium sensu lato*; this is the earliest valid generic name for species with
              spherical zoosporangia]

Cystosiphon canterae (Karling) M. W. Dick   [Dick, 2001*b*]
Cystosiphon closterii (De Wild.) M. W. Dick   [Dick, 2001*b*]
Cystosiphon dictyosporum (Racib.) M. W. Dick   [Dick, 2001*b*]
Cystosiphon pythioides Roze & Cornu **[type species]**
Cystosiphon reducta (de Wild.) M. W. Dick   [Dick, 2001*b*]

DIASPORANGIUM Höhnk

Diasporangium jonesianum Höhnk **[type species; monotypic]**

ENDOSPHAERIUM D'Eliscu [doubtful genus]

Endosphaerium funiculatum D'Eliscu **[type species; monotypic]**

HALOPHYTOPHTHORA H. H. Ho & S. C. Jong

Halophytophthora avicenniae (Gerr.-Corn. & J. A. Simpson) H. H. Ho & S. C. Jong
                                                    [*Index of Fungi* **5**: 358; **6(1)**: 14]
Halophytophthora bahamensis (Fell & Master) H. H. Ho & S. C. Jong   [*Index of Fungi* **4**: 417; **6(1)**: 14]
Halophytophthora batemanensis (Gerr.-Corn. & J. A. Simpson) H. H. Ho & S. C. Jong
                                                    [*Index of Fungi* **5**: 358; **6(1)**: 14]
Halophytophthora epistomium (Fell & Master) H. H. Ho & S. C. Jong   [*Index of Fungi* **4**: 417; **6(1)**: 14]
Halophytophthora exoprolifera H. H. Ho *et al.*
                     [H. H. Ho, A. Nakagiri & S. Y. Newell] [*Index of Fungi* **6(5)**: 269]

Table 4, continued.

Halophytophthora kandeliae H. H. Ho *et al.*
        [H. H. Ho, S. Y. Hsieh & H. S. Chang] [*Index of Fungi* **6(2)**: 76; **6(4)**: 208]
Halophytophthora masteri A. Nakagiri & S. Y. Newell   [*Index of Fungi* **6(10)**: 547]
Halophytophthora mycoparasitica (Fell & Master) H. H. Ho & S. C. Jong
        [*Index of Fungi* **4**: 417; **6(1)**: 14]
Halophytophthora operculata (Pegg & Alcorn) H. H. Ho & S. C. Jong   [*Index of Fungi* **5**: 210; **6(1)**: 14]
Halophytophthora polymorphica (Gerr.-Corn. & J. A. Simpson) H. H. Ho & S. C. Jong
        [*Index of Fungi* **5**: 358; **6(1)**: 14]
Halophytophthora porrigovesica Nakagiri *et al.*
        [K. Nakagiri, Tad.Ito, L. Manoch, & M. Tanticharoen] [*Mycoscience* **42**: 34 (2001)]
Halophytophthora spinosa (Fell & Master) H. H. Ho & S. C. Jong var. *spinosa*
        [*Index of Fungi* **4**: 417; **6(1)**: 14]
Halophytophthora spinosa (Fell & Master) H. H. Ho & S. C. Jong var. *lobata* (Fell & Master) H. H. Ho &
  S. C. Jong   [*Index of Fungi* **4**: 417; **6(1)**: 14]
Halophytophthora tartarea A. Nakagiri & S. Y. Newell   [*Index of Fungi* **6(10)**: 547]
Halophytophthora vesicula (Anastasiou & Churchl.) H. H. Ho & S. C. Jong **[type species]**
        [*Index of Fungi* **3**: 534; **6(1)**: 14]

LAGENIDIUM Zopf

    [NOTE: Retypification of the genus by *L. giganteum* is recommended: the genus would then
                be monotypic. The existing 'type' [see Dick, 1999*b*] species being transferred to
                Myzocytium.]

Lagenidium giganteum Couch **[proposed type species; monotypic]**

MYZOCYTIUM Schenk

Myzocytium megastomum De Wild.
Myzocytium proliferum Schenk **[type species]**
Myzocytium netrii (C. E. Mill.) M. W. Dick   [Dick, 2001*b*]
Myzocytium rabenhorstii Zopf [type species of *Lagenidium*]

PERONOPHYTHORA W. H. Ko *et al.*
        [C. C. Chen *ex* W. H. Ko, H. S. Chang, H. J. Su, C. C. Chen & L. S. Leu]

    [NOTE: Referable to *Phytophthora sensu lato.*]

Peronophythora litchi W. H. Ko *et al.* **[type species; monotypic]**
        [C. C. Chen *ex* W. H. Ko, H. S. Chang, H. J. Su, C. C. Chen & L. S. Leu]

Table 4, continued.

PHYTOPHTHORA de Bary

[NOTES: Species groups in Stamps *et al.* (1990) justified right.
Erwin & Ribeiro (1996) cite *publication authors*, not authorities, after the binomial.

Phytophthora nicotianae and *P. parasitica* re: Erwin & Ribeiro (1996: 391). There is no separate and distinct Code of Nomenclature governing the nomenclature of straminipilous organisms (Chromista) other than the Zoological Code (for heterotrophs without a cell wall) and the Botanical Code. The convention has been to use the Botanical Code for both the photosynthetic organisms and the fungal organisms, but taxa (especially Labyrinthista) have been described under either code. Unilateral action by workers in *Phytophthora* would create, within this kingdom, a chaotic nomenclatural situation, which must be avoided. Therefore the use of *P. nicotianae* should prevail over *P. parasitica* unless or until *P. parasitica* is formally conserved. The extensive use, over many years, of both these names inevitably weakens a submission in support of *P. parasitica*, but a workshop of interested parties could achieve consensus.]

Phytophthora arecae Rosenbaum	**[II]**
Phytophthora boehmeriae Sawada	**[II]**
Phytophthora botryosa Chee	**[II]**
Phytophthora cactorum (Lebert & Cohn) J. Schröt.	**[I]**
Phytophthora cajani K. S. Amin *et al.* [K. S. Amin, Baldev & F. J. Williams] **[VI]**	[*Index of Fungi* **4**: 542]
Phytophthora castanae Katsura & S. Uchida	**[II]** [*Index of Fungi* **4**: 514]
Phytophthora castanae (L. Mangin) I. MacFarl. [*nom. invalid.*]	[*Index of Fungi* **4**: 635]
Phytophthora cinchonae Sawada	[*Index of Fungi* **4**: 71]
Phytophthora cambivora (Petri) Buisman	**[VI]**
Phytophthora capsici Leonian	**[II]**
Phytophthora cinnamomi Rands var. *cinnamomi*	**[VI]**
Phytophthora cinnamomi Rands var. *parvispora* Kröber & R. Marwitz	[*Index of Fungi* **6(6)**: 347]
Phytophthora citricola Sawada	**[III]**
Phytophthora citrophthora (R.E.Sm. & E.H.Sm.) Leonian	**[II]**
Phytophthora clandestina P. A. Taylor *et al.*	
[P. A. Taylor, Pascoe & F. C. Greenh.] **[I]** [*Index of Fungi* **5**: 447]	
Phytophthora colocasiae Racib.	**[IV]**
Phytophthora cryptogea Pethybr. & Laff.	**[VI]**
Phytophthora cryptogea Pethybr. & Laff. forma specialis *begoniae* Kröber	[*Index of Fungi* **5**: 210]
Phytophthora cyperi (Ideta) S. Ito	**[III]**
Phytophthora cyperi-bulbosi Seethal. & K. Ramakr.	**[III]**
Phytophthora drechsleri Tucker var. *drechsleri*	**[VI]**
Phytophthora drechsleri Tucker var. *cajani* M. Pal *et al.*	
[M. Pal, Grewal & A. K. Sarbhoy] [*Index of Fungi* **4**: 71]	
Phytophthora eriugena Clancy & Kavanagh [*nom. invalid.*]	**[IV]** [*Index of Fungi* **4**: 635]
Phytophthora erythroseptica Pethybr. var. *erythroseptica*	**[VI]**
Phytophthora erythroseptica Pethybr. var. *drechsleri* (Tucker) Sarej.	[*Index of Fungi* **4**: 71]
Phytophthora erythroseptica Pethybr. var. *pisi* Bywater & Hickman	
Phytophthora fagi R. Hartig [≡ *Peronospora fagi*, ? = *Phytophthora quercina*]	
Phytophthora formosana Sawada [= *P. nicotianae*]	[*Index of Fungi* **4**: 41]
Phytophthora fragariae Hickman var. *fragariae*	**[V]**
Phytophthora fragariae Hickman var. *oryzobladis* J. S. Wang & J. Y. Lu	[*Index of Fungi* **4**: 542]
Phytophthora fragariae Hickman var. *rubi* W. F. Wilcox & Deacon	[*Index of Fungi* **6(6)**: 347]
Phytophthora gonapodyides (H. E. Petersen) Buisman	**[VI]**
Phytophthora heveae A. W. Thomps.	**[II]**

Table 4, continued.

Phytophthora hibernalis Carne	**[IV]**
Phytophthora humicola W. H. Ko & Ann	**[V]**
Phytophthora idaei D. M. Kennedy	**[I]** [*Index of Fungi* **6(10)**: 560]
Phytophthora ilicis Buddenh. & Roy A. Young	**[IV]**
Phytophthora imperfecta Sarej. var. *imperfecta*	[*Index of Fungi* **4**: 72]

Phytophthora imperfecta Sarej. var. *citrophthora* (R. E. Sm & E. H. Sm.) Sarej.
[*Index of Fungi* **4**: 72, 176]

Phytophthora imperfecta Sarej. var. *nicotianae* (Breda de Haan) Sarej.   [*Index of Fungi* **4**: 72]

Phytophthora inflata Caros. & Tucker var. *infestans* **[type species]**   **[IV]**

Phytophthora infestans (Mont.) de Bary var. *phaseoli* (Thaxt.) Leonian   [*Index of Fungi* **4**: 72]

Phytophthora infestans (Mont.) de Bary forma specialis *mirabilis* E. M. Möller & De Cock
[*Index of Fungi* **6(8)**: 445]

Phytophthora inflata Caros. & Tucker	**[III]**
Phytophthora insolita Ann & W. H. Ko	**[V]** [*Index of Fungi* **5**: 91]
Phytophthora ipomoeae W. G. Flier & N. J. Grünwald	
Phytophthora iranica Ershad	**[I]** [*Index of Fungi* **4**: 72]

Phytophthora italica S. O. Cacciola *et al.*   [S. O. Cacciola, G. Magnano di San Lio & A. Belisario]

Phytophthora japonica Waterh.   **[VI]** [*Index of Fungi* **4**: 316]

   Phytophthora katsurae W. H. Ko & H. S. Chang [≡ *P. castanae*]
[**II**] [*Index of Fungi* **4**: 635; **6(16)**: 868]

Phytophthora lateralis Tucker & Milbrath   **[II]**

Phytophthora leersiae H. H. Ho & H. S. Chang
[Sawada *ex* H. H. Ho & H. S. Chang] [*Index of Fungi* **6(4)**: 221]

Phytophthora lepironiae Sawada	**[III]**
Phytophthora lycopersici Sawada [= *P. nicotianae*]	[*Index of Fungi* **4**: 41]
Phytophthora macrochlamydospora J. A. G. Irwin	**[III/IV]** [*Index of Fungi* **6(4)**: 221]
Phytophthora meadii McRae	**[II]**
Phytophthora medicaginis E. M. Hansen & D. P. Maxwell	**[V]** [*Index of Fungi* **6(3)**: 159]
Phytophthora megakarya Brasier & M. J. Griffin	**[II]** [*Index of Fungi* **4**: 571]
Phytophthora megasperma Drechsler var. *megasperma*	**[V]**

Phytophthora megasperma Drechsler var. *sojae* (Kaufm. & Gerd.) A. A. Hildebr.
[*P. sojae* Kaufm. & Gerd.]

Phytophthora megasperma Drechsler forma specialis *glycines* T.-L. Kuan & Erwin [*nom. illeg.*]
[*Index of Fungi* **5**: 171]

Phytophthora megasperma Drechsler forma specialis *medicaginis* T.-L. Kuan & Erwin [*nom. illeg.*]
[*Index of Fungi* **5**: 171]

Phytophthora megasperma Drechsler forma specialis *trifolii* R. G. Pratt [*nom. illeg.*]
[*Index of Fungi* **5**: 134]

Phytophthora melonis Katsura	**[VI]** [*Index of Fungi* **4**: 514]
Phytophthora mexicana H. H. Hotson & Hartge	**[II]**
Phytophthora mirabilis Galindo & H. R. Hohl	**[IV]** [*Index of Fungi* **5**: 606]
Phytophthora multivesiculata Ilieva *et al.*	

[Ilieva, Man in't Veld, W. Veenb.-Rijksa & R. Pieters] [*Index of Fungi* **6(18)**: 992]

Phytophthora nicotianae Breda de Haan var. *nicotianae*   **[II]**

Phytophthora nicotianae Breda de Haan var. *parasitica* (Dastur) Waterh.

   Phytophthora oryzae (Brizi) K. Hara [= *P. japonica*]   [*Index of Fungi* **4**: 72]

   Phytophthora oryzae (S. Ito & Nagai) Waterh. [later homonym]   [*Index of Fungi* **4**: 176]

Phytophthora palmivora (E. J. Butler) E. J. Butler var. *palmivora*   **[II]**

Phytophthora palmivora (E. J. Butler) E. J. Butler var. *heterocystica* Babacauh
[*Index of Fungi* **5**: 358, 899]

   Phytophthora parasitica Dastur var. *parasitica*   **[II]**

   Phytophthora parasitica Dastur var. *capsici* (Leonian) Sarej.   [*Index of Fungi* **4**: 72]

Table 4, continued.

Phytophthora parasitica Dastur var. *colocasiae* (Racib.) Sarej.	[*Index of Fungi* **4**: 72]
Phytophthora pistaciae Mirabolfathy	[*Mycol. Res.* **105**: 1173 (2001)]
Phytophthora phaseoli Thaxt.	[**IV**]
Phytophthora porri Foister	[**III**]
Phytophthora primulae J. A. Toml.	[**III**]
Phytophthora pseudotsugae Hamm & E. M. Hansen	[**I**] [*Index of Fungi* **5**: 358]
Phytophthora quercina T. Jung	[**I**] [*Index of Fungi* **7**(2): 75]
Phytophthora quininea Crand.	[**V**]
Phytophthora ramorum Werres, De Cock & Man in't Veld	[*Mycol. Res.* **105**: 1164 (2001)]
Phytophthora richardiae Buisman	[**VI**]
Phytophthora ricini Sawada [= *P. nicotianae*]	[*Index of Fungi* **4**: 41]
Phytophthora sinensis Y. N. Yu & W. Y. Zhuang	[**VI**]
Phytophthora sojae Kaufm. & Gerd.	
[= *P. megasperma* Drechsler var. *sojae* (Kaufm. & Gerd.) A. A. Hildebr.]	[**V**]
Phytophthora syringae (Kleb.) Kleb.	[**III**]
Phytophthora tentacula Kröber & R. Marwitz	[**I**] [*Index of Fungi* **6**(6): 347]
Phytophthora trifolii E. M. Hansen & D. P. Maxwell	[**V**] [*Index of Fungi* **6**(3): 159]
Phytophthora undulata (H. E. Petersen) M. W. Dick	[**VI**]
Phytophthora verrucosa Alcock & Foister	[**V**]
Phytophthora vignae Purss	[**VI**]
Phytophthora vignae Purss forma specialis *adzukicola* S. Tsuchiya *et al.*	
[S. Tsuchiya, M. Yanagawa & A. Ogoshi] [*Index of Fungi* **5**: 606]	
Phytophthora vignae Purss forma specialis *medicaginis* S. Tsuchiya *et al.*	
[S. Tsuchiya, M. Yanagawa & A. Ogoshi] [*Index of Fungi* **5**: 606]	

PYTHIUM Pringsh. [*Pythium* sensu lato]

[NOTE: For a complete list of binomials and full citations see Dick (1990*b*); citation
                    information in Plaats-Niterink (1981) is incomplete.]

Pythium acanthicum Drechsler	
Pythium acanthophoron Sideris	
Pythium acrogynum Y. N. Yu	[*Index of Fungi* **4**: 252]
Pythium adhaerens Sparrow	
Pythium amasculinum Y. N. Yu	[*Index of Fungi* **4**: 252]
Pythium anandrum Drechsler	
Pythium angustatum Sparrow	
Pythium aphanidermatum (Edson) Fitzp.	
Pythium apleroticum Tokun.	
Pythium aquatile Höhnk	
Pythium aristosporum Vanterp.	
Pythium arrhenomanes Drechsler	[*Index of Fungi* **6**(14): 772]
Pythium australe M. W. Dick [*nomen nudum*]	[*Index of Fungi* **6**(3): 162]
Pythium betae M. Takah.	[*Index of Fungi* **4**: 183]
Pythium boreale R. L. Duan	[*Index of Fungi* **5**: 451]
Pythium buismaniae Plaäts-Nit.	[*Index of Fungi* **5**: 137]
Pythium capillosum B. Paul var. *capillosum*	[*Index of Fungi* **5**: 695]
Pythium capillosum B. Paul var. *helicoides* B. Paul	[*Index of Fungi* **5**: 779]
Pythium catenulatum V. D. Matthews	
Pythium caudatum (G. L. Barron) M. W. Dick	[*Index of Fungi* **4**: 411; Dick, 2001*b*]
Pythium chamaehyphon Sideris	
Pythium chondricola De Cock	[*Index of Fungi* **5**: 499]

Table 4, continued.

Pythium coloratum Vaartaja	
Pythium conidiophorum Jokl	
Pythium connatum Y. N. Yu	[*Index of Fungi* **4**: 252]
Pythium contiguanum B.Paul [≡ *Pythium drechsleri* B. Paul]	
Pythium cucumerinum Bakhariev	[*Index of Fungi* **5**: 404]
Pythium cryptogynum B. Paul	[*Index of Fungi* **6(2)**: 94]
Pythium cylindrosporum B. Paul	[*Index of Fungi* **6(5)**: 288]
Pythium debaryanum R. Hesse	
Pythium deliense Meurs	
Pythium diclinum Tokun.	
Pythium dictyosporum Racib.	
Pythium dimorphum F. F. Hendrix & W. A. Campb.	[*Index of Fungi* **4**: 106]
Pythium dissimile Vaartaja	
Pythium dissotocum Drechsler	
Pythium drechsleri B. Paul [later homonym - see *Pythium contiguanum*]	
Pythium drechsleri S. Rajgopalan & K. Ramakr.	[*Index of Fungi* **4**: 252]
Pythium echinulatum V. D. Matthews	
Pythium erinaceus G. I. Robertson [probably = *Pythium echinulatum*]	[*Index of Fungi* **4**: 636]
Pythium flevoense Plaäts-Nit.	[*Index of Fungi* **4**: 224]
Pythium fluminum D. Park var. *fluminum*	[*Index of Fungi* **4**: 515]
Pythium fluminum D. Park var. *flavum* D. Park	[*Index of Fungi* **4**: 515]
Pythium folliculosum B. Paul	[*Index of Fungi* **6(4)**: 225]
Pythium graminicola Subraman.	
Pythium grandisporangium Fell & Master	[*Index of Fungi* **4**: 419]
Pythium helicandrum Drechsler	
Pythium helicoides Drechsler	
Pythium helicum T. Itô	
Pythium hemmianum M. Takah.	
Pythium heterothallicum W. A. Campb. & F. F. Hendrix	
Pythium hydnosporum (Mont.) J. Schröt.	
Pythium hypoandrum Y. N. Yu & Y. L. Wang	[*Index of Fungi* **5**: 779]
Pythium hypogynum Middleton	
Pythium indigoferae E. J. Butler	
Pythium inflatum V. D. Matthews	
Pythium insidiosum De Cock *et al.*	
[De Cock, L. Mend., A. A. Padhye, Ajello & Kaufman]	[*Index of Fungi* **5**: 695]
Pythium intermedium de Bary	
Pythium irregulare Buisman	
Pythium iwayamae S. Ito	
Pythium kunmingense Y. N. Yu	[*Index of Fungi* **4**: 252]
Pythium lobatum S. Rajagopalan & K. Ramakr.	[*Index of Fungi* **4**: 252]
Pythium lucens Ali-Shtayeh	[*Index of Fungi* **5**: 499]
Pythium lutarium Ali-Shtayeh	[*Index of Fungi* **5**: 499]
Pythium macrosporum Vaartaja & Plaäts-Nit.	[*Index of Fungi* **5**: 137]
Pythium mamillatum Meurs	
Pythium marinum Sparrow	
Pythium maritimum Höhnk	
Pythium marsipium Drechsler	
Pythium mastophorum Drechsler	
Pythium megalacanthum de Bary	
Pythium middletonii Sparrow	
Pythium middletonii S. Rajagopalan & K. Ramakr. [later homonym]	[*Index of Fungi* **4**: 252, 386]

Table 4, continued.

Pythium minor Ali-Shtayeh   [*Index of Fungi* **5**: 499]
Pythium monospermum Pringsh. **[type species]**
Pythium multisporum Poitras
Pythium mycoparasiticum Deacon *et al.*
                              [Deacon, S. A. K. Laing & L. A. Berry] [*Index of Fungi* **6(3)**: 162; **6(5)**: 228]
Pythium myriotylum Drechsler
Pythium nagae S. Ito & Tokun.
Pythium nodosum B. Paul *et al.*
                              [B. Paul, D. Galland, T. Bhatnagar & H. Dulieu] [*Index of Fungi* **6(18)**: 994]
Pythium nunn Lifsh. *et al.*   [Lifsh., Stangh. & R. Baker] [*Index of Fungi* **5**: 362]
Pythium oedochilum Drechsler
Pythium okanoganense P. E. Lipps   [*Index of Fungi* **5**: 93]
Pythium oligandrum Drechsler
Pythium opalinum M. W. Dick (*nomen nudum*)   [*Index of Fungi* **6(3)**: 162]
Pythium ornacarpum B. Paul   [*Index of Fungi* **7(2)**: 78]
Pythium ornamentatum B. Paul   [*Index of Fungi* **5**: 779]
Pythium orthogonon C. Ahrens   [*Index of Fungi* **4**: 73]
Pythium ostracodes Drechsler
Pythium pachycaule Ali-Shtayeh var. *pachycaule*   [*Index of Fungi* **5**: 499]
Pythium pachycaule Ali-Shtayeh var. *ramificatum* B. Paul   [*Index of Fungi* **6(6)**: 351]
Pythium paddicum Hirane
Pythium palingenes Drechsler
Pythium papillatum V. D. Matthews
Pythium parasiticum S. Rajagopalan & K. Ramakr.   [*Index of Fungi* **4**: 252]
Pythium paroecandrum Drechsler
Pythium parvum Ali-Shtayeh   [*Index of Fungi* **5**: 499]
Pythium periilum Drechsler
Pythium periplocum Drechsler
Pythium perniciosum Serbinow
Pythium perplexum H. Kouyeas & Theoh.   [*Index of Fungi* **4**: 636]
Pythium pleroticum T. Itô
   Pythium plurisporum G. Abad *et al.* [= *Pythium minor* ?]
                              [G. Abad, H. D. Shew, L. F. Grand & L. T. Lucas] [*Index of Fungi* **6(13)**: 729]
Pythium podbielkowskii (A. Batko) M. W. Dick   [*Index of Fungi* **4**: 215; Dick, 2001*b*]
Pythium polycarpum B. Paul   [*Index of Fungi* **5**: 555]
Pythium polymastum Drechsler
Pythium polypapillatum T. Itô
Pythium polytylum Drechsler
Pythium porphyrae M. Takah. & M. Sasaki   [*Index of Fungi* **4**: 515]
Pythium prolatum F. F. Hendrix & W. A. Campb.
Pythium pulchrum Minden
Pythium pyrilobum Vaartaja
Pythium radiosum B. Paul   [*Index of Fungi* **6(5)**: 288]
Pythium ramificatum B. Paul   [*Index of Fungi* **5**: 695]
Pythium rostratum E. J. Butler
Pythium salinum Höhnk
Pythium salpingophorum Drechsler
Pythium scleroteichum Drechsler
Pythium sinense Y. N. Yu   [*Index of Fungi* **4**: 252]
Pythium spinosum Sawada
Pythium splendens Hans Braun
Pythium sulcatum R. G. Pratt & J. E. Mitch.   [*Index of Fungi* **4**: 224]

Table 4, continued.

Pythium sylvaticum W. A. Campb. & F. F. Hendrix
Pythium tardicrescens Vanterp.
Pythium tenue Gobi
Pythium toruloides B. Paul   [*Index of Fungi* **5**: 555]
Pythium torulosum Coker & P. Patt.
Pythium tracheiphilum A. Matta
Pythium uladhum D. Park   [*Index of Fungi* **4**: 515]
Pythium ultimum Trow var. *ultimum*
Pythium ultimum Trow var. *sporangiiferum* Drechsler
Pythium uncinulatum Plaäts-Nit. & I. Blok   [*Index of Fungi* **4**: 574]
Pythium vanterpoolii V. Kouyeas & H. Kouyeas
Pythium vexans de Bary var. *vexans*
Pythium vexans de Bary var. *minutum* G. S. Mer & Khulbe   [*Index of Fungi* **5**: 310]
Pythium violae Chesters & Hickman
Pythium volutum Vanterp. & Truscott
Pythium zingiberis M. Takah.

TRACHYSPHAERA Tabor & Bunting

[NOTE: Probably referable to *Phytophthora sensu lato*.]

Trachysphaera fructigena Tabor & Bunting **[type species; monotypic]**

TABLE 5.   Pythiales: Pythiogetonaceae

MEDUSOIDES Voglmayr

Medusoides argyrocodium Voglmayr **[type species; monotypic]**
                                                               [*Index of Fungi* **17(1)**: 17]

PYTHIOGETON Minden

Pythiogeton autossytum Drechsler
Pythiogeton dichotomum Tokun.
Pythiogeton nigrescens A. Batko
Pythiogeton ramosum Minden
Pythiogeton transversum Minden
Pythiogeton uniforme A. Lund *
Pythiogeton utriforme Minden **[type species]**
Pythiogeton zeae H. J. Jee *et al.*   [H. J. Jee, H. H. Ho & W. D. Cho]

TABLE 6. Sclerosporales: Sclerosporaceae. For taxonomic reviews of the genera of the *Sclerosporales*, see Dick *et al.* (1984, 1989; Dick, 2001*b*) and Waterhouse (1964).

SCLEROSPORA J. Schröt.

Sclerospora butleri W. Weston
Sclerospora graminicola (Sacc.) J. Schröt. [type species]
Sclerospora iseilmatis Thirum. & Naras.
Sclerospora northii W. Weston
Sclerospora secalina Naumov

PERONOSCLEROSPORA (S. Ito) Hara
 [Hara *in* Shirai, *not* Shirai & Hara (Shaw & Waterhouse, 1980); *not* C. G. Shaw (as in Hawksworth *et al.* (1995), see also type species.]

Peronosclerospora dichanthiicola (Thirum. & Naras.) C. G. Shaw   [*Index of Fungi* **4**: 541, 570; **5**: 444]
Peronosclerospora globosa Kubicek & R. G. Kenneth   [*Index of Fungi* **5**: 399]
Peronosclerospora heteropogonis Siradhana *et al.*
                         [Siradhana, Dange, Rathore & S. D. Singh] [*Index of Fungi* **5**: 20]
Peronosclerospora maydis (Racib.) C. G. Shaw   [*Index of Fungi* **4**: 541; **5**: 444]
Peronosclerospora miscanthi (T. Miyake) C. G. Shaw   [*Index of Fungi* **4**: 541; **5**: 444]
Peronosclerospora noblei (W. Weston) C. G. Shaw   [*Index of Fungi* **5**: 20]
Peronosclerospora philippinensis (W. Weston) C. G. Shaw   [*Index of Fungi* **4**: 541; **5**: 444]
Peronosclerospora sacchari (T. Miyake) Hara [type species]   [*Index of Fungi* **4**: 541, 570; **5**: 445]
Peronosclerospora sorghi (W. Weston & Uppal) C. G. Shaw   [*Index of Fungi* **4**: 541]
Peronosclerospora spontanea (W. Weston) C. G. Shaw   [*Index of Fungi* **4**: 541; **5**: 445]
Peronosclerospora westonii (M. C. Sriniv. *et al.*) C. G. Shaw
                   [M. C. Sriniv., Naras. & Thirum.] [*Index of Fungi* **4**: 541; **5**: 445]
Peronosclerospora zeae Yao [*nom. illeg.*]

TABLE 7. Sclerosporales: Verrucalvaceae

PACHYMETRA B. J. Croft & M. W. Dick

Pachymetra chaunorhiza B. J. Croft & M. W. Dick [type species; monotypic genus][*Index of Fungi* **5**: 897]

SCLEROPHTHORA Thirum. *et al.*   [Thirum., C. G. Shaw & Naras.]

Sclerophthora cryophila W. Jones
Sclerophthora farlowii (Griffiths) R. G. Kenneth
Sclerophthora lolii R. G. Kenneth   [*Index of Fungi* **3**: 285]
Sclerophthora macrospora (Sacc.) Thirum. *et al.* [type species]   [Thirum., C. G. Shaw & Naras.]
Sclerophthora rayssiae R. G. Kenneth *et al.* var. *rayssiae*
                         [R. G. Kenneth, Koltin & Wahl] [*Index of Fungi* **3**: 285]
Sclerophthora rayssiae R. G. Kenneth *et al.* var. *zeae* Payak & Naras.
                         [R. G. Kenneth, Koltin & Wahl] [*Index of Fungi* **3**: 447]

Table 7, continued.

VERRUCALVUS P. Wong & M. W. Dick

Verrucalvus flavofaciens P. Wong & M. W. Dick **[type species; monotypic genus]** *[Index of Fungi* **5**: 458]

2. Acknowledgement

I am greatly indebted to Dr John David, of CABI Bioscience, for scanning these lists, checking them against the database established by IMI (now CABI) and searching original literature to check nomenclatural details.

3. References

Angiosperm Phylogeny Group (APG) (1998) An ordinal classification for the families of flowering plants, *Annals of the Missouri Botanic Garden* **85**, 531-553.

Barreto, R.W., and Dick, M.W. (1991) Monograph of *Basidiophora* (Oomycetes) with the description of a new species, *Botanical Journal of the Linnean Society* **107**, 313-332.

Biga, M.L.B. (1955) Riesaminazione delle specie del genere *Albugo* in base alla morfologia dei conidi, *Sydowia* **9**, 339-358.

Brummitt, R.K., and Powell, C.E. (1992) *Authors of Plant Names. A list of authors of scientific names of plants, with recommended standard forms of their names, including abbreviations*, Royal Botanic Gardens, Kew, UK.

Constantinescu, O. (1979) Revision of *Bremiella* (Peronosporales), *Transactions of the British Mycological Society* **72**, 510-515.

Constantinescu, O. (1989) *Peronospora* complex on Compositae, *Sydowia* **41**, 79-107.

Constantinescu, O. (1991*a*) An annotated list of *Peronospora* names, *Thunbergia* **15**, 1-110.

Constantinescu, O. (1991*b*) *Bremiella sphaerosperma* sp. nov. and *Plasmopara borreriae* comb. nov., *Mycologia* **83**, 473-479.

Constantinescu, O. (1992) The nomenclature of *Plasmopara* parasitic on Umbelliferae, *Mycotaxon* **43**, 471-477.

Constantinescu, O. (1996*a*) *Paraperonospora apiculata* sp. nov., *Sydowia* **48**, 105-110.

Constantinescu, O. (1996*b*) *Peronospora* on *Acaena* (Rosaceae), *Mycotaxon* **58**, 313-318.

Constantinescu, O. (1998) A revision of *Basidiophora* (Chromista, Peronosporales), *Nova Hedwigia* **66**, 251-265.

de Bary, A. (1863) Recherches sur le développement de quelques champignons parasites, *Annales des Sciences Naturelles, Partie Botanique, Paris, Série IV* **20**, 5-148.

Dick, M.W. (1988) Coevolution in the heterokont fungi (with emphasis on the downy mildews and their angiosperm hosts), in K.A. Pirozynski & D.L. Hawksworth (eds.), *Coevolution of Fungi with Plants and Animals*, Academic Press, London, pp. 31-62.

Dick, M.W. (1990*a*) Phylum Oomycota, in L. Margulis, J. O. Corliss, M. Melkonian & D. Chapman (eds.), *Handbook of Protoctista*, Jones & Bartlett, Boston, pp. 661-685.

Dick, M.W. (1990*b*) *Keys to Pythium*, Reading, UK, published by the author [ISBN 0-9516738-0-7].

Dick, M.W. (1995) Sexual reproduction in the Peronosporomycetes (chromistan fungi), *Canadian Journal of Botany, Supplement 1, Sections E-H* **73**, S712-S724.

Dick, M.W. (2001*a*) The Peronosporomycetes, in D. J. McLaughlin & E. McLaughlin (eds.), *The Mycota, Volume VII, Systematics and Evolution*, pp. 39-72. Springer-Verlag.

Dick, M.W. (2001*b*) *Straminipilous Fungi, Systematics of the Peronosporomycetes including Accounts of the Marine Straminipilous Protists, the Plasmodiophorids and Similar Organisms*, Kluwer Academic Publishers, Dordrecht, The Netherlands.

Dick, M.W. (in press) Towards an understanding of the evolution of the downy mildews [this volume].

Dick, M.W., Croft, B.J., Magarey, R.C., Cock, A.W.A.M. de, and Clark, G. (1989) A new genus of the Verrucalvaceae (Oomycetes), *Botanical Journal of the Linnean Society* **99**, 97-113.

Dick, M.W., Vick, M.C., Gibbings, J.G., Hedderson, T.A., and Lopes Lastra, C.C. (1999) 18S rDNA for species of *Leptolegnia* and other Peronosporomycetes: justification for the subclass taxa Saprolegniomycetidae and Peronosporomycetidae and division of the Saprolegniaceae *sensu lato* into the Leptolegniaceae and Saprolegniaceae, *Mycological Research* **103**, 1119-1125.

Dick, M.W., Wong, P.T.W., and Clark, G. (1984) The identity of the oomycete causing 'Kikuyu Yellows', with a reclassification of the downy mildews, *Botanical Journal of the Linnean Society* **89**, 171-197.

Erwin, D.C., and Ribeiro, O.K. (1996) Phytophthora *Diseases Worldwide*, American Phytopathological Society, St Paul, Minnesota, USA.

Farr, D.F., Bills, G.F., Chamuris, G.P., and Rossman, A.Y. (1989) *Fungi on Plants and Plant Products in the United States*, American Phytopathological Society, St Paul, Minnesota, USA.

Gäumann, E. (1923) Beiträge zu einer Monographie der Gattung *Peronospora* Corda, *Beiträge zur Kryptogamenflora der Schweiz* **5 (4)**, 1-360.

Gustavsson, A. (1959*a*) Studies on Nordic Peronosporas. I. Taxonomic revision, *Opera Botanica* **3 (1)**, 1-271.

Gustavsson, A. (1959*b*) Studies on Nordic Peronosporas. II. General account, *Opera Botanica* **3 (2)**, 1-61.

Hall, G.[S.] (1996) Modern approaches to species concepts in downy mildews, *Plant Pathology* **45**, 1009-1026.

Hibbett, D.S., and Donoghue, M.J. (1998) Integrating phylogenetic analysis and classification in fungi, *Mycologia* **90**, 347-356.

Kirk, P.M., and Ansell, A.E. (1992) *Authors of Fungal Names. A list of authors of scientific names of fungi, with recommended standard forms of their names including abbreviations. Index of Fungi, Supplement. 1992*, CAB International, Kew.

Kochman, J., and Majewski, T. (1970) *Grzyby (Mycota). [Flora Polska] Tom IV. Glonowce (Phycomycetes) Wroślikowe (Peronosporales)*, Państwowe Wydawnictwo Naukowe, Warszawa.

Korf, R.P. (1996) Simplified author citations for fungi and some old traps and new complications, *Mycologia* **88**, 146-150.

Plaats-Niterink, A.J. Van der (1981) Monograph of the genus *Pythium, Studies in Mycology, Centraalbureau voor Schimmelcultures, Baarn* **21**, 1-242.

Săvulescu, O. (1962) A systematic study of the genera *Bremia* Regel and *Bremiella* Wilson, *Revue de Biologie. Academia Republicii Populare Romîne, Bucarest* **7**, 43-62.

Săvulescu, T., and Săvulescu, O. (1952) Studiul *Sclerospora, Basidiophora* si *Peronoplasmopara, Buletin ştiinţific. Academia Republicii Populare Romîne*, pp. 327-457.

Shaw, C.G. (1978) *Peronosclerospora* species and other downy mildews of Graminae, *Mycologia* **70**, 594-604.

Shaw, C.G. (1981) Taxonomy and evolution, in D. M. Spencer (ed.), *The Downy Mildews*, Academic Press, London, pp. 17-29.

Shaw, C.G., and Waterhouse, G.M. (1980) *Peronosclerospora* (Ito) Shirai & K. Hara antedates *Peronosclerospora* (Ito) C. G. Shaw, *Mycologia* **72**: 425-426.

Skalický, V. (1964) Beitrag zur infraspezifischen Taxonomie der obligat parasitischen Pilze (Prispevek k vnitrodruhove taxonomii nekterych obligatne parasitickych plisni a hub), *Acta Universitatis Carolinae (Praha) 1964*, 25-89.

Skalický, V. (1966) Taxonomie der Gattungen der Familie Peronosporaceae (Taxonomie rodu celedi Peronosporaceae), *Preslia (Praha)* **38**, 117-129.

Skidmore, D.I., and Ingram, D.S. (1985) Conidial morphology and the specialization of *Bremia lactucae* Regel (Peronosporaceae) on hosts in the family Compositae, *Botanical Journal of the Linnean Society* **91**, 503-522.

Soltis, D.E., Soltis, P.S., Chase, M.W., Albach, D., Mort, M.E., Savolainen, V., and Zanis, M. (1998*a*) Molecular phylogenetics of angiosperms: congruent patterns inferred from three genes. Part II, *American Journal of Botany* **85**, 157 [abstract].

Soltis, P.S., Soltis, D.E., Chase, M.W., Albach, D., Mort, M.E., Savolainen, V., and Zanis, M. (1998*a*) Molecular phylogenetics of angiosperms: congruent patterns inferred from three genes. Part I, *American Journal of Botany* **85**, 157-158 [abstract].

Soltis, P.S., Soltis, D.E. and Chase, M.W. (1999) Angiosperm phylogeny inferred from multiple genes as a tool for comparative biology, *Nature (London)* **402**: 402-404.

Stamps, J., Waterhouse, G.M., Newhook, F.J., and Hall, G.S. (1990) Revised tabular key to the species of *Phytophthora*, *Mycological Papers* **162**, 1-28.

Waterhouse, G.M. (1964) The genus *Sclerospora*. Diagnoses (or descriptions) from the original papers and a key, *Commonwealth Mycological Institute, Commonwealth Agricultural Bureaux, Miscellaneous Publication, Number 17*, 1-30.

Waterhouse, G.M. (1968) The genus *Pythium* Pringsheim. Diagnoses (or descriptions) and figures from the original papers, *Mycological Papers* **110**, 1-71.

Waterhouse, G.M., and Brothers, M.P. (1981) The taxonomy of *Pseudoperonospora*, *Mycological Papers* **148**, 1-28.

INDEX

A

AAA 34
AABA 211
α-aminoadipin acid (AAA) 34
ABC pumps 149
accessions 89, 95, 98, 99, 102, 106
acibenzolar-S-methyl 126, 138, 139
acquired resistance 64
acropetal 122
actin coding region 21
AFLP autoradiograph 187
AFLP markers 187
agronomical practices 152
Albuginaceae 252
Albugo 43, 252-254
Albugo candida 93
alkaloids 34, 36
alternative oxidase 128, 129
alternative respiration 127
amino acid amide carbamates 126, 133
amino butyric acid 126, 140, 141, 207,
    211, 216, 218
ancestral character 15
Angiosperm coevolution 30
Angiosperm evolution 26
Angiosperms 22, 24
antifungal compounds 126
apical growth 132
apomorphous character 15
apoplastic mobility 122
apoptosis 104
appressorium 105
Arabidopsis thaliana 66, 68, 72, 93,
    139, 140, 185
arachidonic acid 140, 141
aromatic hydrocarbons 126, 131
Arthraxon 90
Asteraceae 87
ATP production 126, 128
ATP synthase 127, 128

ATP uncouplers 129
ATR loci 185, 186
author abbreviations 225
authority notation 225
autofluorescence 71, 209
avirulence 185
avirulence genes 63, 70
avoidance 59
Avr genes 63, 67, 70
azoxystrobin 126, 127, 129, 137, 195,
    196, 198, 201, 203-205

B

BABA 65, 72, 141, 207, 209, 210-214,
    216, 219-221
Bacillus 140, 141
Barremian 30
Basidiophora 228
basipetal 122
benalaxyl 124, 126
Benua 228
benzothiadiazoles 138, 140
β-1,3-glucanase 223
β-1,3-glucans 212
binomials 225
biodiversity 88, 90
biotrophic 4
biotrophic habit 32
biotrophic species 163
biotrophs 108
breeding lines 173
Bremia 228, 229
Bremia lactucae 61, 67, 70, 71, 85, 86,
    90, 100, 108, 119, 179-181
Bremiella 229
BTH 65, 66, 138

X

Z